Anonymous

A New and General Biographical Dictionary

Vol. 7

Anonymous

A New and General Biographical Dictionary
Vol. 7

ISBN/EAN: 9783744718288

Printed in Europe, USA, Canada, Australia, Japan

Cover: Foto ©berggeist007 / pixelio.de

More available books at **www.hansebooks.com**

A NEW AND GENERAL BIOGRAPHICAL DICTIONARY;

CONTAINING

AN HISTORICAL AND CRITICAL ACCOUNT

OF THE

LIVES and WRITINGS

OF THE

Moſt Eminent Perſons

IN EVERY NATION;

PARTICULARLY THE BRITISH AND IRISH;

From the Earlieſt Accounts of Time to the preſent Period.

WHEREIN

Their remarkable ACTIONS and SUFFERINGS,

Their VIRTUES, PARTS, and LEARNING,

ARE ACCURATELY DISPLAYED.

With a CATALOGUE of their LITERARY PRODUCTIONS.

A NEW EDITION, IN FIFTEEN VOLUMES,
GREATLY ENLARGED AND IMPROVED.

LONDON:

Printed for G. G. and J. ROBINSON, J. JOHNSON, J. NICHOLS, J. SEWELL,
H. L. GARDNER, F. and C. RIVINGTON, W. OTRIDGE and SON,
G. NICOL, E. NEWBERY, HOOKHAM and CARPENTER,
R. FAULDER, W. CHAPMAN and SON, J. DEIGHTON,
D. WALKER, J. ANDERSON, T. PAYNE, J. LOWNDES,
P. MACQUEEN, J. WALKER, T. EGERTON, T.
CADELL, jun. and W. DAVIES, R. EDWARDS,
VERNOR and HOOD, J. NUNN, MURRAY
and HIGHLEY, T. N. LONGMAN, LEE
and HURST, and J. WHITE.

1798.

BIOGRAPHICAL DICTIONARY.

GESNER (CONRAD), an eminent scholar, physician, and philosopher, was born at Zurich in Switzerland in 1516, where he also received the first rudiments of learning. He discovered great facility of genius; but the circumstances of his father would not allow him to make his son a scholar, and he was about to discontinue his studies, when Ammien, professor of Latin and eloquence at Zurich, took him to his own house, and charged himself with the care of his education. By the death of his father, he was a second time reduced to great extremities, and fell into a dropsical disorder. On the recovery of his health, he resolved to travel, and seek his fortune. He went to Strasburg, where he made some progress in the hebrew language; and, the civil wars of Switzerland having subsided, he was allowed a pension from the academy of Zurich, to enable him to make the tour of France. He accordingly went to Paris, accompanied by John Frisius, the early friend of his youth, and partner of his studies. From Paris he returned to Strasburg, whence he was invited by the university of Zurich, to preside over a school. Here he married, but, finding his appointment inadequate to the maintenance of a family, he was obliged to seek for other resources. From his childhood he had a great propensity to study physic, and he now devoted all the time he could spare from his school to books of medicine. At length he left in disgust his situation at Zurich, and proceeded to Basil, where he employed his time in reading the greek physicians in their own language, till he was made greek professor at Lausanne. This situation afforded him the means

means of attending to his favourite purfuits, and enabled him to go to Montpelier, where having ftudied anatomy and botany for fome time, he returned to Zurich, entered upon his profeffion as a phyfician, and was admitted to a doctor's degree. He was foon afterwards made profeffor of philofophy, a fituation which he adorned for the fpace of twenty-four years, that is, as long as he lived, for in 1565 he was carried off by the plague. He wrote no lefs than fixty-fix pieces, on the fubjects of grammar, botany, medicine, natural hiftory, of which the principal are thefe.

I. An univerfal dictionary, publifhed at Zurich in 1545, or a dictionary of books and authors, which was the firft work of the kind, and has been the model of all fubfequent ones.

II. Hiftory of animals, in 4 vol. folio, Zurich 1551; a great and fplendid work, though not always the moft accurate.

III. A greek and latin lexicon. He was a perfect mafter of both the languages; but, as he himfelf acknowledges in the work, he wrote it for bread, and it is confequently not without faults.

IV. Opera botanica, Nuremburg, folio. We owe to Gefner the having firft diftinguifhed the genera of plants, from a comparifon of their flowers, feeds, and fruits.

In juftice to the memory of Conrad Gefner, it is neceffary to add, that, on account of the variety of his attainments, and the extent of his learning, he was named, by way of diftinction, the german Pliny.

GESNER (SOLOMON), bookfeller at Zurich, in Switzerland, and author of many elegant and admired poems in the german language, born at Zurich, 1730. He was, for feveral years before his death, a member of the fenate of his native city. He was an admirable landfcape-painter as well as poet. The greater part of his pictures were difpofed of in England, where perhaps he has been better known as a painter than a poet. Among his writings, " the death of Abel" is that which is beft known in England. He died March 2, 1788.

GESNER (SOLOMON), a celebrated theologian of the lutheran perfuafion, in the fixteenth century. He wrote a great number of works, fuch as effays, differtations, fermons, &c. He died at Wittenburg in 1605.

GESNER (JOHN-MATTHEW), a profound fcholar, and moft acute critic, was born at a village near Newburg, in Germany, in the year 1691. He was of the family of Conrad Gefner above mentioned. He loft his father at a very early age; but, by the kindnefs of a father-in-law, he was enabled to follow the bent of his natural inclination for learning, and ftudied for eight years under Nicolas Keelerus,

at

at Anspach. In consequence of the recommendation of Buddeus, he was appointed to superintend the public school of Weinheim, in which character he remained eleven years. From Weinheim he was removed to a situation equally honourable, and more lucrative, at Anspach; whence, after some other changes of no great importance in his situation, he finally returned to Gottingen. Here he received the reward of his talents and industry in several advantageous appointments. He was made professor of humanity, public librarian, and inspector of public schools, in the district of Luneburg. He died at Gottingen, universally lamented, and esteemed, in the year 1761.

His works of greatest importance are various editions of the classics, both greek and latin, and, above all, a Thesaurus of the latin tongue, which whoever possesses will probably not require the aid of any other latin lexicon. The editions of the classics which received the correcting hand of Gesner, and which are more popular, are the Horace and the Claudian. The work which he himself valued the most, and which was not published till after his death, is the Argonautics of Orpheus, with the tracts de lapidibus, and the hymns. Many ingenious and learned men have not thought it beneath them to write in recommendation of Gesner's talents and virtues; but our readers will receive more various and particular information from a narrative on this subject written by Ernestus, and addressed to Ruhnkenius. An excellent portrait of Gesner is prefixed to his latin Thesaurus.

GETA (SEPTIMIUS), son of the emperor Severus, and brother of Caracalla who was jealous of his accomplishments, and finally stabbed him in the arms of Julia, their common mother. Geta died at the age of twenty-three years, and 212 after Christ. Caracalla consecrated, to the temple of Serapis, the sword with which, as he boasted, he had slain his brother Geta, who nevertheless received the funeral honour of a Roman emperor, and was placed among the gods. Sit *ditus*, dum non sit *vivu*, was the observation of his brother.

GETHIN (Lady GRACE), an English lady of uncommon parts, was the daughter of Sir George Norton, of Abbots-Leith in Somersetshire, and born in 1676. She had all the advantages of a liberal education, and became the wife of Sir Richard Gethin, of Gethin-grott in Ireland. She was mistress of great accomplishments natural and acquired, but did not live long enough to display them to the world, for she died in her 21st year. She was buried in Westminster-abbey, where a beautiful monument with an inscription is erected over her: and, moreover, for perpetuating her memory, provision was made for a sermon to be preached in Westminster-abbey, yearly, on Ash-wednesday, for ever. She wrote, and left behind her in loose papers, a work, which, soon after her death, was methodized and published under the title of Reliquiæ Gethinianæ; or, some remains of the most ingenious and excellent lady, Grace Lady Gethin, lately deceased; being a Collection of choice Discourses, pleasant Apophthegms, and witty Sentences. Written by her, for the most Part, by Way of Essay, and at spare hours, 1700," 4to. with her picture before it. This work consists of discourses upon Friendship, Love, Gratitude, Death, Speech, Lying, Idleness, The World, Secrecy. Prosperity, Adversity, Children, Cowards, Bad Poets, Indifferency, Censoriousness, Revenge, Boldness, Youth, Age, Custom, Charity, Reading, Beauty, Flattery, Riches, Honour, High Places, Pleasure, Suspicion, Excuses, &c.; and, as it is very scarce, and not easily to be procured, the following extract from it may properly be produced as a specimen of the author's abilities and manner. "Reading," says she, "serves for delight, for ornament, and for ability: it perfects nature, and is perfected by experience, the crafty condemn it, simple admire it, and wise men use it. Some books are to be tasted or swallowed, and some few to be chewed or digested. Reading makes a full man, conference a ready man, and writing an exact man. He that writes little needs a great memory: he that confers little, a present wit: and he that reads little needs much cunning, to make him seem to know that which he does not. History makes men wise, poetry witty, mathematics subtle, philosophy deep, morals grave, logic and rhetoric able to contend; nay, there is no impediment in the wit but may be wrought out by fit study, where every defect of the mind hath its proper receipt." Among Mr. Congreve's poems are to be found, " Verses to the Memory of Grace Lady Gethin, occasioned by reading her Book, intituled, ' Reliquiæ Gethinianæ;" in which the agreeable writer, after speaking of the shortness of life, and the difficulty of attaining knowledge, proceeds thus:

> Whoe'er on this reflects, and then beholds
> With strict attention what this book unfolds,
> With admiration struck, shall question, who
> So very long could live so much to know?
> For so complete the finish'd piece appears,
> That learning seems combin'd with length of years;
> And, both improv'd by purest wit, to reach
> At all that study or that time can teach.
> Put to what height must his amazement rise,
> When, having read the work, he turns his eyes
> Again to view the foremost opening page,
> And there the beauty, sex, and tender age,
> Of her beholds, in whose pure mind arose
> Th' ethereal source, from whence this current flows!

GETHING

GETHING (RICHARD), a curious penman, was, according to Wood, of Herefordshire, but settled at the hand and pen in Fetter-lane, London, as early as 1616, about which time he published a copy-book of various hands, in 26 plates, oblong quarto, well executed, considering the time. In 1645, he published his Chirographia, in 37 plates, wherein he principally aims at the improvement of the italian hand. There is another edition of this book dated 1664, perhaps after his death, as it has this title, "Gething's Redivivus," with his picture in the front. In 1652, his Calligraphotechnia was propagated from the rolling-press—it contains thirty-six folio-plates, with his picture, which has a label round it, inscribing him aged 32. This seems to be a re-publication of his former works, for some of the plates are dated 1615, 1616, and it is dedicated to Sir Fran. Bacon, who died in 1626, which was a long time before this publication in 1652.

GEVARTIUS (JOHN GASPAR), a learned critic, was the son of an eminent lawyer, and born at Antwerp in 1593. Many authors have called him simply John Gaspar, and sometimes he himself was content with doing this; so that, perhaps, he is better known by the name of Gaspar than Gevartius. His first application to letters was in the college of Jesuits at Antwerp, whence he removed to Louvain, and then to Douay. He went to Paris in 1617, and spent some years there in the conversation of the learned. Returning to the Low-countries in 1621, he took the degree of LL.D. in the university of Douay, and afterwards went to Antwerp, where he was made town-clerk, a post he held to the end of his life. He married in 1625, and died in 1666, aged 72. He had always a taste for classical learning, and devoted a great part of his time to literary pursuits. In 1621, he published at Leyden, in 8vo. "Lectionum Papinianarum Libri quinque in Statii Papinii Sylvas;" and, at Paris in 1619, 4to. "Electorum Libri tres, in quibus plurima veterum Scriptorum loco obscura & controversa explicantur, illustrantur, & emendantur." These, though published when he was young, have established his reputation as a critic: but he was also a poet, and gave many specimens of his skill in versifying: witness, amongst others, a Latin poem, published at Paris 1618, on the death of Thuanus, "Historiæ sui temporis scriptoris incomparabilis," as he justly calls him. He kept a constant correspondence with the learned of his time, and some of his letters have been printed: there are 12 to Nicholas Heinsius, in the "Sylloge Epistolarum," by Burman. Our Bentley mentions Gaspar Gevartius as a man famous in his day; and tells us, that "he undertook an edition

tion of the poet Manilius, but was prevented by death" from executing it.

GHILINI (JEROME), an Italian writer, born at Monza, in Milan, 1589, was educated by the Jesuits at Milan in polite literature and philosophy. He went afterwards to Parma, where he began to apply himself to the civil and canon law; but was obliged to desist on account of ill health. He returned home, and upon the death of his father married: but losing his wife, he became an ecclesiastic, and resumed the study of the canon law, of which he was made doctor. He lived to be 80 years of age, and was the author of several works; the most considerable of which, and for which he is at present chiefly known, is his "Theatro d' Huomini Letterati." The first part of this was printed at Milan, 1633, in 8vo, but it was enlarged and reprinted in 2 vols. 4to. at Venice, 1647. Baillet says, that this work is esteemed for its exactness, and for the diligence which the author has shewn, in recording the principal acts and writings of those he treats of, but this is not the opinion of M. Monnoye his annotator, nor of the learned in general. It is pretty well agreed, that, excepting a few articles, where more than ordinary pains seem to have been taken, Ghilini is a very injudicious author, deals in general and insipid panegyric, and is, to the last degree, careless in the matter of dates. This work, however, for want of a better, has been made much use of, and is even quoted at this day by those who know its imperfections.

GHIRLANDAIO (DOMENICO), a Florentine painter, born in 1449, was at first intended for the profession of a goldsmith, but followed his more prevailing inclinations to painting with such success, that he is ranked among the first masters of his time. Nevertheless, his manner was gothic and very dry; and his reputation is not so much fixed by his own works, as by his having had Michael Angelo for his disciple. He died at 44 years of age, and left three sons, David, Benedict, and Rhandolph, who were all of them painters.

GIANNONI (PETER), born at Naples, in 1680, died in Piedmont in 1748. He wrote a history of Naples, which so offended the court of Rome, that, to avoid persecution, he was compelled to take refuge in the territories of the king of Sardinia. His work has been translated into French, by Desmanceaux, and is admired for its purity of style.

GIBBON (EDWARD), author of the "History of the Decline and Fall of the Roman Empire." He was born at Putney in 1737, and was sent at a very early age to the grammar school at Kingston, from which he was removed to Westminster-school; from this seminary he went to Magdalen-College,

College, Oxford; and from Oxford to Lausanne. At Oxford he imbibed the principles of popery; his friends, alarmed at this, sent him to the Sage of Ferney, for a cure.—It wrought an effectual one, for he came home a confirmed infidel. He was, at one period of life, a member of parliament and a lord of trade; but when his friend, lord North, ceased to be minister, Mr. Gibbon retired to Switzerland, where he proposed to spend the remainder of his life in literary retirement. His "History of the Decline and Fall of the Roman Empire" requires no remark nor eulogium from us. The following is given as his character as a writer, by Mr. Porson, in his "Letters to Archdeacon Travis."

"An impartial judge, I think, must allow that Mr. Gibbon's history is one of the ablest performances of its kind that has ever appeared. His industry is indefatigable; his accuracy scrupulous; his reading, which indeed is sometimes ostentatiously displayed, immense; his attention always awake; his memory extensive; his periods harmonious.—His reflections are often just and profound; he pleads eloquently for the rights of mankind, and the duty of toleration; nor does his humanity ever slumber, unless when women are ravished, or the christians persecuted. Mr. Gibbon shews, it is true, so strong a dislike to christianity as visibly disqualifies him for that society, of which he has created Ammianus Marcellinus president. I confess that I see nothing wrong in Mr. Gibbon's attack on Christianity; it proceeded, I doubt not, from the purest and most virtuous motive. We can only blame him for carrying on the attack in an insidious manner, and with improper weapons. He often makes, when he cannot easily find, an occasion to insult our religion; which he hates so cordially, that he might seem to revenge some personal injury. Such is his eagerness in the cause, that he stoops to the most aukward perversion of language for the pleasure of turning the Scripture into ribaldry, or of calling Jesus an impostor. Though his style is in general correct and elegant, he sometimes draws out the thread of his verbosity finer than the staple of his argument. In endeavouring to avoid vulgar terms, he too frequently dignifies trifles, and clothes common thoughts in a splendid dress, that would be rich enough for the noblest ideas. In short, we are too often reminded of the great man, Mr. Prig, *the auctioneer, whose manner was so inimitably fine, that he had as much to say upon a ribbon as a Raphael.* Sometimes, in his anxiety to vary his phrase, he becomes obscure; and, instead of calling his personages by their names, defines them by their birth, alliance, office, or other circumstances of their history. Thus an honest gentleman is often described by a circumlocution, lest the same word should be twice repeated in the fame page. Sometimes, at his attempts at elegance, he lofes fight of Englifh, and fometimes of fenfe. A lefs pardonable fault is that rage for indecency, which pervades the whole work; but efpecially the laft volumes. And, to the honour of his confiftency, this is the fame man who is fo prudifh, that he dares not call Belifarius a cuckold; becaufe it is too bad a word for a *decent* hiftorian to ufe. If the hiftory were anonymous, I fhould guefs that thefe difgraceful obfcenities were written by fome debauchee, who, having from age, or accident, or excefs, furvived the practice of luft, ftill indulged himfelf in the luxury of fpeculation, *and expofed the impotent imbecillity, after he had loft the vigour of the paffions.*"

Mr. Gibbon wrote other things befides his hiftory, which will probably laft as long as the Englifh language. His Effay on Hiftory was firft written in French, and afterwards tranflated into Englifh: the Differtation on the Sixth Æneid difplays great acutenefs as well as erudition.; and his letter to Mr. Davis, in anfwer to animadverfions on his hiftory, will ever be confidered as a mafter-piece of writing. We could eafily fill a large portion of our volume with anecdotes of this truly eminent man; but our limits oblige us to be concife. He died in 1794.

GIBBS (JAMES), was born at Aberdeen, in 1683. He ftudied architecture in Italy, and about the year 1720 became the architect moft in vogue. He gave the defign of St. Martin's church, which was finifhed in five years, and coft 32,000*l*. The New-church, at Derby; the New-buildings, at King's College, Cambridge, and the Senate-houfe there; are a part of his works. His, likewife, was St. Mary's in the Strand, one of the fifty new churches, a monument of the piety more than the tafte of the nation. In 1728, he publifhed a large folio of his own defigns, which he fold for 1500*l*. and the plates for 400*l*. more. Died Auguft 5, 1754.

GIBALYN (LE COMPTE DE), author of "Le Monde Primitif," born 1725, died 1784. For this really great work he twice received the prize of 1200 livres given by the french academy to the moft valuable work which has appeared in the courfe of the year.

GIBSON (EDMUND), bifhop of London, fon of Edward Gibfon, of Knipe in Weftmorland, was born there in 1669; and, having laid the foundation of claffical learning at a fchool in that county, became a fervitor of Queen's college, Oxford, in 1686. The ftudy of the Northern languages being then particularly cultivated in this univerfity, Gibfon applied himfelf vigoroufly to that branch of literature, wherein he was affifted by Dr. Hickes. The quick proficiency that he made appeared in a new edition of William

liam Drummond's "Polemo-Middiana," and James V. of Scotland's "Cantilena Ruſtica:" theſe he publiſhed at Oxford, 1691, in 4to. with notes. His obſervations on thoſe facetious tracts ſtand as a monument of his abilities in the witty way; and the ſingular learning ſhewn in the annotations is really valuable. But his inclination led him to more ſolid ſtudies; and, in a ſhort time after, he tranſlated into Latin the "Chronicon Saxonicum," and publiſhed it, together with the Saxon original, and his own notes, at Oxford, 1692, in 4to. This work he undertook by the advice of Dr. Mill, the learned Editor of the "Greek Teſtament," in folio; and it is allowed by the learned to be the beſt remains extant of Saxon antiquity. The ſame year appeared a treatiſe, intituled, "Librorum Manuſcriptorum in duabus inſignibus Bibliothecis, altera Teniſoniana Londoni, altera Dugdaliana Oxonii, Catalogus. Edidit E. G. Oxon, 1692," 4to. The former part of this catalogue, conſiſting of ſome ſhare of Sir James Ware's manuſcript collection, was dedicated to Dr. Thomas Teniſon, then biſhop of Lincoln, as at that time placed in his library. He had a natural inclination to ſearch into the antiquities of his country; and, having laid a neceſſary foundation in the knowledge of its original languages, he applied himſelf to them for ſome years, with great diligence. He publiſhed Camden's "Britannia," and other works, which may be ſeen in a note p. 11; and concluded, in this branch of learning, with "Reliquiæ Spelmannianæ, or the Poſthumous Works of Sir Henry Spelman, relating to the Laws and Antiquities of England," which, with his own life of the author, he publiſhed at Oxford, 1698, folio. This he likewiſe dedicated to Dr. Teniſon, then Abp. of Canterbury; and probably, about that time, he was taken as domeſtic chaplain into the archbiſhop's family: nor was it long after, that we find him both rector of Lambeth, and archdeacon of Surrey.

Teniſon dying Dec. 14, 1715, Wake, biſhop of Lincoln, ſucceeded him; and Gibſon was appointed to the ſee of Lincoln. After this advancement, he went on indefatigably in defence of the government and diſcipline of the Church of England: and on the death of Robinſon, in 1720, was promoted to the biſhoprick of London. Gibſon's talents ſeem to have been perfectly ſuited to the particular duties of this important ſtation; upon the right management of which ſo much depends, in reſpect to the peace and good order of the civil, as well as the eccleſiaſtical, ſtate of the nation. It is well known, that he had a very particular genius for buſineſs, which he happily tranſacted, by means of a moſt exact method that he uſed on all occaſions: and this he purſued

sued with great advantage, not only in the affairs of his own diocese in England, which he governed with the most exact regularity, but in promoting the spiritual affairs of the church of England colonies, in the West-Indies. The minister, at this time, were so sensible of his great abilities in transacting business, that there was committed to him a sort of ecclesiastical ministry for several years; and more especially from the long decline of health in Abp. Wake, when almost every thing that concerned the church was in a great measure left to the care of the bishop of London.

The writer of his life, among many instances which he declares might be assigned of his making a proper use of that spiritual ministry he was honoured with, specifies some few of a more eminent kind. One was his occasional recommendation of several worthy and learned persons to the favour of the secular ministry, for preferments suited to their merits, as he had frequently the disposal of the highest dignities in the church. Another, that of procuring an ample endowment from the crown, for the regular performance of divine service in the Royal-chapel, at Whitehall, by a succession of ministers, selected out of both universities, with proper salaries. A third, that he constantly guarded against the repeated attempts of certain persons to procure a repeal of the corporation and test acts. By baffling the attacks made on those fences of the church, he thought he secured the whole ecclesiastical institution: for, it was his fixed opinion, that it would be an unjustifiable piece of presumption to arm those hands with power, that might possibly employ it, as was done in the days of our fathers, against the ecclesiastical constitution itself. He was entirely persuaded, that there ought always to be a legal establishment of the church, to a conformity with which some peculiar advantages might be reasonably annexed: and at the same time, with great moderation and temper, he approved of a toleration of protestant dissenters; especially as long as they keep within the just limits of conscience, and attempt nothing that is highly prejudicial to, or destructive of, the rights of the establishment in the church. But he was as hearty an enemy to persecution, in matters of religion, as those that have most popularly declaimed against it.

Lastly, one more service to the church and clergy, done by the bishop of London, well claims their grateful acknowledgements; namely, his distinguished zeal (after he had animated his brethren on the bench to concur with him) in timely apprizing the clergy of the bold schemes that were formed by the quakers, in order to deprive the clergy of their legal maintenance by tithes; and in advising them to avert so great a blow to religion, as well as so much injustice to themselves,

by

by their early application to the legislature, to preserve them in the possession of their known rights and properties. But, though the designs of their adversaries were happily defeated, yet it ought ever to be remembered, in honour of the memory of the bishop of London, that such umbrage was taken by the then great minister, on occasion of the advice given by him and his brethren to the clergy in that critical juncture, as in fact soon terminated in the visible diminution of his interest and authority.

However, no discouragements, he met with, were able to break his firm and steady attention to the duties of his office; in writing and printing pastoral letters to the clergy and laity, in opposition to infidelity and enthusiasm; in visitation-charges, as well as occasional sermons, besides less pieces of a mixt nature, and some particular tracts against the prevailing immoralities of the age [A].

[A] For the reader's satisfaction we shall insert here a catalogue of his works as follows: An edition of Drummond's "Polema-middiana, &c. 1691," 4to. has been already mentioned, as also the "Chronicon Saxonicum, 1692," 4to. and his "Librorum Manuscriptorum Catalogus," printed the same year, all three at Oxford: where he likewise published "Julii Cæsaris Portus Iccius Illustratus," a tract of W. Somner, with a dissertation of his own, 1694. An edition of "Quintilian de Arte Oratoria, with Notes," Oxon. 1693, 4to. A translation of Camden's "Britannia into English, 1695," fol. and again with large editions in 1722, and 1772, two vols. fol. "Vita Thomæ Bodleii Equitis Aurati, & Historia Bibliothecæ Bodleianæ," prefixed to a book, intituled, "Catalogi Librorum Manuscriptorum in Anglia & Hibernia in unum collecti Oxon. 1697," in 2 vols. folio. "Reliquiæ Spelmannianæ, &c." mentioned above, 1695, fol. "Codex Juris Ecclesiastici Anglicani, &c. 1713," fol. "A Short State of some present Questions in Convocation, 1700," 4to. "A Letter to a Friend in the Country, concerning the Proceedings in Convocation, in the years 1700 and 1701, 1701," 4to. "The Right of the Archbishop to continue or prorogue the whole Convocation. A Summary of the Arguments in Favor of the said Right." "Synodus Anglicana, &c. 1702." "A Parallel between a Presbyterian Assembly, and the new Model of an English Provincial Synod," 4to. "Reflections upon a Paper, intituled, "The Expedient proposed," 4to. "The Schedule of Prorogation reviewed," 4to. "The pretended Independence of the Lower-House upon the Upper-House a groundless notion, 1703," 4to. "The Marks of a defenceless Cause, in the Proceedings and Writings of the Lower House of Convocation," 4to. "An Account of the Proceedings in Convocation in a Cause of Contumacy, upon the Prolocutor's going into the Country without the Leave of the Archbishop, commenced April 10, 1707." All these upon the disputes in Convocation. except the "Synodus Anglicana," &c. are printed without his name, but generally ascribed to him. "Visitations parochial and general, with a Sermon, and some other Tracts, 1717," 8vo. "Five Pastoral Letters, &c. Directions to the Clergy, and Visitation Charges, &c" 8vo. Family Devotion; A Treatise against Intemperance; Admonition against Swearing. Advice to Persons who have been Sick; Trust in God; Sinfulness of neglecting the Lord's Day; Against Lukewarmness in Religion; Several occasional Sermons; Remarks on Part of a Bill brought into the House of Lords by the Earl of Nottingham, in 1721, intituled, "A Bill for the more effectual Suppression of Blasphemy and Profaneness" is also ascribed to the bishop, as is also, "The Case of addressing the Earl of Nottingham, for his Treatise on the Trinity," published about the same time. Lastly, "A Collection of the principal Treatises against Popery, in the Papal Controversy, digested into proper Heads and Titles, with some Prefaces of his own. Lond. 1738." 3 vols. fol.

He was very senfible of his decay for fome time before his death, in which he complained of a langour that hung about him. As, indeed, he had made free with his conftitution by incredible induftry, in a long courfe of ftudy and bufinefs of various kinds; he had well nigh exhaufted his fpirits, and worn out a conftituion which was naturally fo vigorous, that life might, otherwife, have probably been protracted to more than 79; towards the end of which year of his age, namely, September 6, 1748, he died with true chriftian fortitude, an apparent fenfe of his approaching diffolution, and in a perfect tranquillity of mind, during the intervals of his laft fatal indifpofition at Bath, after a very fhort continuance there. His lordfhip was married, and left feveral children of each fex, who were all handfomely provided for by him

GIBSON (RICHARD), commonly called the dwarf, was an eminent englifh painter, in the time of Sir Peter Lely, to whofe manner he devoted himfelf, and whofe pictures he copied to admiration. He was originally fervant to a lady at Mortlake, who, obferving that his genius led him to painting, put him to De Cleyn, to be inftructed in the rudiments of that art. De Cleyn was mafter of the tapeftry-works at Mortlake, and famous for the cuts which he defigned for fome of Ogilby's works, and for Sandys's tranflation of Ovid. Gibfon's paintings in water-colours were well efteemed; but the copies he made of Lely's portraits gained him the greateft reputation. He was greatly in favour with Charles I. to whom he was page of the back-ftairs; and he alfo drew Oliver Cromwell feveral times. He had the honour to inftruct in drawing queen Mary and queen Anne, when they were princeffes, and he went over to Holland to wait on the former for that purpofe. He was himfelf a dwarf; and he married one Mrs. Anne Shepherd, who was alfo a dwarf. Charles I. was pleafed, out of curiofity or pleafantry, to honour their marriage with his prefence, and to give away the bride. Waller wrote a poem on this occafion, "of the marriage of the dwarfs," which begins thus:

> Defign or chance makes others wive,
> But Nature did this match contrive.
> Eve might as well have Adam fled,
> As fhe deny'd her little bed
> To him, for whom Heaven feem'd to frame
> And meafure out this only dame, &c.

Fenton, in his notes on this poem, tells us, that he had feen this couple painted by Sir Peter Lely; and that they appeared to have been of an equal ftature, each of them meafuring three feet ten inches. They had, however, nine children; five of
which

which attained to maturity, and were well-proportioned to the usual standard of mankind. To recompense the shortness of their stature, nature gave them an equivalent in length of days; for Gibson died in Covent-Garden, in his 75th year; and his wife, surviving him almost 20 years, died in 1709, aged 89.

GIBSON (WILLIAM), nephew to the above Richard, was instructed in the art of painting both by him and Sir Peter Lely, and became also eminent. His excellence, like his uncle's, lay in copying after Sir Peter Lely; although he was a good limner, and drew portraits for persons of the first rank. His great industry was much to be commended, not only for purchasing Sir Peter Lely's collection after his death, but likewise for procuring from beyond sea a great variety of valuable things in their kind; insomuch, that his collection of prints and drawings was not inferior to any persons of his time. He died of a lethargy in 1702, aged 58.

GIBSON (EDWARD), William's kinsman, was instructed by him, and first painted portraits in oil; but afterwards finding more encouragement in crayons, and his genius lying that way, he applied himself to them. He was in the way of becoming a master, but died when he was young.

GIBSON (WILLIAM), was a self-taught mathematician, born at Boulton, near Appleby, in Westmorland, in 1720, died in 1791. His knowledge of the art of navigation, the principles of mechanics, the doctrine of motion, of falling bodies, and the elements of optics, though not evinced by any publications on those subjects, was so notorious to his countrymen, and so frequently as well as usefully exercised, that it deserves thus to be recorded.

GIBSON (THOMAS), an eminent painter, practised in London and Oxford, died April 28, 1751, aged about 71.

GIBSON (THOMAS), a native of Morpeth in Northumberland, was famous in his time for the studies of physic, divinity, history, and botany, in which studies he made considerable progress. Bale bears witness to his character as a physician, by saying, that he performed almost incredible cures. He was a friend to the reformation, and wrote some pieces in defence of that cause. He was a fugitive for his religion, in the reign of queen Mary; but, on the accession of Elizabeth, returned, and died in London in 1562. He wrote many pieces, the titles of all which are very verbose, and may be seen in Tanner. See also Aikin's Biographical Memoirs of Medicine.

GIFANIUS (HUBERTUS, or OBERTUS), a learned critic and great civilian, was born at Buren in Guelderland in 1534. He

He studied at Louvain and at Paris, and was the first who erected the library of the german nation at Orleans. He took the degree of doctor of civil law there in 1567; and went thence to Italy in the retinue of the french ambassador. Afterwards he removed to Germany, where he taught the civil law with high repute. He taught it first at Strasburg, where he was likewise professor of philosophy; then in the university of Altdorf, and at last at Ingoldstadt. He forsook the protestant religion to embrace the roman-catholic. He was invited to the imperial court, and honoured with the office of counsellor to the emperor Rodolph. He died at Prague in 1609, if we believe some authors; but Thuanus, who is more to be depended on, places his death in 1604. Besides notes and comments upon authors of antiquity, he wrote several pieces relating to civil law.

As to his literary character, he has been accused of a notorious breach of trust, with regard to the MSS. of Fruterius. Fruterius was a great genius, and had collected a quantity of critical observations; but died at Paris in 1566, when he was only 25. He left them to Gifanius, to be published, who acted fraudulently, and suppressed them as far as he was able; for which he is severely treated by Janus Douza in his satires and elsewhere. The fact is also mentioned by Thuanus. He was charged with plagiarism, and had quarrels with Lambin upon this head. Gifanius, it seems, had inserted in his edition of Lucretius all the best notes of Lambin, without acknowledging to whom he was obliged; and with some contempt of Lambin; for which, however, Lambin, in a third edition of that author, has loaded him with all the hard names he could think of. He calls him "audacem, arrogantem, impudentem, ingratum, petulantem, insidiosum, fallacem, infidum, nigrum." He had, also, another quarrel with Scioppius, about a MS. of Symmachus; which Scioppius, it is said, had taken away, and used without his knowledge. These quarrels are not worth relating. It is pity, that polite literature will not restrain the passions, and civilize the manners of its possessors; but experience has shewn, that it will not: which gives us reason to conclude, that human nature will be human nature still, and that its depravity will appear under some mode or other, in spite of all applications to correct it.

GILBERT (WILLIAM), a learned physician, who first discovered several of the properties of the load-stone, was born at Colchester, where his father was recorder, in 1540; and, after an education at a grammar-school, was sent to Cambridge. Having studied physic for some time, he went abroad for his farther improvement; and, in one of of the foreign universities, had the degree conferred upon him of M. D. He returned to England with a considerable reputation for his learning in general, and had especially the character of being deeply skilled in philosophy and chemistry; and, resolving to make his knowledge useful to his country by practising in this faculty, he presented himself a candidate to the college of physicians in London, and was elected a fellow of that society about 1573. Thus, every way qualified for it, he practised in this metropolis with great success and applause; which being observed by queen Elizabeth, whose talent it was to distinguish persons of superior merit, she sent for him to court, and appointed him her physician in ordinary; and gave him, besides, an annual pension to encourage him in his studies. In these, as much as his extensive business in his profession would give him leave, he applied himself chiefly to consider and examine the various properties of the load-stone; and proceeding in the experimental way, a method not much used at that time, he discovered and established several qualities of it not observed before. This occasioned much discourse; and, spreading his fame into foreign countries, great expectations were raised from his treatise on that subject, which were abundantly fulfilled when it appeared in public.

He printed it, in 1600, under the following title, "De Magnete, magneticisque Corporibus & de magno Magnete tellure, Physiologia nova:" i. e. "Of the Magnet (or Loadstone) and magnetical Bodies, and of that great Magnet the Earth." It contains the history of all that had been written on that subject before his time [B], and is the first regular system on this curious subject, and may not unjustly be styled the parent of all the improvements that have been made therein since. In this piece our author shews the use of the declination of the magnet, which had been discovered by Norman in finding out the latitude, for which purpose also he contrived two instruments for the sea. This invention was published by Thomas Blandeville, in a book intituled, "Theoriques of the Planets, together with the making of two Instruments for Seamen, for finding out the Latitude without Sun, Moon, or Stars, invented by Dr. Gilbert, 1624." But the hopes from this property, however pro-

[B] Among such writers are Harriot, Huer, Wright, Kendal, Barlow, and Norman, which shews Wood's observation to be uncandid at least when he tells us, that Barlow had knowledge in the magnet 20 years before Gilbert's book came out; and, whatever was the intention of the antiquary's remark, it is certain, from his own account, that Gilbert first improved this knowledge to that degree of perfection, as to be fit for public view and use, since Barlow did not publish his magnetical advertisement till 1616. Ath. Oxon. Vol. I. See also the article BARLOW (WILLIAM), in Biog. Brit.

mifing at firſt, have by a longer experience been found to be deceitful.

After the death of Elizabeth, the doctor was continued as chief phyſician to James I. but he enjoyed that honour only a ſhort time, paying his laſt debt to nature, Nov. 30, 1603. His corpſe was interred in Trinity church at Colcheſter, where he was born, and where there is a handſome monument raiſed to his memory; a print of which is to be ſeen in the Hiſtory and Antiquities of Colcheſter, by Morant. By a picture of him in the ſchool gallery of Oxford, he appears to have been tall of ſtature, and of a chearful countenance. All that is left us of his character has been ſaid on the occaſion of his famous book; on which account we have the higheſt encomiums of him, ſuch as are uſually made by one author upon another. Thus Carpenter tells us, that he had trodden out a new path to philoſophy. Sir Kenelm Digby compares him with Harvey, the diſcoverer of the circulation. Barrow ranks him with Galileo, Gaſſendus, Merſennus, and Des Cartes; whom he repreſents as men reſembling the ancients in ſagacity and acuteneſs of genius. Theſe atteſtations of his high merit are indeed given him by his countrymen; but, that they may not be ſuſpected of partiality [c], there is good reaſon to believe, that his fame was ſtill more celebrated among foreigners; of which this is one very ſtrong confirmation, that the famous Peireſc often lamented, that when he was in England he was not acquainted with our philoſopher.

Beſides his principal work printed in his life-time, he left another treatiſe in MS. which coming into the hands of Sir William Boſwell was from that copy printed at Amſterdam, in 1651, 4to. under this title, " De mundi noſtro ſublunari Philoſophia nova." As he was never married, he gave by his laſt will all his library, conſiſting of books, globes, inſtruments, &c. and a cabinet of minerals, to the college of phyſicians; and this part was punctually performed by his brothers, who inherited his eſtate, which muſt have been ſomewhat conſiderable. Wood obſerves, he was the chief perſon in his pariſh at Colcheſter.

GILBERT (THOMAS, B. D.) He was educated in Edmund-hall, and ordained miniſter at Eggmond, in Shropſhire, where he continued till he was rejected for noncon-

[c] This remark of lord Bacon is the leaſt free from that cenſure. He frequently mentions Gilbert's book with applauſe; and in one place particularly ſtyles it a painful and experimental work, Advancement of Learning, L. i. c. 13. words, in his lordſhip's mouth, of ſingular force and extent of meaning, and which are handſomely illuſtrated by the compliment of Mr. Wright prefixed to the book; by which it appears, that our author ſpent no leſs than 18 years in bringing it to perfection.

formity

formity 1662. He afterwards returned to Oxford, where he lived privately, and was much refpected by the principal men in the univerfity. He difputed with Dr. South concerning predeftination, and made the latter a convert to his doctrines. He died July 14, 1694, aged 83. He wrote feveral theological treatifes, particularly one againft Dr. Owen, entituled, "The Poffibility of Salvation, without Satisfaction."

GILBERT (SIR HUMPHREY), an able navigator, related to Sir Walter Raleigh. He gained a confiderable reputation in Ireland, in the military capacity; and was one of thofe gallant adventurers who improved our navigation. He took poffeffion of Newfoundland, in the name of Queen Elizabeth; but was unfuccefsful in his attempt to fettle a colony on the continent of America. He wrote a book to prove the exiftence of a N. W. paffage to the Indies. Died 1583.

GILBERT (JEFFERY), barrifter at law, and afterwards lord chief baron of the exchequer, firft in Ireland, and then in England. This gentleman (among other things) was author of "an Abridgment of Mr. Lock's Effay on Human Underftanding," publifhed in 1750, by Dr. Dodd, and of an excellent tranflation of the 12th ode of the fecond book of Horace, printed (without a name) in "the Wits Horace," p. 67.

GILDAS. He is the moft antient Britifh writer extant, for his famous epiftle was written 560, about twelve years after the Romans evacuated this ifland. Bifhop Nicolfon calls him a monk of Bangor, which is denied by Lloyd and Stillingfleet. If he ever was a monk of Bangor, it muft have been after he wrote his epiftle; for he tells us that he refided near the wall of Severus, which, running from fea to fea, divided the Caledonians from the Britains. He even tells us that he faw the Caledonians pull down part of the wall; and fays, that they had more hair on their faces, than cloaths on their bodies. It is therefore plain, that he was a native of Valentia, which includes, at prefent, none of the North of England or South of Scotland. His epiftle has been printed in latin, and fome time in the reign of Charles II. tranflated into englifh.

GILDON (CHARLES), an englifh critic, was born at Gillingham, in Dorfetfhire, about 1666: his father was a member of Gray's-inn, and had fuffered much by his adherence to Charles I. Gildon had the firft rudiments of his education at the place of his nativity, whence his relations, who were Roman-catholics, fent him to the Englifh-college, at Douay, to make him a prieft: but, after fome time, he found his inclination tending another way. He returned to England in 1685; and, as foon as he was grown up, he came to London. Here he spent the greatest part of his paternal estate; and married a woman with no fortune, at the age of 23. During the reign of James II. he employed himself in reading the controversies of those times; and declared, that it cost him above seven years study, before he could overcome the prejudices of his education. Necessity constraining him, as he himself owns, he made his first attempt in the dramatic way in his 23d year; and, at length, produced three plays, none of which, however, had any success. He was the author of many other things, as Letters, Essays, Poems, &c. and, as he affected criticism above all things, published several works in that way. Among the rest, were " The Complete Art of Poetry," and " The Laws of Poetry, as laid down by the Duke of Buckingham in his Essay on Poetry, by the Earl of Roscommon in his Essay on Translated Verse, and by Lord Landsdown on unnatural Flights in Poetry, illustrated and explained." He was also an author in the religious or philosophical way, and published in 1705, " The Deist's Manual, or Rational Enquiry into the Christian Religion, with some animadversions on Hobbes, Spinoza, The Oracles of Reason, Second Thoughts, &c." as he had in 1695, published, " The Miscellaneous Works of Charles Blount, Esq. to which he had prefixed the Life of that Gentleman, together with an Account and Vindication of his Death. By these publications we may be convinced that, however difficult he might find it, he certainly got rid of his popish prejudices. Gildon had been concerned in some plot against Pope, which procured him a place in the Dunciad:

" Ah Dennis! Gildon ah! what ill stars'd rage
Divides a friendship, long confirm'd by age?
Blockheads with reason wicked wits abhor,
But fool with fool is barbarous civil war, &c."

GILES (JOHN), in latin, JOANNES ÆGIDIUS, or de SANCHO ÆGIDIO, was a native of St. Alban's, and flourished in the 13th century. He received his education at Paris, and was made physician in ordinary to Philip, king of France, and professor of medicine in the universities of Paris and Montpellier. According to the custom of those times, he was made a doctor of divinity, and was the first Englishman upon record who entered among the Dominicans, with whom he became a noted preacher. In his old age he was famous for his divinity, lectures, and physical receipts.

GILL (ALEXANDER), born 1564, and admitted at Christ's College, Oxford, in 1608. In 1608 he was made headmaster of St. Paul's school, and trained up many persons of note,

note, both in church and state, till the time of his death, in 1635. He published only two or three theological tracts, and lies buried in Mercer's-Chapel.

GILL (ALEXANDER), his son, born in London, and admitted of Trinity College, Oxford, 1612. He served his father and Thomas Farnaby, in the quality of an usher; and, after many changes, rambles, and some imprisonment, he succeeded his father in St. Paul's School, September 1635, whence he removed in 1640, and kept a private school in Aldersgate-street till his death, which happened in 1642. Wood, who censures his conduct, accounts him one of our best latin poets, Ath. Ox.

GILL (JOHN, D. D.). He was born at Kettering, in Northamptonshire, November 19, 1697. His parents were not in affluent circumstances; but they supported themselves above want. This, their son, was put to a Grammar-school, at a very early age, and made such an amazing progress in latin and greek as is seldom to be found in one so young as he was. He afterwards studied logic, rhetoric, metaphysics, theology, and all the other branches of learning; and in 1716 was admitted member of the Anabaptist Church, at Kettering; and in 1718 he accepted of the Meeting at Higham-Ferrars, in Northamptonshire, where he was much followed. In 1721 he accepted of an invitation to be minister of the Meeting at Horslydown, whence he removed to Tooley-street, where he officiated as pastor, till his death, October 13, 1771, in the 74th year of his age. He was a learned orientalist, a rigid calvinist, a voluminous writer, and endowed with an excellent memory, which he improved by extensive reading and study.

GILLESPIE (GEORGE), a minister of the church of Scotland, and a staunch defender of the presbyterian rights, was a noted preacher before the year 1638, to the time of his death, December 17, 1648. He was one of those four divines who were sent as commissioners from the church of Scotland to the Westminster assembly in 1643, to forward the covenanted work of reformation. His works are, Aaron's Rod blossoming; Miscellany Questions, first printed 1649; English Popish Ceremonies, &c.

GILPIN (BERNARD), an english divine, was descended from an ancient family in Westmorland, and born at Kentmire, in that county, 1517. After passing through a grammar-school, he was sent to Oxford, and admitted a scholar on the foundation of Queen's college, in 1533. Here he diligently pursued his studies, and made himself master of Erasmus's works, which were then in vogue; at the same time,

time, cultivating logic and philosophy, he became a diſtinguiſhed diſputant in the ſchools. To theſe acquiſitions he added a ſingular knowledge in the greek and hebrew tongues; in which laſt he was inſtructed by Thomas Neale, then fellow of New college; but who afterwards became hebrew profeſſor. March 1541, he proceeded M.A. having before taken his degree of B. A. at the uſual time. He was now alſo choſen fellow of his college and was much beloved for ſweetneſs of diſpoſition and unaffected ſincerity of manners. At the ſame time, his eminence for learning was ſuch, that he was choſen one of the firſt maſters to ſupply Chriſt's-church college, after the completing of its foundation by Henry VIII.

As he had been bred in the Roman-catholic religion, ſo he had hitherto continued ſteady to that church; and in defence thereof, while he reſided at Oxford, held a diſputation againſt Hooper, afterwards biſhop of Worceſter, and martyr for the proteſtant faith. But in Edward VI's time being prevailed upon to diſpute with Peter Martyr, againſt ſome poſitions maintained by him in his divinity-lecture, at Oxford; and, being ſtaggered a little therein, he began more ſeriouſly to read over the ſcriptures and writings of the fathers, expecting to confirm himſelf in his opinions by ſtronger arguments: on the contrary, the reſult of his enquiries was the cooling of his zeal for popery, and kindling a deſire toward the new religion: in which temper he applied for farther inſtruction to Tonſtall, biſhop of Durham, who was his mother's uncle. That prelate told him, that in the matter of tranſubſtantiation Pope Innocent III. had done unadviſedly in making it an article of faith; and confeſſed, that the pope had alſo committed a great fault, in taking no better care than he had done in the buſineſs of indulgences and other things. After this, he conſulted other private friends; and at the ſame time, continuing his diligence in ſearching the ſcriptures and the fathers, he began to obſerve many abuſes and enormities in popery, and to think reformation neceſſary.

Whilſt he was going on in this courſe, having taken orders, he was over-ruled by his friends to accept, againſt his will, the vicarage of Norton, in the dioceſe of Durham. This was in 1552; and being, a grant from Edward VI. before he went to reſide, he was appointed to preach before his majeſty, who was then at Greenwich. His ſermon was greatly approved and recommended him to the notice of ſir Francis Ruſſel and ſir Robert Dudley, afterwards earls of Bedford and Leiceſter, and to ſecretary Cecil, afterwards lord-treaſurer Burleigh, who obtained for him the king's licence for a general preacher, during his majeſty's life; which, however

however, happened to be not much above half a year after. Thus honoured, he repaired to his parish, but he soon grew uneasy here: for, however resolved he was against popery, he was scarcely settled in some of his religious opinions; and he found the country overspread with popish doctrines, the errors of which he was unable to oppose. In this unhappy state he applied to bishop Tonstall, then in the Tower; who advised him to provide a trusty curate for his parish, and spend a year or two abroad in conversing with some of the most eminent professors on both sides the question. The proposal to travel was quite agreeable to Gilpin; who, after resigning his living, from a scruple of conscience, set out for London, to receive the bishop's last orders, and embark. The bishop promised to support him abroad; and at parting put into his hands a treatise upon the Eucharist, which the times not suiting to be printed here, he desired might be done under his inspection at Paris [A]. With this charge he embarked for Holland, and on landing went immediately to Malin, to visit his brother George, who was then a student there. After a few weeks he went to Louvain, which he fixed on for his residence; proposing to make occasional excursions to Antwerp, Ghent, Brussels, and other places in the Netherlands. Louvain was then a chief place for students in divinity, some of the most eminent divines, on both sides of the question, residing there; and the most important topics of religion were discussed with great freedom. Gilpin made the best use of his time, and soon began to have juster notions of the doctrine of the reformed, when he was alarmed with the news of Edward's death, and the accession of Mary to the throne.

However, this bad news came attended with an agreeable account of Bishop Tonstall's release from the Tower, and re-establishment in his bishopric: but the consequence of this was not so agreeable; for afterwards he received a letter from his brother George, inviting him to Antwerp upon a matter of great importance. Coming thither, he found that the business was a request of the bishop's, to persuade him to accept of a living of considerable value, which was become vacant in his diocese. George used all his endeavours for the purpose, but in vain [B]: Bernard was too well pleased with his

[A] It was written in latin with this title " De veritate corporis & sanguinis Christi Domini in Eucharistia," and contained a defence of the real presence in the gross sense; an opinion which Gilpin, who had a great reverence for his uncle, seems to have imbibed from him, and to have retained ever after.

[B] He succeeded better in a request made afterwards, at the instance of the earls of Bedford and Leicester, to give him in writing an exact account of the progress of his change from the romish religion; which was executed, and is printed in his life by bishop Carleton. George was now at the English court, but employed as a minister from thence in the Low Courtries, where he usually resided.

present

present situation to think of a change, and excused himself to his patron on the same scruple of conscience as before, against taking the profits while another did the duty. "And whereas," concludes he, "I know well your lordship is careful how I should live, if God should call your lordship, being now aged, I desire you, let not that trouble you. For, if I had no other shift, I know that I could get a lectureship, either in this university, or at least in some abbey, where I should not lose my time; and this kind of life, if God be pleased, I desire before any benefice [c]." This letter was dated Nov. 22, 1554. Meanwhile, he was greatly affected with the misfortune of the English from queen Mary's persecution; and not a little pleased to find, that, though unable personally to assist them, yet his large acquaintance in the country furnished him with the means of serving many of them by recommendations. He had been now two years in Flanders, and made himself master of the controversy, as it was there handled. He left Louvain, therefore, and went to Paris, where his first care was to print his patron's book [D]; which he performed entirely to his lordship's satisfaction this same year 1554, and received his thanks for it. Here popery became quite his aversion; for he now saw more of its superstition and craft; the former among the people, the latter among the priests. In this city he met with his old hebrew-master, Neal, of New-college. Neal had always been a favourer of popery, and was now a bigot to it; and he tried his strength upon his quondam pupil, but found him above his match. This was the same Neal who was afterwards chaplain to bishop Bonner, and distinguished himself by vouching the silly story of the Nag's-head consecration.

After three years absence, Gilpin returned to England in 1556, a little before the death of queen Mary; and soon after received from his uncle the archdeaconry of Durham, to which the rectory of Easington was annexed. He immediately repaired to his parish, where, notwithstanding the persecution, which was then in its height, he preached boldly against the vices, errors, and corruptions of the times, especially in the clergy [E]. This was infallibly to draw vengeance upon himself; and, accordingly, a charge consisting of

[c] He was much delighted with his present situation, which was near to a monastery of Minorite friars; and had the use of an excellent library of theirs, and enjoyed the company of the best scholars; nor, says he, was I ever more desirous to learn.

[D] For this purpose, he took lodgings at the house of Vascosan, an eminent printer, to whom he had been recommended by his friends in the Netherlands

[E] He often preached against pluralities, and non-residence; upon which the popish clergy cried out, that all who broached that doctrine, would quickly become heretics; and he was accordingly accused of heresy.

thirteen

thirteen articles was drawn up againſt him, and preſented in form to the biſhop; but Tonſtall found a method of diſmiſſing the cauſe in ſuch a manner as to protect his nephew without endangering himſelf. The malice of his enemies could not, however, reſt: his character at leaſt, was in their power; and they created him ſo much trouble, that, not able to undergo the fatigue of both his places, he begged leave of the biſhop to reſign either the archdeaconry or his pariſh; and the rich living of Houghton le Spring becoming vacant, the biſhop preſented him to it, on his reſignation of the archdeaconry. He now lived retired, and gave no immediate offence to the clergy; the experience he had of their temper made him more cautious not to provoke them. But all his caution availed nothing. He was ſoon formally accuſed to the biſhop a ſecond time; and again protected by him. But his enemies were not yet quieted: enraged at this ſecond defeat, they accuſed him to Bonner, biſhop of London; and here they went the right way to work. Bonner was juſt the reverſe of Tonſtall, and immediately gave orders to apprehend him. Gilpin had no ſooner notice of it, but, being no ſtranger to this prelate's BURNING zeal, he prepared for martyrdom; and commanding his houſe-ſteward to provide him a long garment, that he might go the more comely to the ſtake, he ſet out for London. It is ſaid, that he happened to break his leg in the journey, which delayed him; however that be, it is certain, that the news of queen Mary's death met him on the road, which proved his delivery.

Upon his return to Houghton, he was received by his pariſhioners with the ſincereſt joy; and, though he ſoon after loſt his patron, biſhop Tonſtall, yet he quickly experienced, that worth like his could never be left friendleſs. When the popiſh biſhops were deprived, the earl of Bedford recommended him to the queen for the biſhopric of Carliſle; and took care that a congé d'elire ſhould be ſent down to the dean and chapter for that purpoſe: but Mr. Gilpin declined this promotion. He refuſed alſo an offer the following year, which ſeems to have been more to his taſte. Queen Elizabeth, at her acceſſion to the throne, had procured one Dr. Francis, a proteſtant phyſician, to be choſen provoſt of Queen's college. Francis was received with great reluctance by the fellows, who were attached to popery; and, finding his ſituation uneaſy among them, determined to reſign, and made an offer of the place to Gilpin. But though he loved the univerſity well, and this college in particular, of which he had been fellow, and was aſſured, likewiſe, that the preſent fellows had a very great eſteem for him; yet all was not able to move him from his parſonage. Here he ſpent the remainder of his days in hoſpitality, hofpitality, charity, and all good works. The fame of his hofpitality, in particular, was fo great and fo extenfive, tha lord Burleigh, returning from Scotland, made a vifit to Houghton; and, though he came without any previous notice, yet he was received with his whole retinue, and treated in fo affluent and generous a manner, that he would often afterwards fay, he could hardly have expected more at Lambeth. Towards the latter part of his life, his health was much impaired; and there happened a very unfortunate affair, which entirely deftroyed it. As he was croffing the market-place, at Durham, an ox ran at him, and threw him down with fuch violence, that it was imagined he had received his death's wound. He lay long confined; and though he got abroad again, he never recovered even the little ftrength he had before, and continued lame as long as he lived. He died, 1583, in his 66th year.

GILPIN (RICHARD). He was a native of Cumberland, and educated in Queen's-college, Oxford, whence he took the degree of M. D. but afterwards entered into holy orders, and became minifter of Greyftock, in his own county; but preached with great applaufe in London, and in many other parts of the kingdom; till he was filenced for refufing to comply with the act of uniformity, 1662. He afterwards practifed phyfic in the north of England, particularly at Newcaftle, where he was greatly efteemed by all that knew him, both as a phyfician and a divine—He died 1657. He was the author of feveral treatifes; but his difcourfe on "Satan's Temptations" is moft efteemed.

GIOLITO (DEL FARRARI), a celebrated printer of the fixteenth century. He printed at Venice, and was eminent more for the elegance of his types and qualities of his paper, than the correctnefs of his works He was ennobled by Charles the 5th, and died in 1547, leaving two fons who were printers alfo.

GIOIA (FLAVIO) is deferving of particular notice, becaufe the learned world are generally agreed in afcribing to him the invention of the compafs. He was a Neapolitan, and born about the year 1300. At that time the fovereigns of Naples were younger branches of the royal family of France; and, to mark the circumftance of this invention of the compafs originating with a fubject of Naples, Gioia diftinguifhed the north with a fleur de lis, a particularity which has been adopted by all nations, to whom the ufe of this inftrument is known. Some have pretended that the ancients were not ignorant of the power of the magnet; but it is certain that Pliny, who often fpeaks of the load-ftone, knew nothing of its appropriate direction to the pole. Some authors alfo

GIORGIONE.

also have conferred the honour of this important discovery on the Chinese, and it has by Dr. Wallis been ascribed to the English. However this may be, the territory of Principato, which is part of the kingdom of Naples, and in which place Gioia was born, bears a compass for its arms. If it be only an improvement of an invention, though but partially known, which may be imputed to Gioia, he is without dispute entitled to a distinguished place in the rank of those, who have contributed to the benefit of society.

GIORGIONE, so called from his noble and comely aspect, was an illustrious painter, and born at Castel Franco in Trevisano, a province in the state of Venice, in 1478. Though he was but of an indifferent parentage, yet he had a fine genius and a large soul. He was bred up in Venice, and first applied himself to music; in which he had so excellent a talent, that he became famous for singing and playing on the lute. After this, he devoted himself to painting, and received his first instructions from Giovanni Bellino; but having afterwards studied the works of Leonardo da Vinci, he soon arrived at a manner of painting, superior to them both. He designed with greater freedom, coloured with more strength and beauty, gave a better relievo, more life, and a nobler spirit to his figures; and was the first among the Lombards, who found out the admirable effects of strong lights and shadows. Titian was extremely pleased with his bold and terrible gusto; and intending to make his advantage of it, frequently visited him, under pretence of keeping up the friendship they had contracted at their master Bellino's: but Giorgione, growing jealous of his intentions, contrived to forbid him his house. Upon this, Titian became his rival, and was so careful in copying the life, that he excelled Giorgione in discovering the delicacies of nature. Titian thought, that Giorgione had passed the bounds of truth; and though he imitated in some things the boldness of his colouring, yet he tamed, as one may say, the fierceness of his colours. He tempered them by the variety of tints, that he might make his objects the more natural; but, notwithstanding his efforts to outdo his rival, Giorgione still maintained his character for the greatness of his gusto; and it is allowed, that if Titian has made several painters good colourists, Giorgione first shewed them the way to be so, Giorgione excelled both in history and portraits. The greatest of his performances is at Venice, on the front of the house wherein the german merchants have their meetings, on the side which looks towards the grand canal. He did this piece of painting in competition with Titian, who painted another side of that building; but both these pieces being

being almost entirely ruined by age, it is difficult to form any judgement of them. His most valuable piece in oil is that of our Saviour carrying his crofs, now in the church of San Royo at Venice; where it is held in wonderful esteem and veneration. He worked much out of Venice, as at Castel Franco and Trivisano; and many of his pieces were bought up and carried to foreign parts, to shew that Tuscany alone had not the prize of painting. Some sculptors in his time took occasion to praise sculpture beyond painting, because one might walk round a piece of sculpture, and view it on all sides; whereas a piece of painting could never represent but one side of a body at once. Giorgione hearing this said, that they were extremely mistaken; for that he would undertake to do a piece of painting, which should shew the fore and hind parts, and the two sides, without putting spectators to the trouble of going round it: and he brought it about thus. He drew the picture of a young man naked, shewing his back and shoulders, with a fountain of clear water at his feet, in which there appeared by reflection all his fore parts: on the left side of him, he placed a bright shining armour, which he seemed to have put off, and in the lustre of that all the left side was seen in profile: and on his right he placed a large looking-glass, which reflected his right side to view.

It being too common for men who excel in the fine arts to be subject to the amorous passion, Giorgione was not exempt from it. He fell extremely in love with a young beauty at Venice, who was no less charmed with him, and submitted to be his mistress. She fell ill with the plague: but, not suspecting it to be so, admitted Giorgione to her bed; where, the infection seizing him, they both died in 1511, he being no more than 33.

GIOSEPPINO, an eminent painter, so called by contraction from Giofeppe d' Arpino, a town of Naples, where he was born in 1560. His father was an ordinary painter, who did business for the country people: but he, being carried to Rome very young, and employed by some painters then at work in the Vatican to grind their colours, foon made himself master of the elements of design, and by degrees grew very famous. Having a great deal of wit and genius, he became a favourite with the popes and cardinals. He had particular respect shewn him by Gregory XIII. and was so well received by the french king Lewis XII. that he made him a knight of the order of St. Michael. He has the character of a florid invention, ready hand, and much spirit, in all his works; but yet, having no sure foundation in the study of nature, or the rules of art, and building only upon fantastical ideas formed in his own head, he has run into a

multitude

multitude of errors, and been guilty of many extravagances neceffarily attending thofe, who have no better guides than their own capricious fancy. His battles in the Capitol are moft efteemed of all his pieces. He died at Rome in 1640, aged 80.

GIOTTO, an eminent painter, fculptor, and architect, was born in 1726, at a village near Florence, of parents who were plain country people. When a boy, he was fent out to keep fheep in the fields; and, having a natural inclination for defign, he ufed to amufe himfelf with drawing his flock after the life upon fand, in the beft manner he could. Cimabue travelling once that way found him at this work, and thence conceived fo good an opinion of his genius for painting, that he prevailed with his father to let him go to Florence, and be brought up under him. He had not applied himfelf long to defigning, before he began to fhake off the ftiffnefs of the grecian mafters. He endeavoured to give a finer air to his heads, and more of nature to his colouring, with proper actions to his figures. He attempted likewife to draw after the life, and to exprefs the different paffions of the mind; but could not come up to the livelinefs of the eyes, the tendernefs of the flefh, or the ftrength of the mufcles in naked figures. What he did, however, had not been done in 200 years before, at leaft with any fkill equal to his. Giotto's reputation was fo far extended, that Pope Benedict IX. fent a gentleman of his court into Tufcany, to fee what fort of a man he was; and withal to bring him a defign from each of the florentine painters, being defirous to have fome notion of their fkill and capacities. When he came to Giotto, he told him of the Pope's intentions, which were to employ him in St. Peter's church at Rome; and defired him to fend fome piece of defign by him to his holinefs. Giotto, who was a pleafant ready man, took a fheet of white paper, and fetting his arm clofe to his hip to keep it fteady, he drew with one ftroke of his pencil a circle fo round and fo equal, that " round as Giotto's O" afterwards became proverbial. Then, prefenting it to the gentleman, he told him fmiling, that " there was a piece of defign, which he might carry to his holinefs." The man replied, " I afk for a defign:" Giotto anfwered, " Go, fir, I tell you his holinefs afks nothing elfe of me." The Pope, who underftood fomething of painting, eafily comprehended by this, how much Giotto in ftrength of defign excelled all the other painters of his time; and accordingly fent for him to Rome. Here he painted a great many things, and amongft the reft a fhip of mofaic work, which is over the three gates of the portico, in the entrance to St. Peter's church: which very celebrated piece is

known

known to all painters by the name of Giotto's veffel. Pope Benedict was fucceeded by Clement V. who transferred the papal court to Avignon; whither, likewife, Giotto was obliged to go. After fome ftay there, having perfectly fatisfied the Pope by many fine fpecimens of his art, he was largely rewarded, and returned to Florence full of riches and honour in 1316. He was foon invited to Padua, where he painted a new-built chapel very curiously; thence he went to Verona, and then to Ferrara. At the fame time the poet Dante, hearing that Giotto was at Ferrara, and being himfelf then in exile at Ravenna, got him over to Ravenna, where he wrought feveral things; and perhaps it might be here that he drew Dante's picture, though the friendfhip between the poet and the painter was previous to this. In 1322, he was again invited abroad by Caftruccio Caftrucani, lord of Luca; and, after that, by Robert king of Naples. Giotto painted many things at Naples, and chiefly the chapel, where the king was fo pleafed with him, that he ufed very often to go and fit by him while he was at work: for Giotto was a man of pleafant converfation and wit, as well as ready with his pencil. One day, it being very hot, the king faid to him, " If I were you, Giotto, I would leave off working this hot weather;" " and fo would I, Sir," fays Giotto, " if I were you." He returned from Naples to Rome, and from Rome to Florence, leaving monuments of his art in almoft every place through which he paffed. The number of his works is fo great, that it would be endlefs to recount them There is a picture of his in one of the churches of Florence, reprefenting the death of the bleffed Virgin, with the apoftles about her: the attitudes of which ftory, Michael Angelo ufed to fay, could not be better defigned. Giotto, however, did not confine his genius altogether to painting: he was, as we have faid, a fculptor and architect. In 1327, he formed the defign of a magnificent and beauteous monument for Guido Tarlati, bifhop of Arezzo, who had been the head of the Ghibeline faction in Tufcany: and in 1334 he undertook the famous tower of Sancta Maria del Fiore; for which work, though it was not finifhed, he was made a citizen of Florence, and endowed with a confiderable yearly penfion.

His death happened in 1336: and the city of Florence erected a marble ftatue over his tomb. He had the efteem and friendfhip of moft of the excellent men of the age in which he lived; and among the reft of Dante and Petrarch. He drew, as we have faid, the picture of the former; and the latter mentions him in his will, and in one of his familiar epiftles.

GIRALDI

GIRALDI (LILIO GREGORIO), in latin Gyraldus, an ingenious critic, and one of the moſt learned men modern Italy has produced, was born at Ferrara in 1479, of an ancient and reputable family. He learned the latin tongue and polite literature under Baptiſt Guarini; and afterwards the greek at Milan under Demetrius Chalcondyles. He retired into the neighbourhood of Albert Picus, prince of Carpi, and of John Francis Picus, prince of Mirandula; and, having by their means acceſs to a large and well-furniſhed library, he applied himſelf intenſely to ſtudy. He afterwards went to Modena, and thence to Rome: in which city he was, when it was plundered by the ſoldiers of Charles V. in 1527. He loſt his all in the general ruin; and, what was worſe even than this, he loſt ſoon after his patron cardinal Rangoni, with whom he had lived ſome time. He was then obliged to ſhelter himſelf in the houſe of the prince of Mirandula, not the great Picus, but a relation of the ſame name; but he had the misfortune to loſe this friend and protector in 1533, who was aſſaſſinated by a cabal, of which his nephew was the head. Giraldi was at that time ſo afflicted with the gout, that he had great difficulty to ſave himſelf from the hands of the conſpirators; after having loſt all which he had acquired, ſince the ſacking of Rome. He then returned to his own country, and lived at Ferrara. The gout tormented him ſo for the ſix or ſeven laſt years of his life, that, as he ſpeaks of himſelf, he might be ſaid rather to breathe than to live. He was ſuch a cripple in his hands and feet, that he was incapable of doing the common neceſſaries of life, or even moving himſelf. To this dreadful ſtate was added extreme poverty. All this did not ſo affect him, but that he made what uſe he could of intervals of eaſe, to read, and even write: and many of his books were compoſed in thoſe intervals. He died at length of this dreadful malady in 1552; and was interred in the cathedral of Ferrara, where the following epitaph, compoſed by himſelf, was inſcribed upon his tomb:

D. M.
Quid hoſpes adſtas? tymbion
Vides Gyraldi Lilij,
Fortunæ utriuſque paginam
Qui pertulit, ſed peſſima
Eſt uſus altera, nihil
Opis ferente Apolline.
Nil ſcire refert amplius
Tua aut ſua; in tuam rem abi.

His works conſiſt of 17 productions, which were firſt printed ſeparately; but afterwards collected and publiſhed in 2 vols.

vols. folio, at Basil 1580, and at Leyden 1696. The moſt valued pieces among them are, Hiſtoria de Deis Gentium,—Hiſtoriæ Poetarum tam Græcorum quam Latinorum Dialogi decem,—and, Dialogi duo de Poetis noſtrorum. The firſt of theſe books is one of the laſt he compoſed, and full of the profoundeſt erudition. The other two, which make up the hiſtory of the ancient and modern poets, are written with great exactneſs and judgement. Voſſius ſpeaks highly of this work declaring, that the author has ſhewn great judgement and learning, as well as induſtry, in compoſing it; and obſerves, that though his profeſſed deſign is to collect memoirs concerning their perſons, characters, and writings in general, yet he has occaſionally interſperſed many things, regarding the very art of poetry, which may be uſeful to thoſe who intend more particularly to cultivate it. Joſeph Scaliger, indeed, would perſuade us, that nothing can be more contemptible than the judgement he paſſes on the poets he treats of: but as men who ſpeak from prejudice or paſſion, as Scaliger often did, are apt to contradict themſelves, ſo it is remarkable, that in another place this ſame Scaliger allows all the works of Giraldus to be very good, and that no man knew better how to temper learning with judgement.

There is a work alſo by Giraldus, de annis & menſibus, cæteriſque temporibus partibus, una cum Kalendario Romano & Græco, written with a view to the reformation of the kalendar, which was afterwards effected by Pope Gregory XIII. about 1582. There are likewiſe among his works a few poems, the principal of which is intituled, Epiſtola in qua agitur de incommodis, quæ in direptione Urbana paſſus eſt; ubi item eſt quaſi catalogus ſuorum, amicorum Poetarum, & defleatur interitus Herculis Cardinalis Rangonis." This poem is annexed to the florentine edition of the Two Dialogues concerning his contemporary Poets; and is curious and intereſting, as it contains a kind of literary hiſtory of that time.

The higheſt eulogies have been beſtowed upon Giraldus by authors of the firſt name. Cauſaubon calls him, vir ſolide doctus, & in ſcribendo accuratus, a man ſolidly learned and an accurate writer. Thuanus ſays, that " he was excellently ſkilled in the greek and latin tongues, in polite literature, and in antiquity, which he has illuſtrated in ſeveral works: and that, though highly deſerving a better fate, he ſtruggled all his life with ill health and ill fortune."

GIRALDI (JOHN BAPTIST CINTIO), an italian poet, of the ſame family with Lilio Giraldi, was born at Ferrara in 1504. His father, being a man of letters, took great care of his education; and placed him under Cælio Calcagnini, to

ſtudy

study the languages and philosophy. He made an uncommon progress, and then applied himself to the study of physic: in which faculty he was afterwards a doctor. He must have been a very surprizing person; for he was pitched upon, at 21 years of age, to read public lectures at Ferrara upon physic and polite literature. In 1542, the duke of Ferrara made him his secretary; which office he held till the death of that prince in 1558. He was continued in it by his successor: but envy having done him some ill offices with his master, he was obliged to quit the court. He left the city at the same time, and removed with his family to Mondovi in Piedmont; where he taught the belles lettres publicly for three years. Then he went to Turin; but the air there not agreeing with his constitution, he accepted the professorship of rhetoric at Pavia: which the senate of Milan, hearing of his being about to remove, and apprized of his great merit, freely offered him. This post he filled with great repute; and afterwards obtained a place in the academy of that town. It was here he got the name of Cintio, which he retained ever after, and put in the title-page of his books. The gout, which was hereditary in his family, beginning to attack him severely, he returned to Ferrara; thinking that his native air might afford him relief. But he was hardly settled there, when he grew extremely ill; and, after languishing about three months, died in 1573.

His works are all written in Italian, except some orations, spoken upon extraordinary occasions, in Latin. They consist chiefly of tragedies: a collection of which was published at Venice 1583, in 8vo. by his son Celso Giraldi; who, in his dedication to the duke of Ferrara, takes occasion to observe, that he was the youngest of five sons, and the only one who survived his father. There are also some prose works of Giraldi: one particularly upon comedy, tragedy, and other kinds of poetry, which was printed at Venice by himself in 1554, 4to. As little as this Giraldi seems to be known, some make no scruple to rank him among the best tragic writers that Italy has produced.

GIRALDUS (SILVESTER), a very learned and eloquent man in his time, was born of noble parents, at the castle of Mainarpir, near Pembroke in South Wales, in 1145. Discovering an early inclination for the service of the church, his uncle, who was bishop of St. David's, took care of his education. When he had made a proper advancement, he was sent to France, and studied theology at Paris under Peter Comestor; for, theology, it seems, was then in its most flourishing state in that city. Having finished his own pursuits, he thought himself capable of reading lectures to others;

and accordingly did so, upon the belles lettres and rhetoric in the english college there. He returned to England about 1172, and brought with him so high a reputation for learning and so much zeal for the church, that Richard, abp. of Canterbury, and the Pope's legate, pitched upon him in 1175, to collect some neglected tithes, and reform some abuses, in the principality of Wales. He was invested with an extraordinary commission; and he exerted himself so vigorously, that, in the course of his progress, he suspended an archdeacon for keeping a concubine. In 1176, the bishop of St David's dying, he was named with three others, to be presented to the king, but declined it. The same year he went to Paris, in order to study the canon law. He spent three years upon it; and with so much success, that he was offered the professorship in the university there: but this he refused to accept, designing to go to Bologna to perfect himself in that science. He returned to England in 1180; and, in 1184, became known to Hen. II. who, perceiving his great merit and abilities, sent him the year after, as secretary, with his son prince John into Ireland. John returned with his army the same year; but Giraldus stayed some months longer in Ireland, to search for antiquities, and to make a typographical description of the isle; for which purpose he travelled all over it, and did not pass over to Wales till 1186. He afterwards spent some time in composing his own memoirs, and then went to Oxford; where he employed three whole days in reciting them publicly. The bishopric of St. David's becoming vacant in 1198, he was elected a second time; but a dispute arose about it, for the settling of which he himself went to Rome in 1200. He did not succeed, having a rich competitor to vie with: " erant tum enim omnia venalia Romæ." He lived about 70 years, and was the author of many works; some of which have been printed, some remain in MS. He was a great enemy to the monks, whom he has treated very severely; and it was a common saying with him, " à monachorum malitia libera nos, Domine," from the malice of monks, good Lord, deliver us. Tanner makes it almost a matter of wonder, that a man in such a dark and ignorant age, could be so universally learned, and withal so eloquent, as Giraldus. However, he had other qualities in common with his neighbours; for he was credulous and superstitious in the highest degree; and there were no dreams or visions so senseless and extravagant, which he did not believe to be divine revelations.

The only works of his, which a reader can have any curiosity to see, are his Topographia Hiberniæ, sive de Mirabilibus & Habitatoribus Hiberniæ libri tres, ad Henricum II.—Expugnatio Hiberniæ, sive Historia Vaticinalis de expugnata ab Anglis

Angliis Hibernia." "Itinerarium Cambriæ." "Defcriptio Cambriæ." Thefe are all to be found in a collection publifhed by Camden at Frankfort, 1602, in folio, under the title of "Anglica, Normannica, & Cambrica, a veteribus fcripta." His three books, "De Rebus à fe geftis," together with other pieces, are publifhed by Wharton, in the fecond volume of "Anglia Sacra:" and in the Lambeth and Cotton libraries there are ftill extant from MS. as among others, "Liber Carminum & Epigrammatum," and "De Principis Inftructione Diftinctiones tres;" which laft, Cave tells us, is a long work, but well deferving to be read.

GIRALDUS (CAMBRENSIS). This antient Britifh writer lived in the reign of Henry II. and was nearly cotemporary with Geoffery of Monmouth: He wrote the Hiftory of the World in monkifh latin, but his Account of Britifh Affairs is nearly copied from Geoffery. There are fome things, however, in his hiftory relating to ecclefiaftical affairs, which are extremely valuable; for he gives us an account of the ftate of the monks in his time, from which we learn, that, although they were then extremely ignorant, yet they were more fimple in their manners than thofe who lived in latter times. He died at St. David's, about the latter end of the twelfth century.

GIRON (D. PIERRE) duke of Offona, a noble Spaniard, whom we are induced to mention principally on this account: When he was viceroy of Naples, the famous confpiracy againft Venice was difcovered by means of Jaffier one of the confpirators, and which the duke of Offona fomented and affifted. This has formed a plot for one of the moft popular tragedies on the Englifh ftage. The duke of Offona, a proud, imperious, and intriguing character, died in prifon in 1624, aged 49.

GIRY (LOUIS), a native of Paris, and one of the firft members of the French Academy; he was a man of great integrity and of refpectable accomplifhments. He tranflated 'the Apology' of Tertullian; the Sacred Hiftory of Sulpicius Severus; St. Auguftin's Tract de Civitate Dei, with fome portion of Cicero's works. He died at the age of 70, in the year 1665.

GISELINUS, a native of Bruges, born in 1743, died in 1551, publifhed a correct and good edition of Prudentius, at Antwerp. He was a phyfician, and affifted in the work above mentioned, by Pulmannus.

GLAIN (N. SAINT), a name that would not be worth preferving, but for the fingularity of the anecdote which happens to be connected with it. This perfon was born at Limoges about 1620, and retired into Holland for the fake thought he should do a service to the public if he made it more accessible. With this view he translated into French the famous "Tractatus Theologico-Politicus" of Spinoza; and published it, at first, under the title of "The Key of the Sanctuary." The work making a great noise, he publishe it a second time, in order to spread it farther, with the title of "A Treatise on the superstitious Ceremonies of the Jews." And, lastly, in a third publication, he intituled it, "Curious and disinterested Reflections upon Points the most important to Salvation." This was printed at Cologne in 1678, 12mo.

GLANDORP (MATTHIAS), a German physician, was born in 1595, at Cologne, where his father was a surgeon. His first application to letters was at Bremen; whence he returned to Cologne, and devoted himself to philosophy, physic, and chirurgery. He studied four years under Peter Holtzem, who was the elector's physician, and professor in this city; and he learned the practical part of surgery from his father. To perfect himself in these sciences, he went afterwards into Italy, and made some stay at Padua; where he greatly benefited himself by attending the lectures of Jerome Fabricius ab Aquapendente, Adrian Spigelius, and Sanctorius. He was here made M.D. After having visited the principal towns of Italy, he returned to his country in 1618, and settled at Bremen; where he practised physic and chirurgery with so much success, that the archbishop of this place made him his physician in 1628. He was also made physician of the republic of Bremen. The time of his death is not precisely known; but the dedication of his last work is dated Oct. 8, 1652, so that he could not be dead before, as some Journalists have asserted, though it is probable he was soon after. He published, at Bremen, " Speculum Chirurgorum," in 1619;" "Methodus Medendæ Paronychiæ," in 1133; "Tractatus de Polypo Narium affectu gravissimo," in 1628; and "Gazophylacium Polypusium Fonticulorum & Setonum Refaratum,"in 1633 These four pieces were collected and published, with emendations, under the title of his works. at London, in 1729, 4to. with his life prefixed: and it must needs suggest an high opinion of this young physician, that, though he died a young man, yet his works should be thought worthy of a republication 100 years after; when such prodigious improvements have been made in philosophy, physic, and sciences of all kinds, of which he had not the benefit.

GLANVIL (JOSEPH), a distinguished writer, was born in 1636, at Plymouth in Devonshire, where he probably received the first rudiments of his education, and was entered at Exeter-college, Oxford, April 19, 1652. He was placed under Samuel Conant, an eminent tutor, and having made great proficiency in his studie, he proceeded B.A. Oct. 11, 1655. The following year, he removed to Lincoln-college, probably

bably upon some view of preferment. Taking the degree of M. A. June 29, 1658, he assumed the priestly office [F], and became chaplain to Francis Rouse, esq; then made provost of Eton-college, by Oliver Cromwell, and designed for one of his upper house [of Lords]. Had this patron lived a little longer, Glanvil's expectations would, no doubt, have been fully answered; since he entirely complied with the principles of the then prevailing party, to whom his very prompt pen must needs have been serviceable. But, Rouse dying the same year, he returned to his college in Oxford, and pursued his studies there during the subsequent distractions in the state. About this time, he became acquainted with Mr. Richard Baxter, who entertained a great opinion of his genius, and continued his respect for him after the Restoration, when he renounced his principles. The friendship was also still kept on Glanville's side, who, Sept. 3, 1661, addressed an epistle to his friend, professing himself to be an admirer of his preaching and writings; he also offered to write something in his defence, but yielded to his advice, not to sacrifice his views of preferment to their friendship [G].

Accordingly, he had the prudence to take a different method; and turning his thoughts to a subject not only inoffensive in itself, but entirely popular at that time, viz. a defence of experimental philosophy, against the notional way of Aristotle and the schools, he published it this year, under the title of " The Vanity of Dogmatizing, or Confidence in Opinions, manifested in a Discourse of the Shortness and Uncertainty of our Knowledge and its Causes, with some Reflections on Peripateticism, and an Apology for Philosophy, 1661," 8vo. These meetings, which gave rise to the Royal Society, were much frequented at this time [H], and encouraged by learned men of all persuasions; so that this small discourse introduced him to the knowledge of the literary world in a favourable light. He had an opportunity of improving by the weakness of an antagonist, whom he answered in an appendix to a piece called, " Scepsis Scientifica, or confessed Ignorance the Way to Science, in an Essay on the Vanity of Dogmatizing, and confident Opinion, 1665," 4to. Our

[F] Assumed it, that is, without any kind of ordination, according to the principles of the sectaries at that time, of which his patron Rouse was an eminent leader. This added to Wood's silence about his having any orders, and his taking orders in the Church of England after the Restoration, is the ground of the conjecture, that he assumed the priesthood.

[G] Baxter's true Defence of the meer Nonconformists, c. v. Lond. 1681. Kennett's Regist. p. 629.

[H] Birch's History of the Royal Society, Vol. I. In the Introduction, Wood says, he reflected with regret upon his university-education, and wished he had been sent to Cambridge, where he should have had a free method of philosophizing. Athen. Oxon. Vol. II. col. 664. This points evidently to Dr. Henry Moore, as will appear hereafter.

author dedicated this piece to the Royal Society, in terms of the highest respect for that institution; and the society being then in a state of infancy, and having many enemies, as might be expected in so novel an undertaking, which seemed to threaten the ruin of the old way of philosophizing in the schools, the "Scepsis" was presented to the council by lord Brereton, at a meeting, Dec. 7, 1664; when his lordship also proposed the author for a member, and he was elected accordingly in that month.

In 1663, the house of John Mumpesson of Tedworth, in Wiltshire, being disturbed by the beating of a drum invisibly every night, our author turned his thoughts to that subject, and in 1666 printed, in 4to. "Some philosophical Considerations, touching the being of Witches and Witchcraft." In this piece he defended the possibility of witchcraft, which drew him into a controversy that ended only with his life: during the course of it, he proposed to confirm his opinion by a collection of several narratives relating to it. Whereupon, as he held then a correspondence with Mr. Boyle, that gentleman, observing with how much warmth the dispute was carried on, gave him many just cautions about his managing so tender a subject; and hinted to him, that the credit of religion might suffer by weak arguments upon such topics. In answer to which, Glanvil professes himself much obliged for those kind admonitions, and promises to be exceeding careful in the choice of his relations: however, he made a shift to pick out no less than twenty-six modern relations, besides that of Mr Mumpesson's drummer [1].

His defence of the Royal Society procured him many friends, some of whom obtained for him the rectory of the Abbey-church at Bath, into which he was inducted June the same year, 1666. From this time he fixed his residence in that city; and, continuing on all occasions to testify his zeal for the new philosophy, by exploding Aristotle, he was desired to make a visit to Mr. Robert Crosse, vicar of Chew, near Pensford in Somersetshire, a great zealot for the old established way of teaching in the schools. Our author accepted the invitation, and, going to Pensford in 1677, happened to come into the room just as the vicar was entertaining his company

[1] These relations were not printed till after his death, in a piece intituled, "Sadducismus Triumphans, in two Parts, 1681," 8vo.; and again in 168., with large editions, by Dr. Henry More, the editor of both editions; to whom our author had addressed a letter on the subject: and in an appendix to the first part concerning the possibility of apparitions, there is added, an account of the nature of a spirit, translated by our author, from the two last chapters of More's "Enchiridion Metaphysicum." This confirms our observation concerning Mr. Glanvil's Moraism; and we shall venture another remark, by way of conjecture, that the famous story of Mumpesson's drummer probably gave birth to Addison's comedy called "The Drummer."

with the praises of Aristotle and his philosophy. After their first civilities were paid, he went on with his discourse, and, applying himself to Mr. Glanvil, treated the Royal Society and modern philosophers with some contempt. Glanvil, not expecting so sudden an attack, was in some measure surprized, and did not answer with that quickness and facility as he otherwise might probably have done. But afterwards, both in conversation and by letters, he attacked his antagonist's assertion, that Aristotle had more advantages for knowledge than the Royal Society, or all the present age had or could have, because, "totam peragravit Asiam," he traveled over all Asia [K].

Neither did Glanvil let the matter rest there, but laid the plan of a farther defence of the Royal Society; but, bishop Sprat's history of it being then in the press, he waited to see how far that treatise should anticipate his design. Upon its publication in 1667, finding there was room left for him, he pursued his resolution [L]; and printed his piece the following year, with this title, expressing the motives of writing it, " Plùs Ultra, or the Progress and Advancement of Knowledge since the Days of Aristotle, in an Account of some of the most remarkable late Improvements of practical useful Learning, to encourage Philosophical Endeavours, occasioned by a Conference with one of the notional Way, 1668." 12mo. In some parts of this piece he treated the Somersetshire vicar with rough raillery [M], which

in

[K] Wool tells us, that Crosse had been fellow of Lincoln-college, and was preferred by the parliament to this rich vicarage; where, leaving his fellowship, he settled in 1655, and was constituted an assistant to the commissioners for ejecting ignorant, &c. ministers. At the Restoration he conformed, and so held his living. While in the university, he was accounted a noted philosopher and divine, an able preacher, and well versed in the fathers and schoolmen. Athen. Oxon.

[L] After Sprat's MS. was read to the Royal Society, Oct. 1664, Mr. Oldenburg, in a letter to Mr. Boyle, dated Nov. 24, following, remarked that he knew not whether there was enough said in it of particulars; and in another letter, dated Oct. 1, 1667, after that history was printed, and ready for publication, he wrote as follows.—" There is a certain gentleman, a florid writer, one of our own royal collegiates, who intends to print shortly some paralipomena relating to the history of our society; wherein he means to take notice of the performances of some eminent members thereof, more than has been done by Mr. Sprat; and farther to recommend and vindicate the modern experimental philosophy, by representing the advantages of this way of trials, both for light and use, above that of former times. It had been extant, I find by some letters, ere this, but that he staid for Mr. Sprat, to see what room he had left for his thoughts; and finding now that he has not throughout prevented him, he seems resolved to pursue his design, though it will not make above half a dozen sheets, and therein to acknowledge some grand contributions to philosophy, that have been omitted by the other. This is but just, and has therefore received encouragement from me, together with the suggestion of some particulars, which this author could not be acquainted with so well as the suggester." Boyle's Works, Vol. V. What the author here intimated was evidently Mr. Joseph Glanvil's, and the book his " Plus Ultra." Birch's Hist. of the Royal Society, Vol. II. p. 197.

[M] The vicar returned the language in a piece, which was denied the press both at Oxford and at London, for its scurrillity.

in return brought him into a very scurrilous dispute with Henry Stubbe, physician at Warwick. In this petulant way, however, of managing the controversy, Glanvil appeared, if not superior to his opponents, at least he had the last blow in it [N]. But when Dr. Meric Casaubon entered the lists, and managed the argument with more candor and greater knowledge, he chose to be silent; because not willing to appear in a controversy with a person, as he says, of fame and learning, who had treated him with so much civility, and in a way so different from that of his other assailants [o]. While he was thus pleading the cause of the institution in general, he shewed himself no unuseful member in respect to the particular business of it. The Society having given out some queries to be made about mines, our author communicated a paper in relation to

scurrillity. However, Glanvil somehow obtaining the contents, got them printed at London, with proper remarks of his own, under the title of "The Chew-Gazette;" but of these there were only 100 taken off, and those dispersed into private hands, to the end, as Glanvil said, that Crosse's shame might not be made public, &c. After this letter was abroad, Crosse wrote ballads against our author and the Royal Society; while other wags at Oxford, pleased with the controversy, made a doggrel ballad on them both, which began thus:
Two gospel knights,
Both learned wights,
And Somerset's renown-a,
The one in village of the shire,
But vicarage too great I fear,
The other lives in town-a.
Glanvil tells us, that Crosse wrote a book called "Biographia," containing rules how lives are to be written, &c. Athen. Oxon.

[N] Stubbe was then, as Wood observes, a summer-practitioner at Bath; and, bearing no good-will to the conceited proceedings of Glanvil, took Crosse's part, and encouraged him to write against the virtuosi, and at the same time entered the lists himself, and the following pamphlets passed between them. 1. "The Plus Ultra reduced to a Nonplus, &c. 1670," 4to. Stubbe. 2. "A prefatory Answer to Mr. Henry Stubbe, the doctor of Warwick, wherein the malignity, &c. of his Animadversions are discovered, 1671," 12mo. Glanvil. 3. "A Preface against Ecebolius Glanvil, F. R. S. subjoined to his Reply, &c. Oxford, 1671," 4to. Stubbe. The doctor also fell upon his antagonist,

in his "Epistolary Discourse concerning Phlebotomy, 1671," 4to; upon which Glanvil immediately published "A farther Discovery of Mr. Stubbe, in a brief Reply to his last Pamphlet, 1671," 8vo. to which was added, "Ad clerum Somersetensem Epistola ΠΡΟΣΦΩΝΗΣΙΣ." And the doctor, among other things, having censured the new philosophy, as tending to encourage atheism, our author published his "Philosophia Pia, &c. 1671, 8vo." This closed the controversy.

[o] Dr. Casaubon's Animadvesions were published in "A Letter to Peter du Moulin, D. D. concerning natural and experimental Philosophy, &c. Cambridge, 1663." The doctor observes, that Mr. Glanvil does not want words to set out his matter to the best advantage, and closes his letter in the following candid style. "What I had to except against the book you brought me, I have told you; I must now thank you for it; for, in very truth, his divinity at the end, which is somewhat mystical, (I hope I do not understand it,) and those two particulars; his contempt of Aristotle, and his censuring all other learning, besides experimental philosophy, and what tendeth to it, as useless, and meer wrangling and disputing excepted; I have read the rest, wherein he doth give us an exact account of late discoveries, with much pleasure, &c." This piece is mentioned, by our author, in the close of his "Prefatory Answer to Stubbe," where he tells us, he had answered the strictures in a particular discourse which he thought to publish next, when he reckoned with Stubbe: but he afterwards changed his mind.

those

those of Mendip hills, and such as respect the Bath, which was well received, ordered to be registered, and afterwards printed in their transactions [P].

In the mean time he was far from neglecting the duties of his ministerial function: on the contrary, he distinguished himself so remarkably by his discourses from the pulpit, that he was frequently desired to preach upon public and extraordinary occasions, and several of these sermons were printed in a collection after his death. But, in justice to his memory, we must not omit to mention one which was never printed. His old antagonist, Stubbe, going from Bath on a visit to Bristol, had the misfortune, on his return, to fall from his horse into a river, which, though shallow, proved sufficient to drown him: his corpse being interred in the abbey-church, our rector paid an honourable tribute to his memory, in a funeral sermon on the occasion. He also wrote an "Essay concerning Preaching," for the use of a young divine; to which he added, "A seasonable Defence of Preaching, and the plain Way of it." This was chiefly leveled against that affectation of wit and fine speaking which began then to be fashionable. This Essay was published in 1678; and the same year he was collated by his majesty to a prebend in the church of Worcester. This promotion was procured by the marquis of Worcester, to whom his wife was something related; and it was the more easily obtained, as he had been chaplain to the king ever since 1672: in which year he exchanged the vicarage of Frome for the rectory of Street, with the chapel of Walton annexed, in Somersetshire. This commodious exchange was easily accomplished, since both the livings were in the patronage of Sir James Thynne.

He published a great number of Tracts besides what have been mentioned, a list of which may be seen below [Q]. As he

[P] The first of these was registered Oct. 10, 1667, and printed in the Phil. Transf. No. 28, and the two others in No. 39 and 49. In this account of the Bath water, he supposes it to be a mixture from several springs of mineral water of a different nature; to confirm which, he observes, "That in 1659, the hot bath was much impaired by the breaking-out of a spring, which the workmen at last found again and restored; that in digging they came to a firm foundation of factitious matter, which had holes in it like a pumice-stone, through which the water played, so that," says he, "it is like the springs which are brought together by art;" which probably was the necromancy the people of antient times believed and reported to have contrived and made these baths; as in a very ancient MS. I find these words; "When Lud Hudibras was dead, Bladud his son, a great necromancer, was made king, and he made the wonder of the hot bath by necromancy, and he reigned 21 years, and after he died, and lies at the new Troy." And in another old chronicle, it is said, "that king Bladud sent to Athens for necromancers to effect this great business; who, 'tis like, were no other than cunning artificers, well skilled in architecture and mechanics."

[Q] These are: 1. "A Blow at modern Sadducism, &c. 1668," to which was added, 1. "A Relation of the fancied Disturbances at the house of Mr. Mumpesson:" as also, 3. "Reflections he had a lively imagination, and a flowing style, these came from him very easily, and he continued the exercise of his pen to the last; the press having scarcely finished his piece, entituled, "The zealous and impartial Protestant, &c. 1680," when he was attacked by a fever, which, baffling the physician's skill, cut him off in the vigour of his age. He died at Bath, Nov. 4th of the same year, about the age of 44. Mr. Joseph Pleydal, archdeacon of Chichester, preached his funeral sermon [R], when his corpse was interred in his own parish-church, where a decent monument and inscription was afterwards dedicated to his memory by Margaret his widow, sprung from the Selwins of Gloucestershire. She was his second wife; but he had no issue by either.

Soon after his decease, several of his sermons, and other pieces, were collected and published with the title of, "Some Discourses, Sermons, and Remains, 1681," 4to. by Dr. Henry Horneck, who tells us, that death snatched him away, when the learned world expected some of his greatest attempts and enterprizes.

GLAPTHORNE (HENRY), lived in the reign of Charles the First, and published several plays. He is called by Winstanley, "one of the choicest dramatic poets of this age." Langbaine, however, speaks of him with more temperate commendation. Glapthorpe also wrote a book of poems, addressed to his mistress, under the name of "Lucinda."

GLASS (JOHN, M. A.). He was born at Dundee, 1638, and educated in the New College, at St. Andrew's, where he took his degrees; and was settled minister of a Country Church, near the place of his nativity. In 1727 he published a treatise to prove that the civil establishment of religion was inconsistent with christianity; for which he was deposed, and became the father of a new sect, called, in Scotland, Glassites; and, in England, Sandemonians. His notions, however, joined to the rigidness of the discipline, deprived him of popularity, so that his followers are far from being numerous. He wrote a great number of controversial tracts, which have been published at Edinburgh, in 4 vols. 8vo. He died at Dundee, 1773, aged 75.

on Drollery and Atheism." 4. "Falpable Evidence of Spirits and Witchcraft, &c. 1668." 5. "A Whip for the Droll Fidler to the Atheist, 1668." 6. "Essays on several important subjects in Philosophy and Religion, 1676," 4to. 7. "An Essay concerning Preaching, 1678," 8vo. to which was added, 8. "A seasonable Defence of Preaching, and the plain Way of it." 9. "Letters to the Dutchess of Newcastle." 10. Three sing'g Sermons, besides four printed together, under the title of "Seasonable Reflections and Discourses, in order to the Conviction and Cure of the scoffing Infidelity of a degenerate Age"

[R] It was afterwards printed: in the close of it he says, he had once thought to have given the audience his character, but was not ashamed to tell them, he found himself not able to do it worthily.

GLASS (JOHN), son of the above, was born at Dundee, 1725, and brought up a surgeon, in which capacity he went several voyages to the West Indies. But, not liking his profession, he accepted the command of a merchant's ship belonging to London, and engaged in the trade to the Brazils. Being a man of considerable abilities, he published, in 1 vol. 4to. "A Description of Teneriffe, with the Manners and Customs of the Portuguese, who are settled there." In 1763 he went over to the Brazils, taking along with him his wife and daughter; and, in 1765, set sail for London, bringing along with him, all his property: but, just when the ship came within sight of the coast of Ireland, four of the seamen entered into a conspiracy, murdered the captain of the ship (Captain Glass), his wife, daughter, the mate, one seaman, and two boys. These miscreants, having loaded their boat with dollars, sunk the ship, and landed at Ross, whence they proceeded to Dublin, where they were apprehended and executed Oct. 1764.

GLAPHYRA, a mistress of Mark Anthony, very celebrated for her beauty, and who, being a native of Cappadocia, obtained from the Roman general the kingdom of that country, for her two sons, Sisinna and Archelaus. The jealousy which this attachment on the part of Anthony excited in Fulvia, his lawful wife, is commemorated in a pointed but obscene epigram, really written by Augustus, but published in the works of Martial.

GLASER (CHRISTOPHER), apothecary to Louis XIV. famous for a treatise on chemistry, which has been translated into English. It is concise, but clear and satisfactory.

GLAUBERT (RODOLPHUS), a German, who applied himself to the study of chemistry in the seventeenth century. His works were published in a volume, entituled, "Glauberus concentratus," this has been translated into English, and published at London, in folio, in 1689.

GLEN (JOHN), a printer and engraver in wood, born at Liege in the middle of the sixteenth century. He published a curious work on ancient and modern dresses, ceremonies, &c. ornamented with a great number of illustrative figures.

GLICAS, or GLYCAS, one of those called the Byzantine Historians. An edition of his works was published in greek and latin, by Labbe, in 1660; the latin translation of which is by Leunclavius.

GLISSON (FRANCIS), an English physician, was son of William Glisson of Rampisham, in Dorsetshire, and grandson of Walter Glisson, of the city of Bristol. Where he learned the first rudiments of his grammar is not known; but he was sent afterwards to Caius College in Cambridge,
apparently

apparently with a view to phyfic. However, as the beft foundation for it, he went through the academical courfes of logic and philofophy, and proceeded in arts, wherein he took both degrees; and, being chofen fellow of his college, was incorporated M A. at Oxford, Oct. 25, 1627 [s]. From this time, applying himfelf particularly to the ftudy of medicine, he took his doctor's degree in that faculty at Cambridge, and was appointed regius profeffor of phyfic in the room of Ralph Winterton; he held this poft forty years, that is, probably as long as he lived. But, not chufing to refide conftantly at Cambridge, he offered himfelf, and was admitted candidate of the college of phyficians, in 1634, and was elected fellow, Sep. 30, the enfuing year.

In the ftudy of his art, he had always fet the immortal Harvey before him as a pattern; and, treading in his fteps, he was diligent to improve phyfic, by anatomical diffections and obfervations. The fuccefs was anfwerable; he was appointed to read Dr. Edward Wall's lecture, in 1639; and, in executing that office, made feveral new difcoveries of principal ufe towards eftablifhing a rational practice of phyfic. He continued to difcharge the duties of this place till the breaking-out of the civil wars, when he retired to Colchefter, and followed the bufinefs of his profeffion with great repute in thofe times of public confufion. He was thus employed during the memorable fiege and furrender of that city to the rebels, 1648; and refided there fome time after.

Amidft his practice he ftill profecuted the improvement of it by anatomical refearches: and in this way publifhed an account of the rickets, in 1650, wherein he fhewed how the vifcera of fuch as had died of that diforder were affected [T]. This was the more curious, as the rickets had but then lately appeared in England; being firft difcovered in the counties of Dorfet and Somerfet, about fifteen years before. In this treatife he had the affiftance of two of his colleagues, Dr. George Bate, and Dr. Ahafuerus Regemorter; and thefe, with other fellows of the college, joining in a requeft to him to communicate to the public fome of his ana-

[s] Wood's Fafti Ox. Vol. I. col. 238. General Dict. and Goodall's account of the college of phyficians.

[T] The title of it is, "De Rachitide; five morbo puerili qui vulgo the Rickets dicitur, Lond. 1650." But though this difeafe was then of fuch modern extraction, yet a treatife had been publifhed, before this of our author, in 1645, 8vo. by Dr. Whiftler, afterwards prefident of the college, with the title of "Pædofplanchnofteocace,"

from the vifcera being judged to be the parts principally affected. In which opinion he was followed by our author, who likewife copied this original, in fhewing what was found præternatural in the vifcera of thofe that died thereof. But the caufe and nature of this diforder was better explained afterwards by Dr. John Mayow, in a fmall treatife publifhed upon it in 1668, 12mo. and again 1681.

tomical lectures which had been read before them, he drew those up in a continued discourse, and printed it with this title, "Anatomia Hepatis, Lond. 1654."

This brought him into the highest esteem among the faculty, and he was chosen one of the electors of the college the year following, and was afterwards president for several years. He published other pieces besides those already mentioned [u]; The last of which was a "Treatise of the Stomach and Intestines," printed at Amsterdam in 1677, not long before his death, which happened that year, in the parish of St. Bride, London.

Wood observes, that he died much lamented, as a person to whose learned lucubrations and deep disquisitions in physic, not only Great Britain, but remoter kingdoms, owe a particular respect and veneration : that, for instance, the world is obliged to him for the discovery of the *capsula communis*, or *vagina portæ* ; and that he hath likewise furnished certain marks for the more easy distinguishing the *venæ cava*, *porta*, and *vasa fellea*, in respect to the liver. It is also said, that he gave such an excellent account of sanguification, and supported it with arguments and experiments, that in 1684 few had doubted of the truth thereof. His treatise of the liver is indeed his *chef d'œuvre* ; though, in his last piece on the stomach and guts, there are several ingenious problems proposed and discussed, both philosophical and physical ; as, for instance, the various colours of the *cutis* or *cuticula*, and the hair : the specifical difference of hunger and thirst, from the five other senses : questions concerning rumination in animals, together with the structure, tenacity, and various uses, of the fibres of the parenchyma of the stomach and guts : the manner of deglutition, concoction, distribution of the chyle, secretion, &c. of the differences, causes, and signs, of flatus : with their most proper discutients : of the hypocondriac flatus : of the parts affected in a rheumatism. But his Physiology is not at present in any esteem.

GLOUCESTER (ROBERT of), the most ancient poet in the records of the English History, flourished in the time of Henry II. Mr. Camden esteemed him much, and quoted many of his old english rhymes in praise of his native country. He is valued now more for his history than his poetry. Died in old age, about the beginning of the reign of king John.

[u] These are, 1. "De Lymphæductis nuper repertis, Amst. 1659" with the Anatomica prolegomena & Anatomia Hepatis." 2. " De naturæ substantia energetica, seu de via vitæ naturæ ejusque tribus primis facultatibus, &c. Lond. 1672." 4to.

GLOVER (RICHARD), was originally brought up in the mercantile way; but always difcovered a ftrong genius for poetry. He began his poem, called "Leonidas," when very young; but was certainly advanced in life before he put his finifhing hand to it. It has been much received for its containing certain great beauties, and has been tranflated into french. Mr. Glover wrote alfo two tragedies, Boadicea, Medea, and afterwards a Sequel to Medea. He died greatly efteemed and much lamented in 1785, aged 74.

GMELIN (SAMUEL GOTTLIEB), fon of a phyfician at Tubinger, born in 1745, was eminent for his attainments in the ftudy of natural hiftory. He made feveral voyages, and died in the profecution of his travels in Tartary. The public have from his pen "Travels in Ruffia," publifhed at Peterfburg, in the German language, in four volumes, quarto; the laft volume of which contains his travels in Perfia. He was of a lively and licentious turn of mind; but was certainly a man of genius and fagacity—his life has been written by Pallas.

GMELIN (JOHN GEORGE) was uncle of the former, and is known in the literary world by his Flora Siberica, and his Travels in Siberia, publifhed in french, in two volumes.

GOAR (JAMES), a Dominican fri r, was born at Paris in 1601, and fent on a miffion to the Levant in 1618. He publifhed what he called Græcorum Euchologium, in greek and latin, concerning which the learned reader may confult the Bilioth. Græc. of Fabricius. He publifhed alfo tranflations of fome of the Byzantine hiftorians. He died at the age of fifty-two, highly refpected as a fcholar.

GOBIER (CHARLES), a jefuit of St. Maloes, a man of an active fpirit and of confiderable reputation, as a writer, born in 1644. He wrote the "Hiftory of des Iles Marianes," as well as "Lettres curieufes et édifiantes." The fubject of this laft is the natural hiftory, the geography, and the politics of this country, which the jefuits had explored. He wrote alfo many tracts on the progrefs of true religion in China, and entered warmly into the difputes betwixt the Miffionaries on the worfhip of Confucius.

GOCLENIUS (CONRAD), born in Weftphalia, in 1486. He wrote many learned notes on Cicero's Offices, publifhed an edition of Lucan, and tranflated the Hermotimus of Lucian. Erafmus, who was his intimate friend, highly valued his manners, and refpected his erudition.

GOCLENIUS (RODOLPHUS) we are induced to mention from no other motive but that he wrote a tract on the cure of
wounds

wounds by application of the magnet, which perhaps laid the foundation of the ridiculous doctrine of magnetism. He died, in 1621, at the age of forty-nine.

GODDARD (JONATHAN), an English physician and chymist, and promoter of the Royal Society, was the son of a rich ship-builder at Deptford, and born at Greenwich about 1617. Being industrious and of good parts, he made a quick progress in grammar-learning; and, at 15 years of age, was entered a commoner at Magdalen-hall, Oxford, in 1632. He staid at the university about four years, applying himself to physic; and then left it, without taking a degree, to travel abroad, as was at that time the custom, for farther improvement in his faculty. At his return, not being qualified, according to the statutes, to proceed in physic at Oxford, he went to Cambridge, and took the degree of batchelor in the faculty, as a member of Christ-college: after which, intending to settle in London, without waiting for another degree, he engaged in a formal promise to obey the laws and statutes of the College of Physicians there. Nov. 1640. Having by this means obtained a proper permission, he entered into practice; but however, being sensible of the advantage of election into the college, he took the first opportunity of applying for his doctor's degree at Cambridge, which he obtained, as a member of Catharine-hall, in 1642: and was chosen fellow of the College of Physicians in 1646. In the mean time, he had the preceding year engaged in another society, for improving and cultivating experimental philosophy. This society usually met at or near his lodgings in Wood-street, for the convenience of making experiments; in which the doctor was very assiduous, as the reformation and improvement of physic was one principal branch of this design. In 1647, he was appointed lecturer in anatomy at the college: and it was from these lectures, that his reputation took its rise. As he, with the rest of the assembly which met at his lodgings, had all along sided with the parliament, he was made head-physician in the army, and was taken, in that station, by Cromwell, first to Ireland in 1649, and then to Scotland the following year; and returned thence with his master; who, after the battle of Worcester, rode into London in triumph, Sept. 12, 1651. He was appointed warden of Merton-college, Oxon, Dec. 9th following, and was incorporated M.D. of the university, Jan. 14th the same year. Cromwell was the chancellor; and returning to Scotland, in order to incorporate that kingdom into one commonwealth with England, he appointed our warden, together with four others, to act as his delegates in all matters relating to grants or dispensations that required his assent [w]. This instrument bore date, Oct. 16, 1652. His powerful patron, dissolving the long parliament, called a new one, named the Little Parliament in 1663; wherein the warden of Merton sat sole representative of the university, and was appointed one of the council of state the same year.

A series of honours and favours bestowed by the usurper, whose interest he constantly promoted, could not fail of bringing him under the displeasure of Charles II. who, presently after his return, removed him from his wardenship, by a letter bearing date July 3, 1660; and claiming the right of nomination, during the vacancy of the see of Canterbury, appointed another warden in a manner the most disgraceful to our author. The new warden was Dr. Edw. Reynolds, then king's chaplain, and soon after bishop of Norwich, who was appointed expressly as successor to Sir Nathaniel Brent, no notice being taken of Dr. Goddard [x]. Thus, driven from Oxford, he removed to Gresham-college, where he had been chosen professor of physic on Nov, 7. 1655. Here he continued to frequent those meetings which gave birth to the Royal Society; and, upon their establishment by the royal charter in 1663, was therein nominated one of the first council. This honour they were induced to confer upon him, both in regard to his merit in general as a scholar, and to his particular zeal and abilities in promoting the design of their institution, of which there is full proof in the "Memoirs" of that society by Dr. Birch, where there is scarcely a meeting mentioned in which his name does not occur for some experiment or observation made by him. At the same time he carried on his business as a physician, being continued a fellow of the college by their new charter in 1663. Upon the conflagration in 1666, which consumed the Old Exchange, our professor with the rest of his brethren removed from Gresham, to make room for the merchants to carry on the public affairs of the city: which, however, did not hinder

[w] The others were Dr. Wilkins, warden of Wadham; Dr. Goodwin, president of Magdalen; Dr. Owen, dean of Christ-church; and Cromwell's brother-in-law, Peter French, a canon of the same church. Three of these deputies were a quorum. Wood's Fasti, Vol. II col. 98.

[x] Our author, it is true, was strongly attached to Cromwel; which, no doubt, brought this mark of the king's resentment upon him; otherwise, it was not deserved by his behaviour in the college. For this we have the testimony of Wood, who was bred at Merton, and always mentions Dr. Goddard, as warden, in terms of kindness and respect. He was, indeed, the first patron to that antiquary; who, as such, dedicated his brother's sermons to him, published in 1659, and sent it him to London, bound in blue Turkey with gilt leaves; as we find it carefully set down in the history of his own life, published by Mr. Hearne.

him from going on with his services both to natural philosophy and physic. In this last, he was not only an able but a conscientious practitioner; for which reason he continued still to prepare his own medicines. He was so fully persuaded that this, no less than prescribing them, was the physician's duty, that in 1668, whatever offence it might give the apothecaries, he was not afraid to publish a treatise, recommending it to general use. He observes, that the greatest part of the apothecaries were far from being possessed of that degree of knowledge, which was necessary to fit them for the due execution of their own employment; notwithstanding which, they were very desirous of invading that of the physician, and of prescribing as well as compounding medicines. He expatiates very largely upon this, and shews what prejudicial consequences attend it, with regard to the art of physic, the progress of which it retards; with regard to the credit of the physician, which suffers often by other men's faults; and, lastly, with regard to the patients themselves, who, while they seek to avoid expence, are brought to a condition, that lays them under a necessity of parting with more money, than might have purchased health at first. The remedy he proposes, as only capable of removing all these mischiefs, is, that physicians make their own medicines.

This treatise was received with applause: but as he found the proposal in it attended with such difficulties and discouragements as were likely to defeat it, he pursued that subject the following year, in "A Discourse setting forth the unhappy condition of the Practice of Physic in London, 1669," 4to. But this availed nothing, and when an attempt was made by the College of Physicians, with the same view, thirty years afterwards, it met with no better success. In 1671, he returned to his lodgings at Gresham-college, where he continued prosecuting improvements in philosophy till his death, which was very sudden. He used to meet a select number of friends at the Crown-tavern in Bloomsbury, where they discoursed on philosophic subjects, and in his return thence in the evening of March 24, 1674, he was seized with an apoplectic fit in Cheapside, and dropped down dead.

His memory was preserved by certain drops, which were his invention, and bore his name; but which, like all such sort of nostrums, have been long ago obsolete. The reader will find an account of his other inventions below [Y]. He had several learned treatises dedicated to him as a patron of learning,

[Y] Two of these are printed in Sprat's "History of the Royal Society," p. 193, 293. The first is a proposal for making
improvements have been added since

learning, all made by persons well acquainted with him, and written without any view of interest; where he is particularly recommended for his extensive learning, his skill in his profession, knowledge of public affairs, and generous disposition, for his candour, affability, and benevolence to all good and learned men [z]. Of this last there is one instance worth preserving; and that is, his taking into his apartment, at Gresham, Dr. Worthington, who lodged with him for the conveniency of preparing for the press the works of Mr. Joseph Mede, which he finished and published in 1664. But he more particularly claims a place in these memoirs, if what Dr. Seth Ward [A], bishop of Salisbury, attests of him, be true; namely, that he was the first englishman, who made that noble astronomical instrument the telescope.

GODEAU (ANTHONY), a French bishop, was descended from a good family at Dreux, and born in 1605. Being inclined to poetry from his youth, he applied himself to it; and so cultivated his genius, that he made his fortune by it. He was but twenty-four when he became a member of that society, which met at the house of Mr. Conrart, to confer upon subjects of polite learning, and to communicate their performances in that way. From this society Cardinal Richlieu took the hint, and formed the resolution, of establishing the French Academy for belles lettres; and our author in a few years obtained the patronage of that powerful ecclesiastic. In 1636, he was advanced to the bishopric of Grasse, which he afterwards relinquished for that of Venice. He assisted in several general assemblies of the clergy, held in the years 1645 and 1655; wherein he vigorously maintained the dignity of the episcopal order, and the system of pure morality [B], against those who opposed both. These necessary absences excepted, he constantly resided upon his diocese, where he was perpetually employed in visitations, preaching, reading, writing, or attending upon the ecclesiastical or temporal affairs of his bishopric, till Easter-day, April 17, 1671; when he was seized with a fit of an apoplexy, of which he died the 21st.

making wine from sugar, to which some improvements have been added by Dr. Shaw, in his "Chymical Lectures." 2. "Arcana Goddordiana."— These are some receipts published at the end of the second edition of the "Pharmacopœia Batavan., Lond. 1691." There are two papers of his published in Philosophical Transactions, No. 127, 138; and a great many others in Birch's "History of the Royal Society."

[z] For instance, Mr. Edmund Dickinson in "Delphi Phœnicizantes, Oxon. 1655." 8vo. Dr. Wall's's "Mathesis Universalis, Ox. 1636-7," 4to.

[A] In this piece entituled, "In Hm. Bul i Mr. Astron. Philol. fund in enta Iæquisitio brevii. Oxon. 1653." 4to.

[B] One of his best pieces is upon this subject, and was published in 1709, with the title of "Christian Morals for the Instruction of the Clergy of the Diocese of Venice:" it was afterwards translated into English, by Basil Kennet.

He

He was a very voluminous author, both in profe and verfe [c]; but it may fuffice to mention one in each way. His "Ecclefiaftical Hiftory," 3 vols. fol. The firft of which appeared in 1653, containing the "Hiftory of the firft eight Centuries;" but as he did not finifh the other two, nothing of them was printed. Hereby, however, he obtained this merit, that he was the firft perfon who gave a "Church Hiftory" in the french language. His other performance is a "Tranflation of the Pfalms into french verfe. Thefe were fo well approved, that thofe of the reformed religion have not fcrupled to ufe them at home in their families, inftead of the verfion of Marot, which is adapted and confecrated to the public fervice [D]. However, the Jefuit, Vavaffor, wrote a piece on purpofe, to prove that our author had no true tafte for poetry [E]; and Boileau remarks feveral defects in his poetical performances.

GODFREY (SIR EDMUND BURY), an able magiftrate of a very fair character, who had exerted himfelf in the difcovery of the popifh plot, was found pierced with his own fword, and many marks of violence on his body. His death, which was imputed to the papifts, who were then fuppofed to be the authors of all mifchief, was generally deemed a ftronger evidence of the reality of the plot than any thing that Oates did or could fwear. His funeral was celebrated with the moft folemn pomp; feventy-two clergymen preceded the corpfe, which was followed by a thoufand perfons, moft of whom were of rank and eminence. His funeral fermon was preached by Dr. Wm. Lloyd, dean of Bangor, and afterwards bifhop of Worcefter. He was found murdered the 17th of October, 1678.

GODFREY of Boulogne, an illuftrious character in the Crufades. After the capture of Jerufalem by the chriftian army, Godfrey was elected, by the confederate Francis, king of that city and the adjacent country. From piety, he rejected the title of king, and was contented with being called duke of the holy fepulchre. He conducted himfelf with great gallantry againft the fultan of Egypt, whofe armies he totally deftroyed, and obtained total poffeffion of the Holy-land. After forming an excellent code of laws for his new fubjects, he died, after having enjoyed his new dignities for no more than

[c] Moreri gives the titles of no lefs than fifty; and then concludes thus: "Our author alfo wrote Chriftian eclogues, feveral poems and poetical pieces, which are more commendable for the fentiments of piety which they infpired than for the beauty and harmony of the verfification.

[D] See a critique upon them, in the preface to an "Effay towards a Paraphrafe on the Pfalms, &c by Bafil Kennet, 1709," 8vo.

[E] The title of it is, "Godellus utrum Poeta?"

the space of a year. It is needless to add that this Godfrey is one of the principal figures in the immortal poem of Tasso.

GODIVA, the name of a beautiful lady, sister of Therald de Burgenhall, sheriff of Lincolnshire, and wife of Leofric, earl of Leicester, who was the eldest son of Algar, the great earl of Mercia. This lady, having an extraordinary affection to Coventry, solicited her husband to release and exempt the inhabitants of that city from a grievous burthen laid upon them. He consented, provided she would ride naked through the streets of Coventry, which she submitted to. This adventure was painted in one of the windows of Trinity-church, in Coventry, with these verses,

> I Luric, for the love of thee,
> Do make Coventry toll-free.

GODOLPHIN (John), an eminent civilian of England, third son of John Godolphin, Esq; was descended from an ancient family of his name in Cornwall, and born, 1617, at Godolphin in the island of Scilly. He was sent to Oxford, and entered a commoner of Gloucester-hall, in 1632; and, having laid a good foundation of logic and philosophy, he applied himself particularly to the study of the civil law. He chose this for his profession; and accordingly took his degrees in that faculty, that of batchelor in 1636, and of doctor in 1642-3. He was then observed to be inclined to Puritanism, which afterwards plainly appeared in two treatises of divinity, published by him in 1650, and 1651 [F]. Going to London afterwards, he sided with the anti-monarchical party; and, taking the oath called the Engagement, was by an act passed in Cromwell's convention, or short Parliament, July 1653, constituted judge of the admiralty jointly with William Clarke, LL. D. and Charles George Cock. Esq. [G]. July 1659, upon the death of Clarke, he and Cock received a new commission to the same place, to continue in force no longer than December following.

[F] The titles are, 1. " The Holy Limbec, or an Extraction of the Spirit from the Letter of certain eminent Places in the Holy Scripture." Other copies were printed with this title, "The Holy Limbec, or a Semicentury of Spiritual Extractions, &c." 2. " The Holy Harbour, containing the whole Body of Divinity, or, the Sum and Substance of the Christian Religion."

[G] This person, who was a counsellor of the Inner Temple, Wood says, was a great anti-monarchist, and in some measure contributed to the death of Charles I. He was one of those twenty-one persons appointed to consult of a Reformation in the law, in 1651: one of the commissioners of the Prerogative-court, and one of the High-court of Justice, in 1651: and author of a canting whimsical book, intituled, " English Law; or, a summary Survey of the Houshold of God on Earth, &c. 1651." To which is added, " An Essay of Christian Government, under the Regimen of our Lord and King, the one immortal, invisible, &c. Prince of Peace, Emanuel." This shews him to be a fifth-monarchy-man.

Notwithstanding

Notwithstanding these compliances with the powers then in being, he was much esteemed for his knowledge in the civil law, which obtained him the post of king's advocate at the restoration: after which, he published several books in his own faculty then in good esteem, as, "A View of the Admiral's Jurisdiction, 1661," 8vo. wherein is printed a translation, by him, of Grasias, or Ferrand's "Extract of the ancient Laws of Oleron [H];" "The Orphan's Legacy, &c. treating of last Wills and Testaments, 1674," 4to. And "Repertorium Canonicum, &c. 1678," 4to. In this last piece he strenuously and learnedly asserts the king's supremacy, as a power vested in the crown, before the Pope invaded the right and authority, or jurisdiction. He died in 1678.

GODWIN (THOMAS), an english bishop, was born, in 1517, at Ockingham in Berkshire; and, being put to the grammar-school there, quickly made such a progress as discovered him to be endowed with excellent parts. But, his parents being low in circumstances, he must have lost the advantage of improving them by a suitable education, had they not been noticed by Dr. Richard Layton, archdeacon of Bucks; who, taking him into his house, and instructing him in classical learning, sent him to Oxford, where he was entered of Magdalen-college about 1538. Not long after, he lost his worthy patron; but his merit, now become conspicuous in the university, had procured him other friends; so that he was enabled to take the degree of B. A. which he did July 12, 1543. The same merit released his friends from any farther expence, by obtaining him, the year ensuing, a fellowship of his college; and he proceeded M. A. in 1547. But he did not long enjoy the fruits of his merit in a college life; his patron, the archdeacon, being a zealous reformer, had taken care to breed up Godwin in the same principles. This brought him into the displeasure of some fellows of his college, who, being zealous for the old religion, made him so uneasy, that, the freeschool at Brackley in Northamptonshire becoming vacant in 1549, and being in the gift of the college, he resigned his fellowship, and accepted it. In this station, he married, and lived without any new disturbance as long as Edward VI. was at the helm: but, upon the accession of Mary, his religion exposed him to a fresh persecution, and he was obliged to quit his school. In this exigence he applied himself to the study of physic; and being admitted to his batchelor's degree in that faculty, at Oxford, July 1555, he practised in it for a support till Elizabeth succeeded to the throne.

[H] This is a small island on the coast of France; but these laws are the first original of all our Admiralty Jurisdiction.

From the period of his being at Magdalen-college he had fixed upon divinity for his profession; and the times now favouring his original design, he was resolved to enter into the church. In this he was encouraged by Bullingham, bishop of Lincoln, who gave him orders, and made him his chaplain: his lordship also introduced him to the queen, and obtained him the favour of preaching before her majesty; who was so much pleased with the propriety of his manner, and the grave turn of his oratory, that she appointed him one of her Lent-preachers. He had discharged this duty by an annual appointment, with much satisfaction to her majesty, some years; when he was made dean of Christ-church, Oxford, in 1565, and had also a prebend conferred on him, by his patron, bishop Bullingham. This year also he took his degree of D. D. at Oxford. In 1566, he was promoted to the deanery of Canterbury, being the second dean of that church: and queen Elizabeth making a visit to Oxford the same year, he attended her majesty, and among others kept an exercise in divinity against Dr. Lawrence Humphries, the professor; wherein the famous Dr. Jewel, bishop of Salisbury, was moderator.

He continued 18 years at Canterbury, and was then, in 1584, advanced to the bishopric of Bath and Wells; but soon after fell under his sovereign's displeasure, by entering a second if not a third time into matrimony. This, and its consequences, made the rest of his life uneasy; so that, gradually losing his strength and spirits, he sunk at length into a quartan ague, and died in 1590. Sir

GODWIN (FRANCIS), son of the preceding, was born at Havington in Northamptonshire, 1561; and, after a good foundation of grammar-learning, was sent to Christ-church-college, Oxford, where he was elected a student in 1678 [I]. He proceeded B. A. in 1580, and M. A. in 1583 [K]; about which time he wrote an entertaining piece upon a philosophical subject, where imagination, judgement, and knowledge, keep an equal pace, but this, contradicting certain received notions of his times, he never published. It came out about five years after his death, under the title of The Man in the Moon; or, a Discourse of a Voyage thither. By Domingo Gonsales, 1638, 8vo [L]. He suppressed also another of his inventions at that time, which was the secret of carrying on a correspondence by signals, and in a much quicker way than by letters. He had probably not been long M. A. when he entered into orders; and became in a short time rector of

[I] His father was dean at this time.   [L] It is mentioned by Bp. Wilkins,
[K] Wood's Fasti, Vol. I.   in his discourse upon the same subject.

Samford

Samford Orcais, in Somersetshire, a prebendary in the church of Wilts, canon residentiary there, and vicar of Weston in Zoyland, in the same county; he was also collated to the sub-deanery of Exeter, in 1587. In the mean time, turning his studies to the subject of the antiquities of his own country, he became acquainted with Camden; and accompanied him in his travels to Wales, in 1590, in the search of curiosities. He took great delight in these enquiries, in which he spent his leisure hours for several years; but at length, leaving the pursuit in a general way to Camden, he confined himself to such antiquities as seemed to concern ecclesiastical matters. After some time, finding with regard to ecclesiastical things that he could add little or nothing to Fox's work on that subject, he restrained his enquiries to persons [M; and here he spared no pains, so that he had enough to make a considerable volume in 1594.

He became B. D. in 1593, and D. D. in 1595; in which year, resigning the vicarage of Weston, he was appointed rector of Bishop's Liddiard, in the same county. He still continued assiduous in pursuing the history of ecclesiastical persons; and, having made an handsome addition to his former collections, published the whole in 1601, 4to, under the following title: A Catalogue of the Bishops of England, since the first planting of the christian religion in this Island; together with a brief History of their Lives and memorable Actions, so near as can be gathered of Antiquity [N]. It appears, by the dedication to lord Buckhurst, that our author was at this time chaplain to this nobleman, who, being in high credit with queen Elizabeth, immediately procured him the bishopric of Llandaff. This was said to be a royal reward for his Catalogue, and this success of it encouraged him to proceed. The design was so much approved, that afterwards he found a patron of it in James I. insomuch, that Sir John Harrington, a favourite of prince Henry, wrote a treatise, by way of supplement to it, for that prince's use [O]. Our author therefore devoted all the time he could spare from the duties of his function towards completing and perfecting this Catalogue; and published another edition in 1615, with

[M] Preface to the first edition of his "Catalogus of English Bishops."

[N] This, containing only a catalogue of the bishops of B th and Wel's, was published by Hearne, at the end of Johannes de Wethamstede's chronicle from a MS. in the library of Trinity-college, Cambridge, of our author's own hand-writing, &c.

[O] It was drawn purely for the private use of the prince, without any intention to publish it; which was done afterwards, with the title of "A brief View of the State of the Church of England." It is carried on only to the year 1608 (when it was written) from the close of our author's works.

great

great additions and alterations [P]. But, this being very erroneously printed, by reason of his distance from the press, he resolved to turn that misfortune into an advantage; and accordingly sent it abroad the year after, in a new elegant latin dress; partly for the use of foreigners, but more perhaps to please the king [Q], to whom it was dedicated, and who in return gave him the bishopric of Hereford, to which he was translated in 1617. In the mean time, various reports having been spread to his disadvantage, about his secret of corresponding already mentioned, and the thing coming at length to the ears of king James, he was careful to communicate the secret to his majesty; and, to convince him that it was a fact and not a fiction, he published his treatise under the title of "Nuncius Inanimatus Utopiæ, 1629," 8vo. In 1630, came out the third edition of his "Annals of the Reigns of Henry VIII. Edward VI. and queen Mary," in latin, 4to: as did also a translation of them into english by his son Morgan Godwin: also, the same year, his small treatise, intituled, "A Computation of the Value of the Roman Sesterce and Attic Talent." After this he fell into a low and languishing disorder, and died in April 1633. He married, when a young man, the daughter of Wollton, bishop of Exeter; by whom he had many children.

GODWIN (Dr. THOMAS), a learned english writer, and an excellent schoolmaster, was born in Somersetshire, 1587; and, after a suitable education in grammar learning, was sent to Oxford. He was entered of Magdalen-hall in 1602; and took the two degrees in arts 1606 and 1609. This last year, he removed to Abingdon in Berkshire, having obtained the place of chief master of the free-school there; and in this employ distinguished himself by his industry and abilities so much, that he brought the school into a very flourishing condition; and bred up many youths who proved ornaments to their country, both in church and state. To attain this commendable end, he wrote his "Romanæ Historiæ Anthologia [R]," an english exposition of the roman antiquities, &c. and printed it at Oxford, in 1613, 4to. However, his inclinations leading him to divinity [s], he

[P] To the former title there was now added, "whereunto is prefixed, A Discourse concerning the first Conversion of our Britain unto the Christian Religion."

[Q] It is well known how ridiculously fond James was of being esteemed a latin scholar. The title is, "De Præsulibus Angliæ Commentarius, &c.

1616," 4'o.

[R] The second edition was published in 1623, with considerable additions. He also printed a "Florilegium Phrasicon, or a Survey of the Latin Tongue."

[s] In the preface to his "Anthologia, &c."

entered

entered into orders, and became chaplain to Montague bishop of Bath and Wells. He proceeded B. D. in 1616, in which year he published at Oxford, "Synopsis Antiquitatum Hebraicarum, &c." a collection of hebrew antiquities, in three books, 4to. This he dedicated to his patron; and, obtaining some time after from him the rectory of Brightwell in Berkshire, he resigned his school, the fatigue of which had been long a subject of his complaint [T]. Amidst his parochial duties, he prosecuted the subject of the jewish antiquities; and, in 1625, printed in 4to. "Moses and Aaron, &c." He took his degree of D. D. in 1637, but did not enjoy that honour many years; dying upon his parsonage in 1642-3. and leaving a wife, whom he had married while he taught school at Abingdon.

Besides the pieces already mentioned, he published "Three Arguments to prove Election upon Foresight by Faith;" which coming into the hands of Dr. William Twise, of Newbury in Berkshire, occasioned a controversy between them; wherein our author is said not to have appeared to advantage.

GOEREE (WILLIAM), born at Middleburg in 1635; a very eminent scholar. He published "Jewish Antiquities," in two volumes folio, Utrecht, 1700. He wrote also and published about the same period a "History of the Jewish Church." He was not merely a scholar properly so called, but a man of general taste, and gave the world an "Essay on the Practice of Painting," and another "on Architecture." He died at Amsterdam in 1715.

GOERTZS (JOHN BARON of), a man very memorable in the history of Sweden. He was in the confidence of Charles XII and his life and character are given at considerable length by Voltaire. He possessed the most surprizing intrepidity joined to a restless activity of character, which perpetually involved him in difficulties, and finally occasioned his death on the scaffold One of his exploits was an endeavour to excite an insurrection in England in

---

[T] Ibid. but the degree of his aversion can only be seen in his own way of expressing it, which indeed is somewhat curious; "Miraris forsan & redarguis, quod nondum destiterim ab his elementaribus; quasi vita mihi vitalis foret, in hisce minutus integram meam ætatem eludere, & votum unicum in his prævis studiis senium contrahere. Qui sic sentis, nec me satis notis nec ludi Literarii (pone lenocinium minimis moletrinæ dices) iniquas leges aut miserias quotidianas & omnigenas. Sentio me in pistrinum damnatum, & cogita to hanc anthologiam è pistrino prodeuntem. Si minus placeat, illud dabis puerorum circumstrepentium susurris, inter quos nata est; si placeat, illud debes puerorum crebris interrogatiunculis, quorum Enodationes me vel invitum indies reducunt ad hæc studia, quæ alias quamdudum jussissem suas sibi res habere: sic me amet Theologia, sacratior mihi pagina in votis, cum hæc in manibus, ludo regente."

favour of the Pretender. When his protector and sovereign loft his life, Goertzs was apprehended, and, to appeafe the people, who imputed to him much of what they had fuffered from the tyranny of Charles, he was beheaded in 1719.

GOES (WILLIAM), a native of Leyden, and a refpectable critic. Among other philological works he wrote fome annotations on Petronius, which Burman has fubjoined to his edition of that author. He was fon-in-law to Dan. Heinfius, and died in 1686.

GOEZ (DAMIAN DE), a portuguefe writer, was born at Alanquar near Lifbon, of a noble family, we know not in what year, and brought up at the court of king Emanuel, whofe valet de chambre he was. Having a ftrong paffion for travelling, he contrived to get a public commiffion; and travelled through almoft all the countries of Europe, contracting as he went an acquaintance with all the learned. Thus, at Dantzic, he was familiar with the brothers, John and Claus Magnus; and he fpent five months at Friburg with Erafmus. He afterwards went to Italy, and was at Padua in 1534. He continued four years in this city, ftudying under Lazarus Bonamicus; not, however, without making frequent excurfions into different parts of Italy. Here he got into the good graces of Peter, afterwards cardinal Bembus, of Chriftopher Madrucius, cardinal of Trent, and of James Sadolet. On his return to Louvain in 1538, he had recourfe to Conrad Glocenius and Peter Nannius, whofe inftructions were of great ufe to him. Here he applied himfelf to mufic and poetry; in the former of which he made fo happy a progrefs, that he was qualified to compofe for the churches. He married at Louvain, and his defign was to fettle in this city, in order to enjoy a little repofe after 14 years travelling. He continued here fome time, and compofed fome works; but, a war breaking out between Charles V. and Henry II. of France, Louvain was befieged in 1542. Goez has written the hiftory of this fiege, in which he bore a confiderable part; for he put himfelf at the head of the foldiers, and contributed much to the defence of the town. When he was old, John III. of Portugal, recalled him into his country, in order to write the hiftory of it; but the favours this monarch loaded him with created him fo much envy, that his tranquillity was at an end, and he came to be accufed; and, though he cleared himfelf from all imputations, was confined to the town of Lifbon. Here he was one day found dead in his own houfe; and in fuch a manner, as to make it doubted whether he was ftrangled by his enemies, or died of an apoplexy He wrote, "Fides, Religio, Morefque Æthiopum."—" De Imperio & Rebus Lufitanorum."—" Hifpaniæ."—" Urbis Oliffiponenfis Defcriptio."

criptio."—" Chronica do Rey Dom Emanuel:"—" Hiftoria do Principe Dom Joam;" and other works, which have been often printed, and are much efteemed. Nicholas Antonio favs, that, though he is an exact writer, yet he has not written the portugueſe language in its purity; which, however, is not to be wondered at, confidering how much time he fpent out of his own country.

GOFF (THOMAS), an englifh writer, was born in Effex in 1592, and received his firft learning at Weftminfter-fchool. Thence he removed to Chrift-church-college in Oxford. and took the degree of B. D. before he left that univerfity. In 1623, he was preferred to the living of Eaft-Clandon in Surrey; where, according to Langbaine, he met with a Xantippe of a wife, whofe intolerable tongue and temper fhortened his days fo, that he died in 1627. He wrote feveral pieces on different fubjects, among which are five tragedies; none of which were publifhed till fome years after his death. Philips and Winftanley have afcribed a comedy to this author. called, " Cupid's Whirligig;" but with no appearance of probability; fince the gravity of his temper was fuch, that he does not feem to have been capable of a performance fo ludicrous. In the latter part of his life he forfook the ftage for the pulpit, and inftead of plays wrote fermons, fome of which appeared the year he died. To thefe works may be added, his " Latin Oration at the Funeral of Sir Henry Savile," fpoken and printed at Oxford in 1622; another in Chrift-church cathedral, at the funeral of Dr. Godwin, canon of that church, printed in London 1627.

GOGAVA (ANTONIUS HENNANNUS), a german phyfician; publifhed at Venice in 1592, " Ariftoxeni Harmonicorum Elementorum," libri 5.

GOGUET (ANTONY-YVES), a french writer, and author of a celebrated work, intituled, " L'Origine des Loix, des Arts, des Sciences, & de leur Progrès chez les anciens Peuples, 1758," 3 vols. 4to. His father was an advocate, and he was born at Paris in 1716. He was very unpromifing as to abilities, and reckoned even dull, in his early years; but, his underftanding developing itfelf, he applied to letters, and at length produced the above work. The reputation he gained by it was great; but he enjoyed it a very fmall time, dying the fame year of the fmall-pox, which diforder, it feems, he always miferably dreaded. It is remarkable, that Conrad Fugere, to whom he left his library and MS. was fo deeply affected with the death of his friend, as to die himfelf three days after him.

GOLDAST (MELCHIOR HAIMINSFIELD), a famous civilian and hiftorian, was born at Bifchoffsel in Switzerland in in 1576, and was a Protestant of the confession of Geneva. He studied the civil law at Altorf under Conrade Rittershusius, with whom he boarded; and returned in 1698 to Bischoffsel. Goldast was always poor; and had no other subsistence but what he acquired by the publication of books. His way was, when he published any work, to send copies of it to the magistrates and great people, from whom he usually received something more than the real value; and his condition was such, that his friends imagined they did him vast service, in helping him to carry on this miserable traffic. In 1599, he lived at St. Gal, in the house of a gentleman, who declared himself his patron, and whose name was Schobinger. The same year he went to Geneva, and lived there at the house of professor Lectius, with the sons of Vassan, whose preceptor he was. In 1602, he went to Lausanne, because he could live cheaper there than at Geneva. His patron Schobinger advised him to it; but with this restriction, says he, "that you refrain hereafter from your frequent removals, which are not for your advantage or credit, and have made you suspected of an odd turn of temper among some persons, who lately complained of it to me at Zurich." This passage is taken from the fifth letter of a collection printed at Francfort in 1688, with this title, "Virorum clarissimorum & doctorum ad Melchiorem Goldastum Epistolæ," 4to; and it is from this collection, that these memoirs of him are originally extracted.

Notwithstanding Scobinger's caution, he returned soon after to Geneva; and, upon the recommendation of Lectius, was appointed secretary to the duke of Bouillon. This place he did not keep long; for he was at Francfort in 1603, and had a settlement at Forsteg in 1604. In 1605, he lived at Bischoffsel; where he complained of not being safe on the score of his religion, which rendered him odious even to his relations. He was at Francfort in 1606, where he married and continued till 1610, in very bad circumstances. We do not know what became of him afterwards; only that he lost his wife in 1630, and died himself in 1635. He was a man of capricious temper, and his want of integrity has been complained o ; not that we are to believe all that Scioppius has said against him, as well because Scioppius was very abusive, as because he supposed Goldast to be the man who had furnished Scaliger with materials for compiling the satire, intituled, "Munsterus Hypobolimæus." The greatest part of the writings published by Goldast are not his own productions, but only r duced by him into a body, or published from MSS. in libraries; and by this it appears, that he was one of the most indefatigable men in the world. Conringius
has

has given him a great character in the following passages: "He is a person", says he, "who has deserved so well of his country, by publishing the ancient monuments of Germany, that undoubtedly the Athenians would have maintained him in the Prytaneum, if he had lived in those times." And elsewhere, "When this more valuable and certain kind of learning," meaning the public law of the german empire, "was promoted in Germany at the beginning of this century by Melchior Goldast, who neither had, nor perhaps ever will have, an equal in illustrating the affairs of Germany, and by whose guidance a more exact knowledge of the empire began by degrees to prevail among us, &c."

We omit to transcribe the titles of his works, they being very numerous, very long, and not very interesting to an Englishman; but the curious reader may find them at full length in Niceron's "Mémoires, &c." and long enough to give him an idea of them in Bayle's "Dictionary."

GOLDHAGEN (JOHN EUSTACHIUS), of Magdeburg; was famous as a translator of various greek writers into latin, and in particular of Herodotus, Pausanias, and Xenophon. He was born in 1701, and died in 1772.

GOLDMAN (NICOLAS), born at Breslaw in 1623; was author of many works; those most known are a "Treatise on Military Architecture;" and another, "De Usu Proportionarii Circuli;" both of which have great merit. He died in 1665.

GOLDSMITH or GOULDSMITH (FRANCIS), lived in the reign of Charles I. and translated the latin play of Grotius called Sophompareas, or history of Joseph, into english verse. The author and his translation were both highly commended.

GOLDSMITH (OLIVER), a poet, and one of those, whose wit, instead of diminishing, served rather to increase his misfortunes. He was born at Roscommon, in Ireland, in 1729; and, being a third son of four, was intended by his father for the church. With this view he was trained in the classics, and sent to Trinity-college, Dublin, in June 1744; where he obtained the degree of B. A. in 1749, but afterwards turned his thoughts to physic, and went to Edinburgh in 1751. Here his beneficent disposition, as we are told, soon involved him in difficulties; and he was obliged precipitately to leave Scotland, in consequence of having engaged himself to pay a considerable sum of money for a fellow-student.

In 1754, he arrived at Sunderland near Newcastle, where he was arrested at the suit of a tailor in Edinburgh, to whom he had given security for his friend; but, by the favour of some gentlemen in the college, who probably admired his wit, as much as they pitied his want of wifdom, he was foon delivered from the bailiff's clutches, and paffed over in a dutch fhip to Rotterdam. He proceeded to Bruffels, then vifited a great part of Flanders; and, after fpending fome time at Strafburg and Louvain, where he obtained the degree of M. B. he accompanied an englifh gentleman to Geneva.

It is an undoubted fact, that this ingenious unfortunate made the greateft part of his tour on foot, having left England with very little money; but being of a philofophic turn, and poffeffed with an almoft enthufiaftic paffion for feeing the manners of different countries and people, he was not difcouraged by any apparent difficulties. He had fome knowledge of the french language, and of mufic; he played tolerably well on the german flute, which, from an amufement, became at times a means of fubfiftence. His learning and other attainments procured him an hofpitable reception at moft of the religious houfes; and his mufic made him welcome to the peafants of Flanders and Germany: " whenever I approached a peafant's houfe towards nightfall," he ufed to fay, " I played one of my moft merry tunes; and that generally procured me not only a lodging, but fubfiftence for the next day[υ]." The higher ranks, it feems, had not any tafte for his mufic; " they always thought my performance odious, and never made me any return for my endeavours to pleafe them."

On his arrival at Geneva, he became a travelling tutor to a young man, who was articled to an attorney; but, on unexpectedly receiving a fortune, was determined to fee the world. This wary youth, in the contract with his preceptor, made a provifo, that he fhould be permitted to govern himfelf; and he was a manager of his money to a parfimonious extreme. During Goldfmith's continuance in Switzerland, he affiduoufly cultivated his poetical talent; and thence fent the firft fketch of his epiftle, called "The Traveller,' to his brother, a clergyman in Ireland; who, giving up fame and fortune, had retired early to happinefs and obfcurity (not that thefe always go together) on an income of 40l. a year. From Geneva the preceptor and pupil vifited the fouth of France, where difagreeing (for, Goldfmith had probably too many humours of his own to attend to thofe of other people) they feparated from each other; and our poet was left once more upon the world at large. He traverfed, however, through many difficulties, the greateft part of France; and, bending his courfe at length to England, arrived at Dover in 1758.

[υ] To this he probably alludes in his Traveller:

" How often have I led thy fportive choir,
" With tunelefs pipe befide the murm'ring Loire!" &c.

His finances were so low on his return to England, that he with difficulty got to London; where, though a batchelor of physic, he applied to several apothecaries to be received into their shops as a journeyman. His broad irish accent, and the uncouthness of his appearance, occasioned him to be treated by these gentry with contempt and insult; but, at length, a chemist near Fish-street, struck with the simplicity of his manner, joined to his forlorn condition, took him into his laboratory; where he continued, till he discovered that his old friend Dr. Sleigh was in London. This was one of those gentlemen, who formerly saved him from limbo, and now took him under his care, till some establishment could be procured for him. Shortly he became an assistant in instructing the youths at the academy at Peckham; then a writer in " The Monthly Review;" and afterwards he was employed in " The Public Ledger," in which his " Citizen of the World" originally appeared, under the title of " Chinese Letters."

Fortune seemed now to take some notice of a man she had long neglected. The simplicity of his character, the integrity of his heart, and the merit of his productions, made his company acceptable to the better sort; and he emerged from apartments he had near the Old Bailey, to the politer air of the Temple; where he took handsome chambers, and lived in a genteel style. His " Traveller," his " Vicar of Wakefield," his " Letters on the History of England," his " Good-natured Man, a Comedy," raised him up, and insured success to any thing that should follow; as " The Deserted Village," " She Stoops to Conquer, &c." Notwithstanding the success of these pieces, by which he cleared vast sums, his circumstances were by no means prosperous; and this his biographer imputes to two causes: partly to the liberality of his disposition, which made him give away his money without wit and wisdom; and partly to an unfortunate habit of gaming, the arts of which (as may well be believed) he very little understood.

With all his accomplishments and powers, he does not appear to have been either wise or happy. Of his want of wisdom enough has appeared; and his temperament does not seem to have been fitted for happiness. Though simple, honest, humane, and generous, he was irritable, passionate, peevish, and sullen; and spleen has run so high with him, that he is said to have " often left a party of convivial friends abruptly in the evening, in order to go home, and brood over his misfortunes." Can wretchedness more extreme be conceived? The latter part of his life was embittered by a violent strangury, which, united with other vexations, brought on a

kind

kind of habitual defpondency. In this unhappy ftate he was attacked by a nervous fever, which being improperly treated, and by himfelf too, put at end to his mortality April 1774, in the 45th year of his age.

Goldfmith, like Smollett, Guthrie, and others who fubfifted by their pens, is fuppofed fometimes to have fold his name to works in which he had little or no concern.

GOLIUS (JAMES), profeffor of arabic at Leyden, and of a confiderable family in that city, was born at the Hague in 1596. He was fent to the univerfity at Leyden, where he fuffered no part of learning to efcape his application; and having made himfelf mafter of all the learned languages, he procceded to phyfic and divinity; neither was he ftill fatisfied without the mathematics. His education being finifhed, he took a journey to France with the duchefs de la Tremouille; when, being invited to teach the greek language at Rochelle, he accepted the employ, and would have held it longer, had not that city been reduced again to the dominion of the french king the year following. Upon this change, Golius refolved to return to Holland. He had early taken a liking to Erpenius, the arabic profeffor at Leyden; by the help of whofe lectures, together with his ufual diligence, he had made a great progrefs in the arabic tongue, and contracted an intimate friendfhip with his mafter. In this difpofition, having obtained an opportunity of attending the Dutch embaffador, in 1622, to the court of Morocco, he confulted with Erpenius, and took proper inftructions from him, for the improvement of both in that language; for the profeffor was deficient fo far, that, having never lived in the country where it flourifhes and is fpoken, he met with many words, proverbs, and terms, whofe meaning he rather gueffed at than really knew. He, therefore, directed his pupil to obferve carefully every production, either of nature, art, or cuftom, which were unknown in Europe; and to defcribe them, fetting down the proper name of each, and the derivation of it, if known. He alfo gave him a letter directed to that prince, together with a prefent of a Grand Atlas, and a New Teftament, in arabic. Thefe procured him a moft gracious reception from Muley Zidan, then king of Morocco, who declared a particular fatisfaction in them, and afterwards read them frequently.

In the mean time, Golius made fo good ufe of Erpenius's advice, that he attained a perfect fkill in the arabic tongue; while the fame curiofity, that led him into the knowledge of the cuftoms and learning of that country, made him very agreeable to the doctors and courtiers. By this means, he became particularly ferviceable to the ambaffador, who,

growing

growing uneasy becaufe his affairs were not difpatched, was advifed to prefent to his majefty a petition, written by Golius in the arabic character and language, and in the chriftian ftyle; a thing very extraordinary in that country. The king was aftonifhed at the beauty of the petition, with respect both to the writing and the ftyle; and fending for the Talips, or fecretaries, fhewed them the petition, which they admired. Whereupon he immediately fent for the ambaffador to know who drew it up; and, being informed it was done by Golius, defired to fee him. At the audience, the king fpeaking to him in arabic, Golius anfwered in fpanifh, that he underftood his majefty very well, but could not anfwer him in arabic, by reafon of its guttural pronunciation, to which his throat was not fufficiently inured. This excufe was accepted by the king, who granted the ambaffador's requeft, and difpatched him immediately. Golius arrived in Holland, with feveral books unknown in Europe; and among others, "The Annals of the Ancient Kingdom of Fez and Morocco," which he refolved to tranflate. He communicated every thing to Erpenius, who well knew the value of them, but did not live long enough to enjoy the treafure; that profeffor dying in Nov. 1624, after recommending this his beft-beloved fcholar to the curators of the univerfity for his fucceffor. The requeft was complied with, and Golius faw himfelf immediately in the arabic chair, which he filled with fo much fufficiency, that the great Erpenius was not miffed.

A mind lefs inflamed with the defire of knowledge would have fet down fatisfied here; but Golius ftill thirfted after farther perfections: and, being perfuaded that this could only be had from the fountain-head, he applied to his fuperiors for leave to take a jouney to the Levant; and obtained letters patent from the prince of Orange, dated Nov. 35, 1625. He fet out immediately for Aleppo, where he continued fifteen months; after which, making excurfions into Arabia, towards Mefopotamia, he went by land to Conftantinople, in company with Cornelius Hago, embaffador from Holland to the Porte. Here the governor of the coaft of Propontis gave him the ufe of his pleafant gardens and curious library· in which retirement, he applied himfelf wholly to the reading of the Arabic hiftorians and geographers, whofe writings were till then either unknown to, or had not been perufed by, him. Upon his return to the city, difcovering occafionally, in converfation with the great men there, a prodigious memory of what he had read, he excited fuch admiration, that a principal officer of the empire treated with him, upon going with
the

the Grand Signor's commission, and viewing the whole empire, in order to describe the situation of places with more exactness than was done in the then present maps. He excused himself on pretence of the oath which he had taken to the States, but in reality on account of the danger of such an undertaking. Here also he found his skill in physic of infinite service, in procuring him the favour and respect of the grandees; from whom, as he would take no fees, he received many valuable and rich presents. Nor was this all, several more costly favours were conferred upon him, with a view of soliciting his stay. He lived four years among them, in the enjoyment of these munificent caresses; and, having in a great measure satisfied his thirst of Eastern learning, and made himself absolute master of the Turkish, Persian, and Arabic, tongues, he returned in 1629, laden with curious MSS. which have been ever since the glory of the university-library at Leyden.

He did not intend, however, that they should continue locked up from the world. On the contrary, as soon as he was settled at home, he began to think of making the best use of them, by communicating them to the public; and, to facilitate the reading of them, he printed an "Arabic Lexicon," and a new edition of "Erpenius's Grammar, enlarged with Notes and Editions;" to which also, he subjoined several pieces of poetry, extracted from the Arabian writers, particularly Tograi and Ababella. But his views were not limited within the bounds of Europe: he had been an eye-witness of the wretched state of christianity in the Mahometan countries, and saw it with the compassion of a fellow-christian. He resolved, therefore, to make his skill in their language serviceable to them, and herein his zeal was very remarkable. Nobody ever solicited so strongly for great offices of state, and in the prosecution of their views, as he did to procure an edition of the "New Testament" in their original language; with a translation into the vulgar greek by an Archmandrite, which he prevailed with the States to present to the Greek church, groaning under the Mahometan tyranny; and, as some of these christians use the arabic tongue in divine service, he took care to have dispersed among them an arabic translation of the Confession of the reformed Protestants, together with the Catechism and Liturgy [w].

[w] For this purpose he employed an Armenian, who understood the vulgar arabic, as well as the phrases consecrated to religion; and could accommodate Golius's style to the capacity of every body; otherwise his expression might probably have been too sublime and abstruse. Golius kept this Armenian two years and a half at his house; and promised him the same pension that the States had granted to the Archmandrite, who translated the New Testament into vulgar greek. Yet he did not know whether the States would be at the expence. He did not propose the matter to them, till the work was finish'd; however, they agreed to his proposal, and likewise made a handsome present to himself.

However,

However, intent as he was upon the services of religion and learning abroad, he did not neglect his duty at home, which was now become double to what it had been before his last journey to the East; for, the curators, during his absence, had honoured him with an additional employ of a very different nature from the former, viz. the professorship of mathematics, to which he was chosen in 1626. He discharged the functions of both, with the highest applause for forty years. He was also appointed interpreter in ordinary to the States, for the arabic, turkish, persian, and other eastern languages; for which he had an annual pension, and a present of a chain of gold with a very beautiful medal, which he wore as a badge of his office. He went through the fatigue of all these posts with the less difficulty, as he always enjoyed a good state of health, which, however, he was careful to preserve, by temperance in diet, and abstinence from enfeebling pleasures. By this means his constitution was so firm, that, at the age of seventy, he travelled on foot all the way from the Meuse to the Wahal, a journey of fourteen hours. This was in 1666; and he died Sept. 28, 1667; having passed through all academic honours, and made himself as much respected for his virtue and piety, as for his learning.

Though he may well be called an universal scholar, yet his chief excellence lay in philology and the languages; for which he had so great a natural talent, that, though he did not begin seriously to study the Persian language till he was fifty-four, he made himself so perfectly a master of it as to write a large dictionary in it, which was printed at London. He could have done as much for the Turkish language: and he made such a progress in the Chinese, that he was able to read and understand their books; though he began late to learn this language, of which to know the characters only is no slight matter, since they amount to the number of 8000. Besides the books which he finished and printed, he left several MSS. of others, which would have been no ways inferior to them, had he lived to complete them. He had begun a Geographical and Historical Dictionary for the Eastern countries; wherein the names of men and places, throughout the East, were explained. He had long given expectations of a new edition of the "Koran," with a translation and confutation of it.

Amidst all this profound literature, his religion was plain, easy, and practical. He lamented and abhorred the factions and disputes, especially about indifferent matters, which disgraced christianity: he could not endure to have divinity looked on as a science: he thought the truth exposed to danger,

danger, even by men of knowledge and learning; who thus introduced philosophy into divinity merely for the sake of disputing.

He married a lady of a very good family, and well allied, with whom he lived twenty-four years, and who survived him, together with two sons, who studied the civil law at Leyden, and became considerable men in Holland. See Funebr. Orat. Jac. Gotii à Gronov. & Swert Athen. Belgsc.

GOLIUS (PETER), brother of the preceding, born at Leyden. He went to Aleppo in the character of a missionary. He was of the order of Barefooted Carmelites, and established a monastery of his order on the summit of Mount Libanus. He was an excellent arabian scholar, and published different works both in arabic and latin.

GOLTZIUS (HENRY), a famous painter and graver, was born in 1658, at Mulbrec in the duchy of Juliers; and learned his art at Haerlem, where he married. Falling into a bad state of health, which was attended with a shortness of breath and spitting of blood, he resolved to travel in Italy. His friends remonstrated against this, but he answered, that " he had rather die learning something than live in such a languishing state." Accordingly, he passed through most of the chief cities of Germany, where he visited the painters, and the curious; and went to Rome and Naples, where he studied the works of the best masters, and designed an infinite number of pieces after them. To prevent his being known, he passed for his man's servant; pretending, that he was maintained and kept by him for his skill in painting: and by this stratagem he came to hear what was said of his works, without being known, which was a high pleasure to him. His disguise, his diversion, the exercise of travelling, and the different air of the countries through which he travelled, had such an effect upon his constitution, that he recovered his former health and vigour. He relapsed, however, some time after, and died at Haerlem in 1617. Mr. Evelyn has given the following testimony of his merit as a graver: " Henry Goltzius," says he, " was a Hollander, and wanted only a good and judicious choice, to have rendered him comparable to the profoundest masters that ever handled the burin; for never did any exceed this rare workman: witness those things of his after Gasporo Celio, &c.—and, in particular, his incomparable imitations after Lucas Van Leyden, in The Passion, the Christus Mortuus, or Pieta; and those other six pieces, in each of which he so accurately pursues Durer, Lucas, and some others of the old masters, as make it almost impossible to discern the ingenious fraud. He was likewise an excellent painter.

GOLTZIUS (HUBERT), a german writer, was born at Venloo, in the duchy of Gueldres in 1526. His father was a painter, and he was himself bred up in this art, learning the principles of it from Lambert Lombard. But he did little at painting, and seems to have quitted it early in life; for he had a particular turn to antiquity, and especially to the study of medals, to which he entirely devoted himself. He considered medals as the very foundation of true history; and travelled through France, Germany, and Italy, in order to make collections, and to draw from them what lights he could. His reputation was high in this respect, so that the cabinets of the curious were every where open to him; and on this account it was, that he was honoured with the freedom of the city of Rome in 1567. He was the author of several excellent works, as, "Imperatorum fere omnium vivæ imagines à J. Cæsare ad Carolum V. ex veteribus numismatibus."—"Fasti Magistratuum, & triumphorum Romanorum ab U. C. usque ad Augusti obitum."—"De Origine & Statu Populi Romani."—"Vitæ & res gestæ J. Cæsaris & Augusti Cæsaris, ex Nummis & Inscriptionibus Antiquis," and other treatises; in all which he applies medals to the clearing up of ancient history. He was so nice and accurate in publishing them, that he had them printed in his own house, and corrected them himself: nay, he even went so far as to engrave the plates for the medals with his own hands. Accordingly, his books were admired all over Europe, and thought an ornament to any library. The learned bestowed the highest eulogies upon them. Lipsius, speaking of the "Fasti Consulares," says, that "he knows not which to admire most, his diligence in seeking so many coins, his happiness in finding, or his skill in engraving them." Scaliger spoke as well of this work, as his great soul could condescend to speak, when he says, "Goltzius nihil me docet, scio omnia illa; sed est bonus liber pro tyronibus;" that is, Goltzius teaches me nothing; I know all those things: but it is a good book for beginners. His books, however, though they abound with erudition and curious knowledge, must be read with some caution; for, there are many false medals in them, which Goltzius adopted for real antiques. It could not be, but that many errors of this nature must be committed by a man, whose love and veneration for Roman antiquities was such, that he gave to all his children nothing but Roman names, such as Julius, Marcellus, &c. so that he might easily receive for antiques what were not so, out of pure fondness for any thing of that kind. Upon this principle, it is probable, that he took, for his second wife, the widow of the antiquary Martinius Smetius; whom he married more for the sake of Smetius's medals and inscriptions than for any thing belonging to herself. However, she was even with him if he did; for she was very ill natured, and plagued him in such a manner as to shorten his days. He died at Bruges in 1583, aged 57.

GOMAR (FRANCIS), native of Bruges, a famous and strenuous defender of the calviniftic doctrines, against Arminius and his followers. He was a very learned man, particularly in the oriental languages. Several treatises, which he wrote at different times, were collected into a volume, and printed at Amsterdam, in 1645. He died at Groningen, where he was first divinity and then hebrew profeffor in 1641.

GOMBAULD (JOHN OGIER DE), a french poet, was born in 1567, at St. Juft de Luffac, near Brouage in Saintongue. He was a gentleman by birth, and his breeding was suitable to it. After a foundation of grammar-learning, he finished his studies at Bourdeaux; and having gone through most of the liberal sciences, under the best masters of his time, he betook himself to Paris, in the view of making the most of his parts; for, being the cadet of a fourth marriage by his father, his patrimonial finances were a little short. At Paris he soon introduced himself to the knowledge of the polite world, by sonnets, epigrams, and other small poetical pieces, which were generally applauded: but, reaping no other benefit for the present, he was obliged to use the strictest œconomy, to support a tolerable figure at court, till the affassination of the king by Ravillac, in 1610. This extraordinary incident provoked every muse in France. The subject was to the last degree interesting, and furnished our poet with one of those opportunities, which are said to fall in every man's way once in his life of making his fortune. He did not let it slip, but exerted his talent to the utmost on the occasion; and the verses he made pleased the queen-regent, Mary de Medicis, so highly, that she rewarded him with a pension of 1200 crowns; nor was there a man of his condition, that had more free access to her, or was more kindly received by her. He was also in the same favour with the succeeding regent, Anne of Auftria, during the minority of Lewis XIV.

In the mean time, he was constantly seen at that delicious meeting-place of all the persons of quality and merit, the house of Mad. Rambouillet. This was like a small choice court, less numerous indeed than that of the Louvre, but, to say the truth, more excellent; since nothing approached this Temple of Honour, where Virtue itself was worshipped under the name of the incomparable Artenice, but what deserved her approbation and esteem. Such was that mansion of politeness, which entirely engaged the heart of Gombauld; and he frequented it with great pleasure, as well as with more assiduity than any other, the Louvre not excepted. Thus he passed his time in a way the most agreeable to a poet, and at length devoted himself entirely to the belles lettres. He published several things which were so many proofs of excellence in this way [z]; so that he grew to be one of those choice spirits, who make up the ministry in the republic of letters, and form the schemes of its advancement. In this employ we find him among those few men of wit, whose meetings in 1626 gave rise to the Academy of Belles Lettres, founded by cardinal Richelieu [A]; and, accordingly, he became a member of that society at its first institution. He was one of the three who was appointed to examine the statutes of the new academy in 1643, and he afterwards finished memoirs for completing them. March 12, 1635, he read a discourse before the academy upon "Je ne sçai quoi," which was the sixth of those that for some years were pronounced at their meetings the first day of every week.

He lived many years in the enjoyment of these honours, and, what is more essential, with good finances, which yet were increased with an additional pension from M. Seguire, chancellor of France. These marks of esteem set his merit in the most conspicuous light; especially when it is considered that he openly professed the reformed religion, and was indeed a zealous Huguenot: but he preserved himself from any ill effects of this by a degree of prudence, very uncommon in men of his profession. He had always enjoyed very good health; but, as he was one day walking in his room, which was customary with him, his foot slipped; and, falling down, he hurt himself so, that he was obliged almost constantly to keep his bed to the end of his life, which lasted near a century. However, in 1657, when at the age of 90, he published a large collection of epigrams; and, many years after, a tragedy called "Danaïdes." This was some time before his death; which did not happen till 1666, in his 92d year.

In his person he is represented tall and well shaped, of a graceful aspect, and with the air of a man of quality; in his

[z] Of these the most admired was his "Endymion," a romance in prose. It was printed in 1624. 2. "Amarantha, a Pastoral." 3. A Volume of "Poems." 4. A Volume of "Letters," all published before 1652. Peliffon's Hist. de l' Acad. Fran. p. 3. 39. Paris, 1672, 12mo.

[A] These meetings were held at the house of Mr. Conrart, who is said to be the author of the preface to Gombauld's treatises and letters upon religion. Colomies Bibl. Choisie, 155. 2d edit.

manners he was modest and regular, sincere in his piety, and proof against all temptations. His mind was as noble as his person was agreeable; he had an upright soul, and was naturally virtuous. His genius was elevated, but more judicious than fanciful. He was of a hot and hasty temper, much inclined to anger, though he had a grave and reserved countenance. His posthumous works were printed in Holland in 1678, with this title, " Traités & Lettres de Monsieur Gombauld sur la Religion." They contain religious discourses, and were most esteemed of all his works by himself; he composed them from a principle of charity, with a design to convert the catholics, and confirm the protestants in their faith.

GOMERSAL (ROBERT), lived in the reign of Charles the First, and was of some eminence as a poet. He was a student of Christ-church, Oxford, where he took his batchelor's and master's degrees; and, in 1627, went out bachelor of divinity. He has left several sermons and poems, both of which have been commended. His best piece is called " The Levites Revenge, containing Poetical Meditations on the 19th and 20th chapter of Judges." He died in 1646.

GOMEZ (DE CIVIDAD, near ALVAREZ), a latin poet of Guadalaxara, in the district of Toledo. His compositions were well received in Spain. Among the most popular of his publications were " The Proverbs of Solomon in verse.— The Epistles of St. Paul, in elegiac verse," and a poem on the " Golden Fleece." He died in 1538.

GOMEZ (DE CASTRO ALVAREZ), was born near Toledo, and was respected by many for his great learning. He wrote " The History of Cardinal Ximenes."

GOMEZ (MAGDELINE ANGELICA POISSON DE), a french lady, who obtained some celebrity as a writer of romances and theatrical pieces. Her compositions are very numerous, but in no very high estimation.

GONDI (JOHN PAUL), afterwards cardinal de Retz, was born in 1613, and died in 1679. He was a doctor of the Sorbonne, and afterwards coadjutor to his uncle the archbishop of Paris; and at length, after many intrigues, in which his restless and unbounded ambition engaged him, became a cardinal. This extraordinary man has drawn his own character in his memoirs, which are written with such an air of grandeur, impetuosity of genius, and inequality, as gives us a very strong representation of his conduct. He was a man who, from the greatest degree of debauchery, and still languishing under its consequences, preached to the people, and made himself adored by them. He breathed nothing
but

but the spirit of faction and sedition. At the age of twenty-three, he had been at the head of a conspiracy against the life of cardinal Richelieu. Voltaire says, that he was the first bishop who carried on a war without the mask of religion: however, his schemes turned out so ill at the long run, that he was obliged to go from France. He went into Spain and Italy, and assisted at the conclave at Rome, which raised Alexander VII. to the pontificate. This pontiff not making good his promises to the cardinal, he left Italy; and went into Germany, then into Holland and England. After having spent the life of an exile and vagabond for five or six years, he obtained leave upon certain terms to return to his own country: which now he could do with safety, his friend cardinal Mazarine being dead in 1661. He was afterwards at Rome, and assisted in the conclave which chose Clement IX; but, upon his return to France, retired from the world, and ended his life like a philosopher: which made Voltaire say, that " in his youth he lived like Catiline, and like Atticus in his old age." In this retreat he wrote his memoirs, " several parts of which " says the same Voltaire, "are worthy of Sallust, but the whole is not equal." They are supposed, however, to be written with impartiality, the author having every where spoken with the same freedom of his own infirmities and vices as any other writer could have done. Some friends, with whom he entrusted the original MS. fixed a mark on those passages, where they thought the cardinal had dishonoured himself, in order to have them omitted, as they were in the first edition: but they have since been restored. The best edition of these memoirs is that of Amsterdam, 1719, in 4 vols. 12mo. This cardinal was the author of other pieces; but these, being of a temporary kind, written as party pamphlets to serve particular occasions and purposes, are not now regarded.

GONGORA (Lewis de), a spanish poet, was born at Cordova, in 1562, of a very distinguished family. He studied at Salamanca, and was known to have a talent for poetry, though he never could be prevailed on to publish any thing. Going into orders, he was made chaplain to the king, and prebendary of the church of Cordova: in which station he died, in 1627. His works are all posthumous, and consist of sonnets, elegies, heroic verses, a comedy, a tragedy, &c. and have been published several times. The spaniards have a very high idea of this poet, even so as to entitle him prince. of the poets of their own nation.

Notes and commentaries have been written on his works, and he has been decked out in form like a variorum classic. Some have found great fault with him, charging him with affectation

affectation in the use of figures, with a false sublime, with obscurity and an embarrassed diction: however, there have not been wanting persons to undertake his defence, and to free him from all such invidious imputations.

GONDRIN (LOUIS ARTOINE), we are induced to mention only as an accomplished courtier, and particular favourite of Louis XIV. The monarch condescended to sleep at the duke's country-house; he complained that he was disgusted by a grove of old trees before his window. In the morning they were no longer to be seen. The prince in his walk was incommoded by an extensive wood, which obstructed his view. He walked there a second time, and repeated his complaint: "Your majesty has only to say you wish it to be removed, and it will immediately disappear." "If that be so," said Louis, "I wish it were away." The matter had previously been prepared, and twelve hundred men in a moment levelled with the ground the whole extent of the wood. "What," said the dutchess of Burgundy, who was present, "if the king had wished our heads thus to disappear, the duke, I fear, would have had no hesitation in gratifying his sovereign!"

GONET (JOHN BAPTIST), a Dominican frier, was a doctor of the university of Bourdeaux, where he taught divinity. He published several works, particularly "A system of Theology, in five volumes, folio." Bayle, in his way, sneers at the spaniards; who say, as he observes, that it was too short, and calls it a pretty compendium of divinity. He died in 1681.

GONNELLI (JOHN), or the blind man of Combassi. He gave extraordinary hopes of his talents as an artist; when at the age of twenty he lost his sight. After this accident he became a sculptor, and by the sense of touch alone obtained a wonderful perfection. He even attempted portraits, and with no mean success; and was happy in obtaining the likeness of Pope Urban the eighth, and Cosmo the first, great duke of Tuscany.

GONSALVA, of Cordova, surnamed the Great Captain. He was of one of the most noble families of Spain, and at first distinguished himself as a warrior against the Portuguese. Afterwards, in the reign of Ferdinand and Isabella, he assisted at the conquest of Grenada. He secured also the kingdom of Naples and the throne of Spain against all the exertions of the french. He was indeed an extraordinary character, and many well-attested facts are recorded of his valour, his generosity, and accomplishments. Florian has made him the hero of an historic romance, which has been translated into english, and well received: it is certainly an elegant as well

well as interesting performance. Gonsalva died in Grenada, which his arms conquered in 1515.

GONTHIER, a latin poet of the thirteenth century. He wrote the history of Constantinople 1203.

GONTHIER (John and Leonard), painters on glass and eminent for their skill, both in their figures and decorations; their works have been highly esteemed, and are yet to be found in the cabinets of the curious.

GONZAGA (Lucretia), an illustrious lady of the 16th century, as remarkable for her wit, learning, and style, as for high birth. She wrote such beautiful letters, that the utmost care was taken to preserve them; and a collection of them was printed at Venice in 1552. There is no learning in her letters, but yet we perceive from them that she was learned; for she declares, in a letter to Robortellus, that his commentaries had led her into a true sense of several obscure passages in Aristotle and Æschylus. All the wits of her time did not fail to commend her highly; and Hortensio Lando, besides singing her praises most zealously, dedicated to her a piece, "Upon moderating the passions of the soul," written in Italian. There was a correspondence between them: and she wrote above thirty letters to him, which have all been printed. In one of them, she blames him for grieving at his poverty: "I wonder," says she, "that you, who are a learned man, and so well acquainted with the affairs of this world, should yet be so strangely vexed at being poor: as though you did not know, that a poor man's life is like sailing near the coast, whereas that of a rich man does not differ from the condition of those who are in the main sea. The former can easily throw a cable on the shore, and bring their ship safe into an harbour; whereas the latter cannot do it without great difficulty, &c." We learn from these letters, that her marriage with John Paul Manfrone was unhappy. She was married to him when she was not fourteen; and his conduct afterwards gave her infinite uneasiness. He engaged in a conspiracy against the duke of Ferrara; was detected and imprisoned by him; but, though condemned, not put to death. She did all in her power to obtain his enlargement; applied to all the powers in christendom to intercede for him; and even solicited the Grand Signior to make himself master of the castle, where her husband was kept. What made her more active, she was not permitted to visit him; and they could only write to each other. But all her endeavours were vain: for he died in prison, having shewn such an impatience under his misfortunes as made it imagined he lost his senses. She never would listen afterwards to any proposals of marriage, though several were made her. Of four children, which she had,

had, there were but two daughters left, whom she put into nunneries. All that came from her pen was so much esteemed, that a collection was made even of the notes she wrote to her servants: several of which are to be met with in the edition of her letters.

GONZALEZ (THYRSUS), a Spaniard, and general of the Jesuits; died at Rome in 1705. He wrote several tracts, which were received with different degrees of satisfaction by the world.

GOOL (JOHN VAN), a dutch painter, born at the Hague in 1685. He was eminent both for the firmness and elegance of 1. pencil. He was also a writer, and published an account of the lives and works of the flemish painters.

GOODALL (WALTER). This learned antiquarian was born in the county of Angus 1689, and educated in King's College Aberdeen, where he took his degrees, and was afterwards appointed deputy-keeper of the advocate's library in Edinburgh. He was at the same time employed as an assistant to the learned Mr. Ruddiman; and in 1736 wrote an introduction in latin to Fordon's Chronicle. In 1751 he published two volumes in vindication of the unfortunate queen Mary, which have been well received by the public, notwithstanding the author's strong and partial attachment to jacobitical principles. He was a very learned philologist, but sacrificed rather too often at the shrine of Bacchus. He died at Edinburgh 1758, aged 71.

GOODWIN (JOHN). He was one of the most extraordinary persons that lived during the last century; and, as appears from some of his writings, a most acute and subtle disputant. He was educated in Queen's College, Cambridge, and in 1633 obtained the living of Coleman-street, London. In 1645 he was turned out of his living, because he refused to administer the sacrament to his people promiscuously. He was such a violent republican, that he wrote a vindication of the death of Charles I. which, at the Restoration, was burnt by the hands of the common hangman. He was excepted out of the act of indemnity, and died soon after lamented by few, for he lived at enmity with all who knew him. His works are numerous, but mostly in support of arminian doctrines.

GOODWIN (THOMAS). He was born at Rolseby in Norfolk, October 5, 1600, and received his education at Cambridge. During his younger years he had wise notions of religion, and his mind was filled with ambition; but, going occasionally to hear Dr. Preston, he was struck with a pious turn of mind, which induced him to join himself to the Puritans; for adhering to their principles he he suffered much; and, in 1630, to avoid the fury of their persecution, he went over to Holland, and settled as pastor of the English church at Arnhiem When the civil wars broke out, he returned to England, and was chosen pastor of a church in London. He was chosen one of the assembly of divines at Westminster; and, in 1649, Oliver Cromwell advanced him to be president of Magdalen College, Oxford. After the ejectment in 1662, he came to London, where he formed a church on the plan of the independents, and continued to preach till the time of his death, in 1680, aged 80.

GOODWIN (THOMAS), was one of the assembly of divines that sat at Westminster, and president of Magdalen College in Oxford. Mr. Wood styles him and Dr. Owen the "two atlasses and patriarchs of independency." He was a man of great reading, but by no means equal to Dr. Owen, and was much farther gone in fanaticism. His works, which consist of sermons and expositions, have been much read. He attended Cromwell upon his death-bed, and was very sure that he would not die, from a supposed revelation communicated to him in a prayer but a few minutes before his death. When he found himself mistaken, he exclaimed in a subsequent address to God, "thou hast deceived us, and we were deceived." He is by Mr. Granger supposed to be the independent minister and head of a college mentioned in No. 494 of the Spectator. Died Feb. 23, 1679.

GORDIANUS (the elder) was of one of the most illustrious families of the Roman senate, descended on the father's side from the Gracchi, on the mother's from Trajan. He had a great estate, an elegant taste, and a beneficent temper. He was twice consul, to which office he was appointed first by Caracalla, and afterwards by Alexander, for he possessed the uncommon talent of acquiring the esteem of virtuous princes without alarming the jealousy of tyrants. He lived at Rome in the ingenuous pursuit of letters, till the voice of the senate, and the approbation of the emperor, named him proconsul of Africa. He was finally, and at the age of fourscore, made emperor. His son, who was his lieutenant in Africa, was declared emperor with him; his manners were less pure than his father's; but his character was equally amiable. The Roman people acknowledged in the features of the younger Gordian the resemblance of Scipio Africanus. They enjoyed their dignities, however, but for a very short period The son was slain in battle in a conflict with some barbarians of Mauritania; and the father on hearing the intelligence put an end to his life.

GORDON (THOMAS), a native of Scotland, greatly distinguished by his writings on political and religious subjects, was born at Kircudbright in Galloway. He had an university education,

education, and went through the common course of academical studies; but whether at Aberdeen or St. Andrew's is uncertain. When a young man, he came to London, and supported himself by teaching the languages. His head was much turned to political and public affairs, and he was employed by the earl of Oxford in queen Anne's time; but we know not in what capacity. He first distinguished himself in the Bangorian controversy by two pamphlets in defence of the bishop; which recommended him to Mr. Trenchard, who took him into his house, at first as his amanuensis, and afterwards into partnership as an author. In 1720, they began to publish, in conjunction, a series of letters, under the name of "Cato," upon various and important subjects relating to the public. About the same time they published another periodical paper, under the title of "The Independent Whig," which was continued some years after Trenchard's death by Gordon alone. The same spirit which appears, with more decent language, in Cato's letters against the administration in the state, shews itself in this work in much more glaring colours against the hierarchy in the church. After Trenchard's death, the minister, Sir Robert Walpole, knowing his popular talents, took him into pay to defend his measures, for which end he wrote several pamphlets. At the time of his death, July 28, 1750, he was first commissioner of the wine-licences, an office which he had enjoyed many years. He was twice married. His second wife was the widow of his great friend, Trenchard; by whom he had children.

He published english translations of Sallust and Tacitus, with additional discourses to each author, which contain much good matter. Two collections of his tracts have been preserved: the first intituled, "A Cordial for Low-spirits," in three volumes; and the second, "The Pillars of Priestcraft and Orthodoxy shaken," in two volumes. But these, like many other posthumous things, had better have been suppressed.

GORDON (ALEXANDER), M. A. a Scotsman, an excellent draughtsman, and a good grecian, who resided many years in Italy, visited most parts of that country, and had also travelled into France, Germany, &c. was secretary to the Society for Encouragement of Learning; and afterwards to the Egyptian club, composed of gentlemen who had visited Egypt (viz. lord Sandwich, Dr. Shaw, Dr. Pococke, &c.). He succeeded Dr. Stukeley as secretary to the Antiquarian Society, which' office he resigned in 1741 to Mr. Joseph Ames. He went to Carolina with governor Glen, where, besides a grant of land, he had several offices, such as register of the province, &c.; and died a justice of the peace, leaving a handsome estate to his family. He published, 1. "Itinerarium
Septentrionale,

Septentrionale, or a Journey through moſt Parts of the Counties of Scotland, in two Parts, with 66 Copper plates, 1726," folio. 2. "Additions and Corrections, by Way of Supplement, to the Itinerarium Septentrionale; containing ſeveral Diſſertations on, and Deſcriptions of, Roman Antiquities, diſcovered in Scotland ſince publiſhing the ſaid Itinerary. Together with Obſervations on other ancient Monuments found in the North of England, never before publiſhed, 1732 [B]," folio. 3. "The Lives of Pope Alexander VI. and his ſon Cæſar Borgia, comprehending the Wars in the Reign of Charles VIII. and Lewis XII. Kings of France; and the chief Tranſactions and Revolutions in Italy, from the Year 1492 to the Year 1516. With an Appendix of original Pieces referred to in the Work, 1729," folio. 4. "A complete Hiſtory of the ancient Amphitheatres, more particularly regarding the Architecture of theſe Buildings, and in particular that of Verona, by the marquis Scipio Maffei; tranſlated from the Italian, 1730," 8vo. afterwards enlarged in a ſecond edition. 5. "An Eſſay towards explaining the Hieroglyphical Figures on the Coffin of the ancient Mummy belonging to Capt. William Lethieullier, 1737," folio, with cuts. 6. "Twenty-five Plates of all the Egyptian Mummies, and other Egyptian Antiquities in England," about 1739, folio.

GORDON (JAMES), a Jeſuit of one of the beſt families of Scotland, who was of deſerved eminence for his knowledge of philoſophy, of theology, and the languages. He taught Hebrew with reputation at Bourdeaux and at Paris. He viſited different parts of Europe, and ſuffered a great deal in behalf of the roman catholic religion. He died at Paris in 1620; he publiſhed a work called "Controverſiarum Chriſtianæ Fidei Epitome."

GORDON (ROBERT), of Stralogh, the author of the "Theatrum Scotiæ," a very excellent work. He died about the middle of the ſeventeenth century. This book contains a deſcription of the whole country of Scotland, with maps of every particular county. It was printed by Janſon Bleaw at Amſterdam, and dedicated to Oliver Cromwell; and to it is added Buchanan's pamphlet, "De Jure Regni apud Scotos."

GORE (THOMAS), originated from an ancient and conſiderable family at Alderton in Wiltſhire. He ſtudied in Oxford, and thence removed to Lincoln's Inn, where continuing ſome years, he retired to his eſtate in Wiltſhire.

[B] A latin edition of the "Itinerarium," including the ſupplement, was printed in Holland, 1731.

He

He died in 1684, and has written several miscellaneous pieces in the latin tongue. Ath. Oxon.

GORELLI, an italian poet, a native of Arezzo. He made Dante his model, and wrote in verse what related to the history of his country, from 1010 to 1384. His work, though not highly to be esteemed as a poem, is useful as a chronicle; and is inserted by Muratori in his character of the italian historians.

GORGIAS (Leontinus), a native of Leontium in Sicily, was a celebrated orator of the school of Empedocles, as was Socrates, and many other distinguished characters. He was deputed by his fellow-citizens to request succour of the Athenians against the people of Syracuse, whom he so charmed with his eloquence that he easily obtained what he required. He also made a display of his eloquence at the olympic and pythian games, and with so much success, that a statue of gold was erected to him at Delphi. He is reputed, according to Quintilian, to be the author and inventor of extemporaneous speaking, in which art he exercised his disciples.

GORGIAS, a renowed person in Epirus, had a remarkable birth. His mother, being near her time, sickened and died; and, as she was carrying to her grave, the bearers and mourners were astonished to hear the cry of an infant in the coffin; whereupon they returned, and opening the coffin, found Gorgias had slipped from the womb in the funeral solemnities of his mother. Her coffin was his cradle, and her death gave a great hero for the service and safety of Epirus. Val. Max.

GORIUS (Antonius Franciscus) of Florence, a respectable historian, critic, and antiquarian. He published an account of greek and latin inscriptions, which have been highly commended by Stoschius and others. The learned world is also indebted to him for many other excellent works on the subject of roman and greek antiquities. He died in 1757. The great elaborate work entitled the "Museum Florentinum, a Description of the Cabinet of the Grand Duke of Tuscany," was the production of this Gorius, or, as he is called in italian, Gorio.

GORLÆUS (Abraham), an eminent antiquary, was born at Antwerp, and gained a reputation by collecting medals and other antiques. He was chiefly fond of the rings and seals of the ancients, of which he published a prodigious number in 1601, under this title, "Dactyliotheca, sive Annulorum Sigillarium, quorum apud priscos tam Græcos quam Romanos usus ex ferro, ære, argento, & auro, Promptuarium." This was the first part of the work: the second was intituled, "Variarum Gemmarum, quibus Antiquitas in signando uti solita, sculpturæ." This work has undergone several editions, the best of which is that of Leyden, 1625: for, it not only contains

tains a vaſt number of cuts, but alſo a ſhort explication of them by Gronovius. In 1608, he publiſhed a collection of medals: which, however, if we may believe the "Scaligerana," it is not ſafe always to truſt. We meet there with the following words: "Gorlæus caſts medals; he ſhewed me ſome, but I found they were not ancient; ſince that time he ſhewed me none but genuine ones: he is a good man." Some have aſſerted, that he never ſtudied the latin tongue; and that the learned preface, prefixed to his "Dactyliotheca," was written by another. Peireſch, as Gaſſendus relates, uſed to ſay, that "though Gorlæus never ſtudied the latin tongue, yet he underſtood all the books written in latin concerning medals and coins." It is a ſign of a good genius to underſtand a latin book, only by the knowledge one has of the ſubject it treats of. Plutarch obſerves ſomewhere, that his ſtudying the roman hiſtory in greek books was the reaſon why he underſtood the language of the latin hiſtorians. But this ſtory of Peireſch cannot be reconciled with what we read in Swertius, who had been familiarly acquainted with Gorlæus, and who relates that he was brought up in the ſame ſchool with Andrew Schottus: where it cannot be ſuppoſed but that he muſt have learned latin. Gorlæus pitched upon Delft for the place of his reſidence, and died there in 1609. His collections of antiques were ſold by his heirs to the prince of Wales.

GORLÆUS (DAVID), a native of Utrecht, lived in the ſeventeeth century. He publiſhed ſome books of philoſophy, in which he departed from the common opinions of the ſchools.

GOROPIUS (JOHN), a phyſician, born in Brabant in 1518; after travelling through great part of Europe, he ſettled at Antwerp. He was a man of whimſical propenſities and very fond of paradox. He wrote and publiſhed "Origines Antverpianæ," which, with every other unaccountable opening on the origin of nations, contains the aſſertion that the flemiſh language was the language of Adam, which poſition he endeavoured to defend from ſome ridiculous etymologies.

GORRÆUS, a proteſtant phyſician of Paris, died in 1572. He was perſecuted for his religion, and, in conſequence of being abruptly apprehended by a party of ſoldiers, loſt his ſenſes. Among other works he publiſhed a tranſlation of Nicander.

GOSSELINI (JULIAN), born at Rome in 1525, was ſecretary to Ferdinand Gonzaga, viceroy of Sicily. The affairs of his ſecretaryſhip, in which he was employed above forty years, did not prevent his publiſhing ſeveral books in italian. He alſo wrote latin verſes and letters, and tranſlated into italian a french book, entituled, "A true Account of Things that have happened in the Netherlands, ſince the Arrival of Don Juan of Auſtria." He died at Milan in 1587.

GODESCHALC, a monk of Orbais, who rendered his name immortal by the controversy which he set on foot concerning predestination and grace. He lived in the ninth century, and was, for his doctrines, thrown into prison, where he languished and died: While in prison, his doctrine gained him followers; his sufferings excited compassion, and both together produced a considerable schism in the church. The death of the persecutor much.considerably diminished the heat of this intricate controversy. The celebrated Maguin published a valuable edition, which is yet extant of all the treatises which were composed on both sides the complicated question. It is in two volumes, quarto, and has this title, " Veterum Auctorum qui nono sæculo de predestinatione & Gratia scripserunt opera et fragmenta, &c."

GOSSELIN (ANTONY), was of Caen, where he was regius professor of history and eloquence, and principal of the college Du Bois. He published the " History of the Ancient Greeks," in latin, in 1636.

GOTHOFRED, the name of a very learned family, originally of France. DENNIS GOTHOFRED, a celebrated lawyer, the son of a counsellor at Paris, was born there in 1549; quitted popery; and retired first to Geneva, then to Germany, where he professed to teach law in some universities. They invited him back to France to fill the chair, which the death of Cujacius vacated in 1590; but calvinism withheld him from accepting it. He died in 1622. What he is now best known by is, an edition of the "Corpus Juris Civilis:" but he left many works upon the subject of law, some of which have been collected and published in Holland, under the title of " Opuscula," in folio.

THEODOSIUS, the eldest son of Dennis, was born at Geneva in 1580, but embraced the catholic religion, which his father had abjured. He became a counsellor of state, and died in 1649 at Munster, where he was assisting the embassy from France for a general peace. He well supported the family-reputation for letters, which his father had begun, by composing many works upon the history, rights, and titles of the kingdom.

JAMES, another son of Dennis, was born in 1587. He persevered in calvinism, and was preferred to the first offices in the republic of Geneva. He was five times Syndic, and died there in 1652. He was a man of very accurate and profound erudition. His works are, 1. An edition of " Philostorgius, in Greek and Latin, 1642," 4to. 2. " Mercure Jesuitique: a Collection of Pieces concerning the Jesuits." 3. " Opuscula Varia: juridica, politica, historica, critica." 4. " De Statu Paganorum sub Imperatoribus Christianis." 5. " Vetus Orbis

bis defcriptio Græci Scriptoris fub Conftantio, &c. Gr. & Lat. cum Notis," 4to. &c. &c.

DENNIS, the fon of Theodofius, and nephew of James, was born at Paris in 1615, and died at Lifle, director of the Chamber of Accounts, in 1681. He inherited his father's tafte for French hiftory, and made great additions to what his father had done. Of this kind are the hiftories of Charles VI. Charles VII. Charles VIII. magnificently printed at the Louvre.

JOHN, fon of the foregoing, had like his father alfo a paffion for the hiftory and antiquities of France. He fucceeded his father as director of the chamber of accounts at Lifle; where he died, very old, in 1732. He gave, 1. An edition of "Philip de Comines." 2. "Journal de Henry III." 3. "Memoires de la Reine Marguerite, &c."

GOTTI (VINCENT LOUIS) was an eminent italian ecclefiaftic, promoted to the office of cardinal, by Benedict XIII. He died in 1742; and was the author of many works, chiefly on fubjects of theology, and in vindication of the doctrines of his church.

GOTTLEBER (JOHN CHRISTOPHER), an excellent scholar and acute critic, was born in 1733. He wrote many learned works; but is particularly remarkable for his animadverfions on different portions of Plato. He died in 1785.

GOUDELIN or GOUDOULI, a favourite poet among his countrymen of Gafcony, who cite his works with great delight and, indeed, admiration. He had much wit and fprightlinefs; his works were publifhed at different times, both at Touloufe and Amfterdam. He died in 1649.

GOUDIMEL (CLAUDIUS), an excellent mufician, was put to death at Lyons for being a proteftant. He flourifhed in the fixteenth century.

GOVEA (MARTIAL), was a good latin poet, and publifhed a grammar of the latin tongue, at Paris, in the fixteenth century.

GOVEA (ANDREW), his younger brother, was a teacher of grammar and philofophy. He was engaged by John III. king of Portugal, to eftablifh a college at Coimbra. He died in 1548.

GOVEA (ANTONY), youngeft brother of the above and the moft famous of them all. Several of his writings have been publifhed both upon philofophy and the civil law. He wrote "Latin Epigrams," with great fuccefs, and publifhed editions both of Virgil and Terence. We have alfo from this eminent man a "Commentary on the Topica of Cicero," and two books of "Various Readings." He is noticed in terms of great refpect by Olivat in the preface to his edition of Cicero. His death happened in 1713, when he was counfellor of ftate at the court of Turin Our countryman, Blount, who speaks of him with much praise, relates that he died in consequence of eating immoderately of cucumbers.

GOUGE (WILLIAM), minister of Blackfriars, London, born in the parish or hamlet of Stratford Le Bow. He was bred in King's College, Cambridge, and is said never to have been absent from public prayers, morning and evening, for nine years together; and to have read fifteen chapters of the bible every day. He never took a journey merely for pleasure all his life. He preached so long till it was a greater difficulty, through age, to get into the pulpit, than to make a sermon. He died seventy-nine years old, leaving an example of humility, faith, and patience, to the imitation of posterity, and was buried in his own church, December 16, 1653. He was a good textuary, as his works, " The whole Armour of God," his " Commentary on the whole Epistles to the Hebrews," his " Exposition to the Lord's Prayer," and his other writings, sufficiently prove. He was one of the assembly of divines, and in esteem with Vossius.

GOUGE (THOMAS), minister of St. Sepulchre's, in London, from the year 1638 to 1662, was son of Dr. W. Gouge of Blackfriars. He was, throughout his life, a man of exemplary piety and benevolence of mind. He caused many thousand copies of the Bible, Catechism, Practice of Piety, and Whole Duty of Man, to be printed in Wales, where he set up upwards of three hundred schools. He was author of several practical books of divinity, which he usually distributed gratis wherever he went. He died in his sleep, with a single groan, in the year 1681, aged 77.

GOUJET (CLAUDE-PETER), a french writer, or rather editor of other people's writings, was born at Paris in 1697, and died there in 1767, after having spent his whole life in literary transactions. He published, 1. " A Supplement to Dupin's Bibliotheque of Ecclesiastical Writers." 2. " Richelet's Dictionary." 3. " An Abridgement of Richelet." 4. " Bibliotheque François," &c. &c.

GOUJON (JOHN), a sculptor and architect of Paris in the reigns of Francis I. and Henry II. He is emphatically stiled by a modern writer the Corregio of sculpture. Many noble works are or were, previous to the revolution, to be seen of this artists, at Paris. He was somtimes incorrect; but always graceful.

GOULART (SIMON), a frenchman, was born near Paris in 1543, and was one of the most indefatigable writers of these latter times. This appears by the great number of works, on which he either wrote notes or summaries of, or translated into french, or composed himself. After he had studied

died theology at Geneva, he was ordained, and succeeded Calvin in the ministry there, which office he held to the time of his death, in 1628. Plutarch's works, translated into french by Amiot, and St. Cyprian's works, are in the list of those on which he wrote notes. Scaliger had a great esteem for him. He made a large collection of very remarkable histories. He has translated into french a great many books; among the rest, the works of Seneca, published at Paris in 1590. He wrote also several treatises of devotion, upon moral subjects and upon the occurrences of his time. D'Aubigné commends these last works; for, having mentioned the titles of some books of that kind, he goes on thus: "To which I shall add the learned pathetic writings, abounding with strong arguments, which Simon Goulart of Senlis published on several occasions; a man worthy to write history, if his character would suffer him to write without partiality." When he did not put his name to his books, he used to mark it by these three initial letters S. G. S. which signified, "Simon Goulart of Senlis." He was remarkably well acquainted with all particulars relating to books and authors: insomuch, that Henry III. sent on purpose to Geneva, to know from him who the author was that assumed the name of Stephanus Junius Brutus, for the sake of publishing some very republican maxims. Goulart was in the secret, but would never reveal it, for fear of hurting those who were concerned in it. The titles of his works may be read in "Niceron's Memoires."

GOULSTON (THEODORE). This medical author was born in Northamptonshire, and became probationer-fellow of Merton College, Oxford, in 1526. In this university he studied physic, and practised for some time with considerable reputation at Wymondeham and its neighbourhood. He took a doctor's degree in 1610, removed to London, and became a fellow of the college of physicians, and afterwards censor. He was many years settled in St. Martin's parish, near Ludgate; and was much esteemed for his classical and theological learning. He died in 1632, and by his will gave 200l. to purchase a rent-charge for the payment of an annual pathological lecture, to be read in the college of physicians, some time between Michaelmas and Easter, by one of the four youngest doctors of the college. Dr. Musgrave has delivered the Goulstonian Lectures with applause. Dr. Goulston left behind him some latin versions and paraphrases of Aristotle and other greek authors. Aikin's Biog. Mem. of Med.

GOULD (ROBERT), a miscellaneous poetical writer, died in 1708. His works were published in 1709, in 2 vols. 8vo.

GOULU (JOHN), a french writer, translated into french Epictetus, Arrian, some tracts of St. Basil, and the works of Diogenes the Areopagite; he was also a controversial writer. He died in 1625, and it was said in his epitaph, that he had restored by his writings the purity of the french tongue. There were other french authors of this name.

GOUPY (JOSEPH), a fine painter in water-colours, and excelled as a copyist. He had the honour to teach her royal highness the princess of Wales. The duke of Chandos gave 300l. for his copies of the cartoons; which, at his death, did not produce 17 guineas. Died 1747.

GOURNAY (MARY DE JARS Lady of), a french female wit, was related to several noble families in Paris, but born, it is said, in Gascony, about 1565 [c]. From her infancy she had a strong turn to literature; and Montaigne publishing his first essays about this time, it was not long before they came to her hands. She read them over with eagerness, was infinitely delighted with them, conceived the highest esteem, and expressed the greatest kindness, for the author. These declarations soon reached the ears of Montaigne, who made many reflections on the occasion in praise of Mademoiselle de Gournay's talents. Hence her esteem grew into a kind of reverential affection for Montaigne, so that, happening to lose her father not long after, she adopted him in his stead, even before she had seen him; and, when he was at Paris in 1588, she made him a visit. She grew intimate with him, and prevailed upon him to accompany her and her mother the lady Gournay, where he passed two or three months. In short, our young devotee to the Muses was so wedded to books of polite literature in general, and Montaigne's Essays in particular, that she resolved never to have any other associate to her happiness. Nor was Montaigne sparing to pay the just tribute of his gratitude. He even foretold, in the second book of his essays, that she would be capable of the first-rate productions. The connexion was carried through the family; Montaigne's daughter, the viscountess de Jamaches, always claimed Mademoiselle de Jars as a sister; and the latter dedicated her piece, "Le Bouqet de Piene," to this sister. Thus she passed many years, blessing and blest in this new alliance, and when she received the melancholy news of Montaigne's death, she crossed almost the whole kingdom of France to mingle her tears and lamentations, which were excessive, with his widow and daughter [D]. Nor did her piety and filial regard stop here. She revised, corrected, and reprinted an edition of his "Essays" in 1634; to which she prefixed a preface, full

[c] Bois Robert, in "Recueil de bons contes, &c." p. 158, Dutch edition. However, Bayle imagines her to be a Parisian. [D] Pasquier's Letters, Vol. II.

of the strongest expressions of esteem and devotion for his memory.

She wrote several things in prose and verse, which were collected into one volume and published by herself in 1636, with this title, "Les Avis, & les presens de la Demoisell deGournai." Thus she took leave of the press, when she was seventy; yet she survived that period many years, not dying till 1645. She died at Paris, and epitaphs were composed for her by Menage, Valois, Patin, La Mothe Vayer, and others.

GOUNVILLE (JOHN HERAULD), was originally valet de chambre to the duke de Rouchfocault: but was afterwards, on account of his talents employed in confidential offices of state. It was on Gounville that Boileau was said to have written this epitaph.

> Ci git justement regretté
> Un savant homme sans science,
> Un gentilhomme sans naissance
> Un tres bon homme sans bonté.

He wrote two volumes of "Memoirs," which contain important anecdotes of the french ministers, from Mazarin to Colbert, and of the reign of Louis XIV. Gounville was was born in 1625, and died in 1705.

GOUSSET (JAMES), a protestant minister of Poitiers, wrote a "Hebrew Dictionary," with other theological works. Died in 1704.

GUTHIERFS (JAMES), a french advocate and man of letters; born at Chaumont, in 1638. The lovers of antiquity are indebted to him for many valuable writings; among which are the following. 1. "De vetere jure Pontificis urbis Romæ II. De officiis Domus Augustæ Publicæ & Privatæ III. De jure Manium, with several other tracts. He wrote also "Latin Verses," with considerable elegance.

GOWER (JOHN), an english poet, contemporary with Chaucer, but older, was descended from an ancient family, and born about 1320. The castle of Swansea, in Glamorganshire, was the paternal estate of Henry Gower, bishop of St. David's, in 1326; and, as this prelate survived till 1347, at which time our prelate must have been twenty-five at least, it is probable he was bred at Oxford, and at Merton-college, whereof his name-sake of St. David's had been a fellow. Some time after leaving the university, he removed to the Middle-Temple; and applied to the law with so much diligence, that he became very eminent in that profession. However, his study did not engross his whole attention; he was well read in polite literature, and had an excellent taste for poetry,

poetry, upon which he spent some of his leisure hours. This part of his character first brought him to an acquaintance with Chaucer, which afterwards grew into a very warm friendship. Many circumstances conduced to unite these two fathers of english poetry; there was a great likeness in their tempers; they were also of the same party. Chaucer had attached himself to John of Gaunt, duke of Lancaster, uncle to Richard II. and Gower had adhered as steadily to Woodstoke, duke of Gloucester, another of the king's uncles. Add to this, that Gower was as much offended with, and censured as freely, the vices of the clergy, as Chaucer did; and it is no wonder, as they were so very intimate, that they conferred together about their works, and sometimes argued warmly without anger, of which Leland speaks with much pleasure, and observes, that the only real dispute between them was, which should honour the other most [E]. Though Gower was born first, yet he outlived Chaucer; and is therefore said, not only to be Chaucer's scholar, but his successor in the laurel.

However, he took care that his inclination and genius for poetry should be no hindrance to the pursuit of his graver studies; on the contrary, while his poetical fame was daily increasing, he was most apt to establish his reputation as a lawyer; and he reaped the advantage of both. In the first character, he became a favourite of his prince, Richard II. insomuch, that one day the king, taking his diversion on the Thames, sent for our poet, who was in a boat near him, into his barge, and commanded him to exert his talent upon some useful subject [F]. He obeyed the royal mandate, and produced his "Confessio Amantis," containing a kind of poetical system of morality; in the conclusion whereof, he gave the king occasionally a great deal of good advice, and upon very delicate subjects, with much dignity and freedom. By this, and other works, he obtained the general opinion of being a good man, and was particularly distinguished by the appellation of the MORAL Gower [G]. In his character as a lawyer, he made so considerable a figure, that he is said to have been raised to the first rank in that profession, and to have sat chief justice of the common pleas. However that be, it is certain he was very eminent for his knowlege this way; and as he was sin-

[E] Leland. Comment. de Scriptor. Britan. Chaucer's Works by Urrey, p. 353. Gower's "Confessio Amantis," fol. 190. edit. 1432.
[F] Prologue to the "Confessio Amantis."
[G] This was first given him by Chaucer, at the close of his "Troilus and Cressida;" in a stanza beginning thus: "O moral Gower, this boke I directe, &c." See it in modern English in Biog. Brit. under our author's article.

gularly

gularly attached to the service of Thomas of Woodstock, first earl of Buckingham, and then duke of Gloucester, it is probable, that he belonged to that prince in the way of his profession. It is well known, that not only the king and prince of Wales, but all the princes of the blood, had their standing counsel learned in the law, who were heard in parliament, in case any bill was read that might be detrimental to their interest; and hence it may be presumed, that Gower was of this prince's counsel. Our lawyer also made his Muse pay the tribute of her tears upon the death of this patron, whose murder at Calais he lamented in a very affecting manner [H].

As his steady attachment to this prince could not but create in him much dislike to the administration of his murderer, he did not spare to lay before king Richard the luxury of his court, the irreligious lives of his clergy, the danger of listening to flatterers, the wickedness of corrupt judges, and the uncertainty of human glory and happiness, even in the most exalted ranks; especially when monarchs (which was his case) gave way to the cruelest oppressions of the people. In these sentiments, as soon as Henry IV. had deposed king Richard, and got possession of the throne, he appeared warmly on the side of the revolution; and added several historical pieces to his chronicle, called, " Vox Clamantis, or, The Voice of one crying in the Wilderness," &c. wherein with one hand he blackened the character of his old master Richard, and with the other blanched that of the new monarch, with the utmost force of his poetical pencil. In the first year of this reign, through the decay of age, being deprived of his eye-sight, he lamented that loss not long after very pathetically in " A Poem of the Commendation of Peace," where he took his leave of the Muses and the world, in such terms as plainly to testify a full sense of his approaching death, which accordingly happened in 1402.

Some short poems of his are printed among those of Chaucer; and there are many more annexed to the first edition of his book, " De Confessione Amantis." And a list of others from the Bodleian, Cotton, and All-Soul's-College libraries, may be seen in Biog. Brit. Where is also an account in Vol. II. of his " Confessio Amantis," printed by Caxton, in 1644, and again in 1554, at London.

GOUYE (JOHN), a jesuit and eminent mathematician. He was member of the Academy of Sciences; and published " Mathematical and Philosophical Observations," in two volumes, 8vo. He is not to be confounded with Gouye Longuemare, who wrote various dissertations and memoirs to illustrate the " History of France."

[H] Both in his " Vox Clamantis," and " Chronica Tripartita."

GRAAF (REGNIER DE), a celebrated phyſician, was born at Schoonhaven, a town in Holland, where his father was the firſt architect, July 30, 1641. After having laid a proper foundation for claſſical learning, he went to ſtudy phyſic at Leyden; in which ſcience he made ſo vaſt a progreſs, that in 1663 he publiſhed a treatiſe "De Succo Pancreatico," which did him the higheſt honour. Two years after he went to France, and was made M. D. at Angers; but returned to Holland the year after, and ſettled at Delft, where he practiſed in his profeſſion ſo ſucceſsfully, that he drew upon himſelf the envy of his brethren. He married in 1672, and died Aug. 17, 1673, when he was only 32 years of age. He publiſhed three pieces upon the organs of generation both in men and women, upon which ſubject he had a controverſy with Swammerdam. His works, with his life prefixed, were publiſhed in 8vo. at Leyden, in 1677 and 1705; they were alſo tranſlated into Flemiſh, and publiſhed at Amſterdam in 1686.

GRABE (JOHN ERNEST), the learned editor of the "Septuagint," from the Alexandrian MS. in the royal library at Buckingham-houſe, was the ſon of Martyn Sylveſter Grabe, profeſſor of divinity and hiſtory in the univerſity of Koningſberg in Pruſſia, where his ſon Erneſt was born, Jan. 10, 1666. He had his education there, and took the degree of M. A. in that univerſity; after which, devoting himſelf to the ſtudy of divinity, he read the works of the fathers with the utmoſt attention. Theſe he took as the beſt maſters and inſtructors upon the important ſubject of religion. He was fond of their principles and cuſtoms, and that fondneſs grew into a kind of unreſerved veneration for their authority. Among theſe he obſerved the uninterrupted ſucceſſion of the ſacred miniſtry to be univerſally laid down as eſſential to the being of a true church: this point, working continually upon his ſpirits, made by degrees ſo deep an impreſſion, that at length he thought himſelf obliged, in conſcience, to quit Lutheraniſm, the eſtabliſhed religion of his country, in which he had been bred, and enter within the pale of the roman church, where that ſucceſſion was preſerved. In this temper he ſaw likewiſe many other particulars in the lutheran faith and practice, not agreeable to that of the fathers, and conſequently abſolutely erroneous, if not heretical.

Being confirmed in this reſolution, he gave in to the electoral college at Sambia in Pruſſia, a memorial, containing the reaſons for his change in 1695; and, leaving Koningſberg, ſet out in order to put it in execution in ſome catholic country. He was in the road to a place called Erfard, in this deſign, when there were preſented to him three tracts in anſwer to his memorial,

memorial, from the elector of Brandenbourg, who had given immediate orders to three Pruſſian divines to write them for the purpoſe [1]. Grabe was entirely diſpoſed to pay all due reſpect to this addreſs from his ſovereign; and, having peruſed the tracts with care, his reſolution for embracing Popery was a little unhinged, inſomuch that he wrote to one of the divines, whoſe name was Spener, to procure him a ſafe-conduct, that he might return to Berlin, to confer with him. This favour being eaſily obtained, he went to that city, where Spener prevailed upon him ſo far as to change his deſign of going among the papiſts, for another. In England, ſays this friend, you will meet with the outward and uninterrupted ſucceſſion which you want: take then your route thither; this ſtep will give much leſs diſſatisfaction to your friends, and at the ſame time equally ſatisfy your conſcience [K]. Our divine yielded to the advice; and, arriving in England, was received with all the reſpect due to his merit, and preſently recommended to king William in ſuch terms, that his majeſty granted him a penſion of 100l. per annum, to enable him to purſue his ſtudies.

He had the warmeſt ſenſe of thoſe favours, and preſently ſhewed himſelf not unworthy of the royal bounty, by the many valuable books which he publiſhed in England; which, from this time, he adopted for his own country; and finding the eccleſiaſtical conſtitution ſo much to his mind, he entered into prieſts' orders in that church, and became a zealous advocate for it, as coming nearer in his opinion to the primitive pattern than any other. In this ſpirit he publiſhed in 1698, and the following year, "Spicilegium SS. Patrum, &c. [L]," or a collection of the leſſer works and fragments, rarely to be met with, of the fathers and hereticks of the three firſt centuries; induced thereto, as he expreſsly declared, by the conſideration, that there could be no better expedient for healing the diviſions of the chriſtian church, than to reflect on the practice and opinions of the primitive fathers [M]. Upon the ſame motive he printed alſo Juſtin Martyr's " Firſt Apo-

[1] The names of theſe divines were Philip James Spener, Bernard Van Sanden, and John William Baier. The firſt was eccleſiaſtical counſellor to the elector, and principal miniſter at Berlin; and the ſecond principal profeſſor at Koningſberg. The three anſwers were printed the ſame year. The firſt at Berlin, the ſecond at Koningſberg, both in 4to, and the third at Jana, in 8vo.

[K] Meneken's "German Dictionary," and "Pfaffii notæ in liturgiam Græcam Grabii."

[L] Both volumes were reprinted at Oxford, in 1700, 8vo.

[M] Some remarks were made upon the firſt volume, in a piece intituled, "A new and full Method of ſettling the canonical Authority of the New Teſtament, by Jer. Jones, 1726," 8vo.

logy"

logy" in 1700 [N]; and the works of Irenæus in 1702 [O]. Upon the acceſſion of queen Anne to the throne this year, our author's affairs grew ſtill better. The very warm affection which that princeſs had for the eccleſiaſtical eſtabliſhment could not but bring ſo remarkable a champion for it into her particular favour. Beſides continuing his penſion her majeſty ſought an occaſion of giving ſome farther proofs of her ſpecial regard for him, and ſhe was not long in finding one.

The "Septuagint" had never been entirely printed from the Alexandrian MS. in St. James's library, partly by reaſon of the great difficulty of performing it, in a manner ſuitable to its real worth, and partly becauſe that worth itſelf had been ſo much diſparaged by the advocates of the roman copy, that it was even grown into ſome neglect. To perform this taſk, and therein to aſſert its ſuperior merit, was an honour marked out for Grabe; and when her majeſty acquainted him with it, ſhe at the ſame time preſented him with a purſe to enable him to go through with it [P]. This was a prodigious undertaking, and he ſpared no pains to complete it. In the mean time, he employed ſuch hours as were neceſſary for refreſhment, in other works of principal eſteem. In 1705, he gave a beautiful edition of biſhop Bull's works, in folio, with notes; for which he received the author's particular thanks [Q]; and he had alſo a hand in preparing for the preſs archdeacon Gregory's pompous edition of the New Teſtament in Greek, which was printed the ſame year at Oxford [R].

From his firſt arrival he had reſided a great part of his time in that univerſity, with which he was exceedingly delighted. Beſides the Bodleian library there, he met with ſeveral perſons of the firſt claſs of learning in his own way,

[N] The works of this father came out in 1722. The editor whereof, in the dedication, obſerves that Dr. Grabe was a good man, and not unlearned, and well verſed in the writings of the fathers; but that he was no critic, nor could be one, not being endowed with genius or judgement, or, to ſpeak the truth, furniſhed with learning ſufficient for that purpoſe. Juſtini Apologia, cum notis Styan Thirlbii. Lond. 1722, fol.—The authors of the "Acta Eruditorum Lipſiæ," in their account of Thirlby's edition of Juſtin Martyr, have animadverted upon him with great ſeverity, on account of that part of his dedication, wherein he has cenſured Grabe.

[O] Several objections were made alſo to this by Rene Maſſuet, a benedictine monk, who publiſhed another edition of Irenæus, at Paris, 1710, folio.

[P] The queen's purſe was 60l. procured by Robert Harley, eſq; and it enabled him to enlarge the prolegomena to the Octateuch. See thoſe prolegomena at the end.

[Q] That learned biſhop on all occaſions, as long as he lived, acknowledged our author's ſingular generoſity as well as learning, in publiſhing his works with ſo much improvement and advantage to the great truths he had defended, and to the learned world.

[R] He reviſed the "Scholia," which Gregory, then dead, had collected from curious authors, and marked the places whence they were taken. Preface to that work.

among

among whom he found that freedom of converse and communication of studies which is inseparable from true scholars, whereby, together with his own application, he was now grown into universal esteem, and every where caressed. The alexandrian MS. was the chief object of his labour. He examined it with his usual diligence, and comparing it with a copy from that of the vatican at Rome, he found it in so many places preferable to the other, that he resolved to print it as soon as possible. With this view, in 1704, he drew up a particular account of the preferences, especially in respect to the book of "Judges," and published it, together with three specimens, containing so many different methods of his intended edition, to be determined in his choice by the learned. This came out in 1705, with proposals for printing it by subscription, in a letter addressed to Dr. Mill, principal of Edmond-hall, Oxford [s]; and that nothing might be wanting which lay in the power of that learned body to promote the work, he was honoured with the degree of D. D. early the following year, upon which occasion Dr. Smalridge, who then officiated as regius professor, delivered two latin speeches, containing the highest compliments upon his merit. The success was abundantly answerable to his fondest wishes; besides the queen's bounty, he received another present from his own sovereign the king of Prussia: and subscriptions from the principal nobility, clergy, and gentry, crouded daily upon him from all parts.

In the midst of these encouragements, the first tome of this important work came out in 1707, at Oxford, in folio and 8vo. This volume contained the Octateuch [T], and his design was to print the rest, according to the tenor of the MS. but, for want of some materials to complete the historical and prophetical books, he chose rather to break that order, and to expedite the work as much as possible [U]. The chief materials for which he waited not yet coming to hand, he was sensible that the world might expect to see the reasons of the delay, and therefore published a dissertation the following year, giving a particular account of it [v].

[s] Among our author's MSS. were found, the alexandrian texts of the "New Testament," and of "St. Clement's Epistles, by Junius, with Notes." But he never discovered his design of printing this work, which would have perfected the whole alexandrian MS. lest he should prejudice the sale of his friend Dr. Mill's "New Testament." This arduous task was reserved for the still superior industry of Dr. Woide.

[T] Prolegom. ad Octateuch.
[U] Some persons were displeased at the preference given by the doctor to the alexandrian MS. above the vatican. Vide Lettre de Th Sal. à Mr. L'Abbe, B. inserted in the supplement to Journal des Sçavans for December 1709.
[v] The title is, " J. Ernest Grabii Dissertatio de variis vitiis lxx. Interpretum ante B. Origenis ævum illatis, & remediis ab ipso Hexaplari ejusdem

In the mean time, he met with the singular misfortune of having his reputation foiled, by the brightness of his own splendor. Mr. William Whiston had not only in private discourses, in order to support his own cause by the strength of our author's character, but also in public writings, plainly intimated, "that the doctor was nearly of his mind about the Constitution of the Apostles, written by St. Clement, and that he owned in general the genuine truth and apostolical antiquity of that collection." This calumny was neglected by our author for some time, till he understood that the story gained credit, and was actually believed by several persons who were acquainted with him. For that reason he thought it necessary to inform the public, that his opinion of the Apostolical Constitutions was quite different, if not opposite, to Mr. Whiston's sentiments about them: this he did in " An Essay upon two Arabic Manuscripts in the Bodleian Library, and that ancient Book called the Doctrine of the Apostles, which is said to be extant in them, wherein Mr. Whiston's mistakes about both are plainly proved [w].

This piece was printed at Oxford, 1711, 8vo. In the dedication, he observes, that it was the first piece which he published in the english tongue, for the service of the church; and it proved in the event to be the last, being prevented in the design he had of publishing many others by his death, which happened Nov. 12, 1712, in the vigour of his age. He was interred in Westminster-abbey, where a marble monument, with his effigy at full length, in a sitting posture, and a suitable inscription underneath, was erected at the ex-

versionis additione adhibitis, deque hujus editionis reliquiis tam manuscriptis tam prælo excusis." The helps he wanted, as above intimated, were a syriac MS. of the historical books of the Old Testament, with Origen's marks upon them; besides two MSS. one belonging to cardinal Chigi, and the other to the college of Lewis le Grand. He received all afterwards, and made collations from them, as also for a volume of annotations upon the whole work, as well as for the prolegomena; all which requiring some time to digest into a proper method, the second volume did not come out till 1700, but was followed by the third the ensuing year.

[w] Grabe was assisted in this piece by Gagnier, who, about ten years before, had come over to the church of England from that of France, and then taught hebrew at Oxford; and, being well skilled in most of the oriental languages, had been appointed the year before, by Sharp, archbishop of York, to assist Grabe in perusing these MSS. having engaged the doctor to write this treatise against Whiston's notion. But as the result of the enquiry was, that the arabic " Didascalia" were nothing else but a translation of the first six entire books of the " Clementine Constitutions," with only the addition of five or six chapters not in the greek, Whiston immediately sent out " Remarks upon Grabe's Essay, &c. 1711;" wherein he claims this MS. for a principal support of his own opinions. He declares, therefore, the doctor could not have served him better than he had done in this essay. Nor has almost, says he, any discovery, I think, happened so fortunate to me, and to that sacred cause I am engaged in from the beginning, as this essay of his before us.

pence

pence of the lord-treasurer Harley [x]. He had so great a zeal for promoting the ancient government and discipline of the church, among all those who had separated themselves from the corruption and superstitions of the church of Rome, that he formed a plan, and made some advances in it, for restoring the episcopal order and office in the territories of the king of Prussia, his sovereign; and he proposed, moreover, to introduce a liturgy much after the model of the english service, into that king's dominions. He recommended likewise the use of the english liturgy itself, by means of some of his friends, to a certain neighbouring court. By these methods, his intention was to unite the two main bodies of Protestants in a more perfect and apostolical Reformation than that upon which either of them then stood, and thereby fortify the common cause of their protestation against the errors of Popery, against which he left several MSS. finished and unfinished; in latin, whereof the titles in english are to be found in Dr. Hickes's account of his MSS. Among which also were several letters, which he wrote with success, to several persons, to prevent their apostacy to the church of Rome, when they were ready to be reconciled to it. In these letters he challenged the priests to meet him in conferences before the persons whom they had led astray; but they knowing, says Dr. Hickes, the Hercules with whom they must have conflicted, wisely declined the challenge.

He left a great number of MSS. behind him, which he bequeathed to Dr. Hickes for his life, and after his decease to Dr. George Smalridge. The former of these divines carefully performed his request of making it known, that he had died in the faith and communion of the church of England, in an account of his life, prefixed to a tract of our author's, which he published with the following title: "Some Instances of the Defect and Omissions in Mr. Whiston's Collections of Testimonies, from the Scriptures and the Fathers, agaist the true Deity of the Holy Ghost, and of misapplying and misinterpreting divers of them, by Dr. Grabe. To which is premised, a Discourse, wherein some Account is given of the learned Doctor, and his MSS. and of this short Treatise found among his English MSS. by George Hickes, D. D. 1712," 8vo. There came out afterwards two more of our author's posthumous pieces. 1. "Liturgia Græca Johannis Ernesti Grabii." This liturgy, drawn up by our author for his own private use, was published by Christopher Matthew Pfaff, at the end of "Irenæi Fragmenta Anec-

[x] It stands against the western wall of the south cross aile, a good height over that of Camden.

dota,"

dota," printed at the Hague, 1715, 8vo. 2. " De Forma Confecrationis Euchariftiæ, hoc eft, Defenfio Ecclefiæ Græcæ, &c." i. e. " A Difcourfe concerning the Form of of Confecration of the Eucharift, or a Defence of the Greek Church againft that of Rome, in the Article of confecrating the Euchariftical Elements, written in Latin, by John Ernett Grabe, and now firft publifhed with an Englifh verfion." To which is added, from the fame author's MSS. fome notes concerning the oblation of the body and blood of Chrift, with the form and effect of the euchariftical confecration, and two fragments of a preface defigned for a new edition of the firft liturgy of Edward VI. with a preface of the editor, fhewing what is the opinion of the church of England concerning the ufe of the fathers, and of its principal members, in regard to the matter defended by Dr. Grabe in this treatife, 1721, 8vo.

GRACIAN (BALTHAZAR), a fpanifh Jefuit, and rector of the college of Tarragon. He wrote feveral works on theological fubjects, and was in great eftimation with his countrymen. He died in 1958.

GRACCHUS (TIBERIUS and CAIUS), fons of Sempronius Gracchus and Cornelia, daughter of Scipio. Tiberius, the elder, was a great patriot, and promoter of the agrarian law. He fell, however, a victim to his zeal; nor did his brother Caius long furvive him, but was killed under fimilar circumftances. To the principles, the conduct, and the fate of thefe men, parallels may be eafily found in the hiftory of many leading men in the french revolution. There were many other individuals of this name diftinguifhed in the annals of Rome.

GRADENIGO (PETER). is celebrated in the hiftory of Europe as having been principally inftrumental in reducing the government of Venice to an ariftocracy, and indeed, to the form which it ftill affumes. He was doge in 1290, and died in 1303.

GRÆME (JOHN), was born at Carnwarth, in Lanarkfhire, in 1748. His father was of the middling clafs of farmers, whofe wealth confifted chiefly in fix children and in his induftry, for which, and his integrity, he was diftinguifhed among his neighbours. He was the youngeft of four fons, and of a conftitution lefs robuft than that of his brothers. Early in life, having difcovered an uncommon proficiency in the learning taught at the fchool of the village, they refolved to difpenfe with his fervices in the bufinefs of the farm, for which he promifed to be unequal, and to educate him in the church; an object of common ambition in that part of the ifland, where the falary of an ecclefiaftic offers no temptation to the rich, and the attainment of a liberal education is within the reach of persons of inferior rank. At the age of fourteen (1763) he was placed at the school of Lanark, under the care of Mr. Robert Thomson [y], a teacher of eminent learning and abilities. Here his progress in grammatical learning was rapid, and, considering his early disadvantages, incredible. His exercises in particular were the admiration of his master; whose discernment construed those eccentricities of imagination, which received his correction, into a presage of future eminence. In 1776 he was removed to the university of Edinburgh. In this justly celebrated seminary his talents found ample scope and encouragement. Accustomed to excel, his desire of excellence found greater excitement, and his industry was equal to his emulation, which prompted him to aim at distinction in the most abstruse and difficult studies, where either a competitor, or applause, could be found. His success was answerable to his assiduity. In classical learning he surpassed the most industrious and accomplished student of his standing. He spoke and composed in latin with a fluency and elegance that had few examples. And, of mathematics, natural philosophy, and metaphysics, his knowledge was considerable. To this was owing a certain proneness to disputation and metaphysical refinement, for which he was remarkable, and which he often indulged to a degree that subjected him to the imputation of imprudence, and (among the unlearned) of free-thinking. His thoughts, full of ardour and vivacity, would often, indeed, make excursions beyond the limits of system, and the narrow views of prejudice, yet were these excursions ever made with modesty; nor was his propensity to argument ever accompanied with arrogance, but was merely the wantonness of conscious talents, and the ebullition of youthful vanity, which abated, and subsided, as he advanced in the study of a more liberal and enlightened philosophy. The belles lettres, a more humanising subject of enquiry, unfolded to his view those attractive beauties to which his mind seemed to have an innate, though hitherto undiscovered, propensity. Recognising, as it were, the standard of excellence congenial to his taste, moral philosophy, history, poetry, and criticism, became his favourite pursuits, and supplanted every inquisitive passion of a less amiable tendency. In tracing the lineaments of humanity, truth, and

[y] This learned and worthy schoolmaster, it is less generally known, was brother-in-law of the celebrated author of "The Seasons." In the memoirs prefixed to his works by Dr. Murdoch, Mrs Thomson should have been added to the two sisters he is said to have left. She died Sept. 3, 1781, and was the last of the poet's three surviving sisters. With a considerable share of his taste, she possessed a large portion of his amiable benevolence.

beauty,

beauty, the feelings of his heart expanded, and his judgement and imagination acquired precision and delicacy. The inchantment of metaphyfical philofophy, the vifions of Malebranche, and the fubtleties of Hume, now loft poffeffion of his admiring fancy. Full of admiration of the inftructive and fublime writings of the moralift, hiftorian, and poet, he forfook the purfuit of an illufive and unfatisfactory philofophy, whofe fophiftry deceives the underftanding, and whofe fcepticifm contracts the heart. His chief delight was to perufe the moft approved delineations of virtue and of nature, and the moft fuccefsful reprefentations of life and of manners; and his higheft ambition to imitate the beft mafters in the different departments of claffical and ornamental learning. His turn for elegant compofition firft appeared in the folution of a philofophic queftion, propofed as a college-exercife, which he chofe to exemplify in the form of a tale, conceived and executed with all the fire and invention of caftern imagination. This happened in 1769; and his firft attempts in poetry are of no earlier date.

About this time, on the recommendation of Alexander Lockhart, efq [z], he was prefented to an exhibition (or burfary, as it is called) in the univerfity of St. Andrew, which he accepted, but found reafon foon after to decline, upon difcovering that it fubjected him to repeat a courfe of languages and philofophy, which the extent of his acquifitions, and the ardour of his ambition, taught him to hold in no great eftimation. This ftep, it may be fuppofed, did not meet with the approbation of his friends; and the only advantage he derived from the event (the moft important in his life), was a view of the venerable city of St. Andrew, which amufed his imagination, and an acquaintance with Dr. Wilkie (author of the Epigoniad), which confirmed him in the purfuit of poetical fame. In 1770, he refumed his ftudies at Edinburgh, and, having finifhed the ufual preparatory courfe, was admitted into the theological clafs: but the ftate of his health, which foon after began to decline, did not allow him to deliver any of the exercifes ufually prefcribed to ftudents in that fociety. It is a confideration mortifying to human genius, that fine talents, and the moft delicate fenfibility, are but too often the predifpofing caufe of an infidious and

---

[z] Dean of the faculty of advocates, and now lord Lovington of the Court of Seffion in Scotland. As an advocate, his learning and eloquence conftitute an æra in the hiftory of the fcottifh bar. He is of the family of Lockhart of Carnwath, fon of the author of the "Memoirs of Scotland," and uncle to General Lockhart (in the auftrian fervice), the prefent reprefentative of the family. The father of Mr. Græme then refided upon the eftate of General Lockhart; as does his eldeft brother, a reputable farmer in the neighbourhood of Carnwath.

fatal

fatal difeafe. In autumn 1771, his ill-health, that had been increafing almoft unperceived, terminated in a deep confumption; the complicated diftrefs of which, aggravated by the indigence of his fituation, he bore with an heroic compofure and magnanimity. Hope, that commonly alleviates the fufferings of the confumptive, he renounced from the beginning: which, at his years, and with his fenfibility, the fires of literary ambition juft kindling, and his wifhes rapt in the trance of fame, required an uncommon union of philofophy and religion. Convinced that his fate was inevitable, and feeling himfelf every day declining, his eafy humour and poetical talent fuffered no confiderable interruption or decay. He continued at intervals to compofe verfes, and to correfpond with his friends, and, after a tedious ftruggle of ten months, expired July 26, 1772, in the 22d year of his age. His poems, confifting of elegies and mifcellaneous pieces, were collected, and printed at Edinburgh, 1773, 8vo.

GRAFFIO, a cafuift of the fixteenth century, born at Capua, wrote two quarto volumes on fubjects of morality.

GRAFIGNY (FRANCES), a french lady of refpectable talents. She wrote the Peruvian letters, which have been tranflated into every European language, and are indeed to be admired for delicacy of fentiment, and elegance of ftyle. She wrote alfo various pieces for the theatre, which were well received. M. Grafigny died at Paris in 1758.

GRAFTON (RICHARD), was born in London, and flourifhed in the reigns of Henry VIII. Edward VI. Mary, and Elizabeth. He publifhed an abridgement of the Chronicles of England, and "A Chronicle, and large meere Hiftory of the Affayers of England, and Kings of the fame, deduced from the Creation of the World."

GRAHAM (GEORGE), clock and watch-maker, was born at Gratwick, a village in the north of Cumberland, in 1675: and, in 1688, came up to London. He was not put apprentice to Tompion, as is generally faid; but, after he had been fome time with another maſter, Tompion received him into his family purely for his merit, and treated him with a kind of parental affection till his death. That Graham was, without competition, the moft eminent of his profeffion, is but a fmall part of his character: he was the beft mechanic of his time, and had a complete knowledge of practical aftronomy; fo that he not only gave to various movements for the menfuration of time, a degree of perfection which had never before been attained, but invented feveral aftronomical inftruments, by which confiderable advances have been made in that fcience: he made great improvements in thofe which had before been in use; and, by a wonderful manual dexterity, constructed them with greater precision and accuracy than any other person in the world.

The great mural arch in the observatory at Greenwich was made for Dr. Halley under his immediate inspection, and divided by his own hand; and, from this incomparable original, the best instruments of the kind in France, Spain, Italy, and the West Indies, are copies, made by english artists. The sector, by which Dr. Bradley first discovered two new motions in the fixed stars, was his invention and fabric. He comprised the whole planetary system within the compass of a small cabinet, from which, as a model, all the modern orreries have been constructed: and when the french academicians were sent to the north, to make observations in order to ascertain the figure of the earth, they thought Graham the fittest person in Europe to furnish them with instruments. They accordingly succeeded, performing their work in one year; so that, by subsequent observations in France, Sir Isaac Newton's theory was confirmed. But the academicians, who went to the south, not taking instruments, were very much embarrassed and retarded.

He was many years a member of the Royal Society, to which he communicated several ingenious and important discoveries, particularly a kind of horary alteration of the magnetic needle; a quicksilver pendulum, and many curious particulars relating to the true length of the simple pendulum, upon which he continued to make experiments till a few years before his death. His temper was not less communicative than his genius was penetrating, and his principal view was not either the accumulation of wealth, or the diffusion of his fame, but the advancement of science, and the benefit of mankind. As he was perfectly sincere, he was without suspicion; as he was above envy, he was candid; and as he had a relish for true pleasure, he was generous. He frequently lent money, but could never be prevailed upon to take any interest; and for that reason he never placed out any money upon government securities. He had bank-notes, which were thirty years old, by him when he died; and his whole property, except his stock in trade, was found in a strong box, which, though less than would have been heaped by avarice, was yet more than would have remained to prodigality.

Nov. 24, 1751, he was carried, with due solemnity and attendance, to Westminster-abbey; and there interred in the same grave with the remains of his predecessor, Tompion.

GRAIN (JOHN BAPTIST LE), a french historian, was born in 1565, and, after a liberal education, became counsellor

sellor and master of the requests to Mary de Medicis, queen of France. He frequented the court in his youth, and devoted himself to the service of Henry IV. by whom he was much esteemed and trusted. Being a man of probity, and no ambition, he did not employ his interest with Henry to obtain dignities, but spent the greatest part of his life in reading and writing. Among other works which he composed, are "The History of Henry IV." and "The History of Lewis XIII. to the Death of the Marshal d'Ancre," in 1617; both which works were published in folio, under the title of "Decades." The former he presented to Lewis XIII. who read it over, and was infinitely charmed with the frankness of the author: but the jesuits, whose policy has never made them fond of free-speakers, found means to have this work castrated in several places. They served "The History of Lewis XIII." worse; for, Le Grain having spoken advantageously therein of the prince of Condé, his protector, they had the cunning and malice to suppress those passages, and to insert others, where they made him speak of him very indecently. Condé was a dupe to this piece of knavery, till Le Grain had time to vindicate himself, by restoring this as well as his former work to their original purity. He died at Paris in 1643, and ordered in his will, that none of his descendants should ever trust the education of their children to the jesuits; which clause, it is said, has been punctually observed by his family.

GRAINDORGE (ANDREW), a native of Caen; a physician, and eminent scholar. He published a treatise on fire, light, and colours; with various other works. He died 1676.

GRAMAYE (JOHN BAPTIST), historiographer of the Low-countries, and provost of Arnheim. He travelled over Germany and Italy, and was going to Spain; but, being intercepted by African corsairs, was carried to Algiers. He returned, some time after, to the Low-countries, and died at Lubeck in 1635. His works are, 1. "Africæ Illustratæ Libri X. 1622," 4to. "An History of Africa," from the earliest Antiquity to his own Time. 2. "Diarium Algeriense." 3. "Peregrinatio Belgica," 8vo. This is reckoned an exact and curious work. 4. "Antiquitates Flandriæ," fol. 5. "Historiæ Namurcensis." Gramaye was also a poet, but his verses are not so good as his prose.

GRAMMOND (GABRIEL, lord of), more respectable as a man of integrity than as a writer. He wrote a history of Louis XIII. He wrote also a history of the wars of Louis XIII. against his protestant subjects, which, though partial, is very curious. He died in 1654

GRAMONT (ANTONY Duke of), who, at a very early age, diſtinguiſhed himſelf as a warrior. He was in great favour with cardinal Richlieu, to whom he was related. For his important military ſervice he was made marſhal of France. He was one of the greateſt ornaments of the court of Louis XIV. and alike accompliſhed in the field and in the cabinet. He wrote two volumes of Memoirs, and died in 1678.

GRAMONT (PHILIBERT, Count of), ſon of the preceding. He ſerved as a volunteer under the prince of Condé and Turenne; came into England about two years after the Reſtoration. He was under a neceſſity of leaving France, as he had the temerity to make his addreſſes to a lady, to whom Lewis XIV. was known to have a tender attachment. He poſſeſſed in a high degree every qualification that could render him agreeable to the engliſh court. He was gay, gallant, and perfectly well-bred, had an inexhauſtible fund of ready wit, and told a ſtory with inimitable grace and humour. Such was his vivacity, that it infuſed life wherever he came, and, what rarely happens, it was ſo inoffenſive, that every one of the company appeared to be as happy as himſelf. He had great ſkill and ſucceſs in play, and ſeems to have been chiefly indebted to it for ſupport. Several of the ladies engaged his attention upon his firſt coming over; but the amiable Mrs. Hamilton, whom he afterwards married, ſeems to have been the only woman who had the entire poſſeſſion of his heart. His elegant "Memoirs" were written from his own information by count Hamilton, and probably in much the ſame language in which they are related.

GRANCOLAS (JOHN), doctor of the Sorbonne, died in 1732. He was author of many works on theological ſubjects, and ſome tranſlations from the fathers. He was a reſpectable ſcholar; but, on the whole, an indifferent writer.

GRAND (ANTONY LE), a Carteſian philoſopher of the laſt century; wrote many works on philoſophical and hiſtorical ſubjects. His moſt eſteemed production is a ſacred hiſtory from the creation to the time of Conſtantine the great, printed in London in 8vo.

GRAND (JOACHIM LE), a french writer on political ſubjects, and indeed a man of general and extenſive accompliſhments. He was in conſiderable eſtimation at the court of Louis XIV. and left many works of conſiderable utility and intereſt to all who are curious in inveſtigating the hiſtory of France. The abbé le Grand tranſlated Lobo's hiſtory of Abyſſinia into french; as well as Ribeyro's hiſtory of the iſland of Ceylon. He died at Paris in 1733, at the age of eighty.

GRAND (MARC ANTONY LE), a french actor and poet, died at Paris in 1728. He wrote a great number of comedies,
ſome

some of which were favourably received, and excelled in different characters as a performer. His works were published in four volumes 12mo. His figure was disagreeable, of which he was not unconscious; for, in one of his addresses to the audience, "Ladies and gentlemen," says he, "it is easier for you to reconcile yourself to my figure than for me to change it.

GRAND (Louis), a french writer, and doctor of the Sorbonne. His writings are admired for their perspicuity and accurate arrangement. His productions are all on theological subjects.

GRANDET (Joseph), a pious and amiable french priest, and accomplished man. He was also an author; but chiefly wrote on subjects of biography, and published several volumes of lives in 12mo.

GRANDIER (Urban), curate and canon of Loudun in France, famous for his intrigues and tragical end, was the son of a notary royal of Sablé, and born at Bouvere near Sablé, we know not in what year. He was a man of reading and good judgement, and a famous preacher; for which the monks of Loudun soon hated him, especially after he had urged the necessity of confessing sins to the curate at Easter. He was a handsome man, of an agreeable conversation, neat in his dress, and cleanly in his person; which made him suspected of loving the fair sex, and of being beloved by them. In 1629, he was accused of having had a criminal conversation with some women in the very church of which he was curate: and the official condemned him to resign all his benefices, and to live in penance. He brought an appeal, this sentence being an encroachment upon the civil power; and, by a decree of the parliament of Paris, he was referred to the presidial of Poitiers, in which he was cleared. Three years after, some ursuline nuns of Loudun were thought, by the vulgar, to be possessed with the devil; and Grandier's enemies, the capuchins of Loudun, charged him with being the author of the possession, that is, with witchcraft. They thought, however, that in order to make the charge succeed according to their wishes, it was very proper to strengthen themselves with the authority of cardinal Richlieu. For this purpose, they wrote to father Joseph, their fellow-capuchin, who had great credit with the cardinal, that Grandier was the author of the piece, intituled, "La Cordonnierre de Loudun;" that is, "The Woman Shoe-maker of Loudon;" which was a severe satire upon the cardinal's person and family. This great minister, among a number of noble perfections, laboured under this defect, that he would prosecute to the utmost the authors of the libels against him; so that, father Joseph having persuaded him that

Grandier was the author of "La Cordonniere de Loudun," though nobody believed him to be so, he wrote immediately to De Laubardemont, counsellor of state, and his creature, to make a diligent enquiry into the affair of the nuns; and gave him sufficiently to understand, that he desired to destroy Grandier. De Laubardemont had him arrested Dec. 1633; and, after he had thoroughly examined the affair, went to meet the cardinal, and to take proper measures with him. July 1634, letters patent were drawn up and sealed, to try Grandier; and were directed to De Laubardemont, and to 12 judges chosen out of the courts in the neighbourhood of Loudun; all men of honour indeed, but very credulous, and on that account chosen by Grandier's enemies. Aug. 18, upon the evidence of Astaroth, the chief of possessing devils; of Easas, of Celsus, of Acaos, of Eudon, &c. that is to say, upon the evidence of the nuns, who asserted that they were possessed with those devils, the commissaries passed judgement, by which Grandier was declared well and duly attained and convicted of the crime of magic, witchcraft, and possession, which by his means happened on the bodies of some ursuline nuns of Loudun, and of some other lay persons, mentioned in his trial; for which crimes he was sentenced to make the *amende honorable*, and to be burnt alive with the magical covenants and characters which were in the register-office, as also with the MS. written by him against the celibacy of priests; and his ashes to be thrown up into the air. Grandier heard this dreadful sentence without any emotion; and, when he went to the place of execution, suffered his punishment with great firmness and courage.

The story of this unhappy person shews how easily an innocent man may be destroyed by the malice of a few, working upon the credulity and superstition of the many: for, Grandier, though certainly a lascivious man, was as certainly innocent of the crimes for which he suffered. Renaudot, a famous physician, and the first author of the french gazette, wrote Grandier's eulogium, which was published at Paris in loose sheets. It was taken from Menage, who openly defends the curate of Loudun, and calls the possession of those nuns chimerical. In 1693, was published at Amsterdam, "Histoire des Diables de Loudun;" from which very curious account it appears, that the pretended possession of the Ursulines was an horrible conspiracy against Grandier's life. Well might Menage affirm, that Grandier "deserves to be " added to Gabriel Naude's Catalogue of great Men, unjustly " charged with Magic."

As to the MS. against the celibacy of priests, mentioned above, Grandier confessed that he composed that work: and
it

it is fuppofed he might write it, although he made that confeffion upon the rack. The funeral oration of Scevola Sammarthanus, which Grandier delivered at Loudun, is printed with Sammarthanus's works.

GRANDIN (MARTIN), doctor of the Sorbonne, wrote a courfe of theology in 6 volumes quarto, which was well received by the public. He died at Paris in 1691.

GRANDUAT (CHARLES), a celebrated french comedian, who, for the fpace of thirty-five years, reprefented the characters of petits maitres in the Paris theatre; neither was he contemptible in tragedy. He was alfo a writer of poetry, and produced fome operas of no defpicable merit.

GRANDIUS (GUIDO), of Cremona, diftinguifhed himfelf as a learned man, and particularly as a mathematician. He wrote various works, and tranflated Euclid into italian; he was born in 1671, and died in 1742.

GRANET (FRANCIS), a french writer of profound and various erudition. The abbé de Fontaine, who was his particular friend, has given him the higheft character for amiable manners and exquifite talents. He tranflated Sir Ifaac Newton's chronology; he wrote remarks on the tragedies of Corneille and Racine, with a great number of other elegant works. He was compelled, contrary to his natural temper, and to the difgrace of his great abilities, to labour as a journalift, an occupation which he hated and defpifed; but fuch undertakings were neceffary to his fupport. He died at Paris in 1741.

GRANGE (JOSEPH DE CHANCEL), a frenchman of great tafte and accomplifhments. He fuffered in early life many fevere hardfhips from his having written fome fatirical verfes againft Philip duke of Orleans. He lived fome years in exile, and not a few in prifon. On the death of his adverfary, he returned to France; and, without referve, indulged the bent of his talents. He died in 1758, leaving many works. The principal of thefe were publifhed in five volumes, and confift of various dramatic pieces and mifcellaneous poems. His tragedies are moft deferving of attention; but all his works are diftinguifhed by a confiderable degree of genius. There were other ingenious frenchmen of this name.

GRANGE (N.), born at Paris in 1738, is known by an edition which he publifhed of the greek antiquities of Le Bos; by a tranflation of Lucretius, with many learned notes; by a tranflation of Seneca, publifhed after his death. Diderot was his friend; and to the laft-mentioned work prefixed a life of Seneca. Grange was diftinguifhed by an intimate acquaintance with both antient and modern authors, by much critical fagacity, and by an excellent and amiable character.

GRANGER or GRAINGER (JAMES, M. D.) author of a tranflation of Tibullus, a poem on the fugar-cane, and feveral medical tracts; was born in Dunfe, a fmall town in the fouth of Scotland, about the year 1723. His fchool-education being finifhed, he was fent to Edinburgh, and placed with Mr. Lawder, a very eminent furgeon there, where he had the opportunity of cultivating his abilities under profeffors who at that time had acquired a great degree of celebrity in the medical world.

The doctor's firft outfet in the line of his profeffion was as furgeon in the army; and, in that capacity, he ferved in Germany under the earl of Stair, till the peace of Aix-la-Chapelle in 1748, after which he fettled in London, and practifed as a phyfician. He was foon taken notice of as a man of genius by the learned of that time; he cultivated the acquaintance of Shenftone, and a great degree of intimacy fubfifted between them till Shenftone's death. Dr. Percy, now bifhop of Dromore, in Ireland, was alfo one of his particular friends.

While in London he publifhed his tranflation of the elegies of Tibullus. This did not meet with all the approbation the Doctor thought it merited; particularly from the late Dr. Smollett, whom Granger conceived to be rather illiberal in his criticifms upon it, waich was the caufe of a long paper war between them, carried on with fuch a degree of warmth, that a reconciliation never could take place.

Whether the practice of phyfic in London anfwered the Doctor's expectation or not, is not certain; but we find that, about the beginning of the war in the late king's reign, he embraced an offer of fettling advantageoufly as phyfician on the ifland of St. Chriftopher. It was on the paffage out, there being a large fleet under convoy to the Weft Indies, that a lady, on-board one of the merchantmen bound for the fame ifland, was taken ill of the fmall-pox, attended with fome alarming fymptoms: a boat was difpatched to the fhip in which Dr. Grainger was a paffenger, foliciting his advice; the Doctor accordingly vifited the lady, and very humanely continued with her during the reft of the voyage. Befides humanity, the Doctor had an inducement to finifh his paffage in this fhip, namely, the company of an agreeable young lady, the daughter of his patient, with whom he became enamoured. It would feem the flame was mutual; they were united in wedlock foon after their arrival in St. Chriftopher's. By his marriage with this lady, whofe name was Burt, he became connected with feveral of the principal families in the ifland. He here practifed phyfic with great fuccefs; but, at the fame time, did not allow his Mufe to lie dormant; for,

during

during his leisure-hours, he wrote his beautiful poem on the culture of the sugar-cane, besides a treatise on the diseases of the West Indies, for the use of planters. On the conclusion of the war, he paid a visit to his native country, and, at the same time, published his Sugar-cane. After a few years residence in Britain, he returned to St. Christopher's, and continued to practice till the beginning of the year 1767, when he was seized with a fever, which then raged in the island, and died on the 9th day of the disease.

Mrs. Grainger and one daughter are all that remain of his family. His daughter inherits a small landed estate in the neighbourhood of Edinburgh.

Dr. Grainger was benevolent in his disposition, engaging in his manners, and an able physician; considered as a poet, he certainly ranks high above the middling class. His Sugar-cane has certainly great poetical merit; the notes are copious, and relate chiefly to the natural history of the island. An Ode to Solitude, and a West-Indian Ballad, the latter published in Dr. Percy's collection, are both much admired. It is to be regretted, that his poetical works have never been collected and published together: they would undoubtedly be very acceptable to the public.

GRANT (FRANCIS), lord Cullen, an eminent lawyer and judge in Scotland, was descended from a younger branch of the antient family of the Grants, of Grant in that kingdom; his ancestor, in a direct line, being Sir John Grant of Grant, who married lady Margaret Stuart, daughter of the earl of Athol. He was born about 1660, and received the first part of his education at Aberdeen; but, being intended for the profession of the law, was sent to finish his studies at Leyden, under the celebrated Voet, with whom he became so great a favourite, by his singular application, that many years afterwards the professor mentioned him to his pupils, as one that had done honour to the university, and recommended his example to them. On his return to Scotland, he passed through the examination requisite to his being admitted advocate, with such abilities as to attract the particular notice of Sir George Mackenzie, then king's advocate, one of the most ingenious men, as well as one of the ablest and most eminent lawyers of that age.

Being thus qualified for practice, he soon got into full employ, by the distinguishing figure which he made at the Revolution in 1688. He was then only 21 years of age; but, as the measures of the preceding reign had led him to study the constitutional points of law, he discovered a masterly knowledge therein, when the Convention of Estates met to debate that important affair concerning the vacancy of the throne,

throne, upon the departure of king James to France. Some of the old lawyers, in pursuance of the principles in which they had been bred, argued warmly against those upon which the Revolution, which had taken place in England, was founded; and particularly insisted on the inability of the Convention of Estates to make any disposition of the crown. Grant opposed these notions with great strength and spirit, and about that time published a treatise, in which he undertook, by the principles of law, to prove that a king might forfeit his crown for himself and his descendants; and that in such a case the States had a power to dispose of it, and to establish and limit a legal succession, concluding with the warmest recommendations of the prince of Orange to the regal dignity.

This piece, being generally read, was thought to have had considerable influence on the public resolutions, and certainly recommended him to both parties in the way of his profession. Those who differed from him in opinion admired his courage, and were desirous of making use of his abilities; as on the other hand, those who were friends to the Revolution were likewise so to him, which brought him into great business, and procured him, by special commissions, frequent employment from the crown. In all which he acquitted himself with so much honour, that, as soon as the union of the two kingdoms came to be seriously considered in the english court, queen Anne unexpectedly, as well as without application, created him a baronet in 1705, in the view of securing his interest towards completing that design; and upon the same principle her majesty about a year after appointed him one of the judges, or (as they are styled in Scotland) one of the senators of the college of justice.

From this time, according to the custom of Scotland, he was styled, from the name of his estate, lord Cullen, and the same good qualities which had recommended him to this post were very conspicuous in the discharge of it; in which he continued for 20 ears with the highest reputation, when a period was put to his life, by an illness which lasted but three days; and, though no violent symptoms appeared, yet his physicians clearly discerned that his dissolution was at hand. They acquainted him therewith, and he received the message not only calmly but chearfully; declaring that he had followed the dictates of his conscience, and was not afraid of death. He took a tender farewel of his children and friends, recommended to them earnestly a steady and constant attachment to the faith and duty of Christians, and assured them that true religion was the only thing that could bring a man peace at the last. He expired soon after quietly, and without any agony, March 16, 1726, in his 66th year.

He was so true a lover of learning, and was so much addicted to his studies, that, notwithstanding the multiplicity of his business while at the bar, and his great attention to his charge when a judge, he nevertheless found time to write various treatises, on very different yet important subjects; some political, which were remarkably well-timed, and highly serviceable to the government; others of a most extensive nature, such as his essays on law, religion, and education, which were dedicated to his late majesty when prince of Wales, by whose command, his then secretary, Mr. Samuel Molyneux, wrote him a letter of thanks in which were many gracious expressions, as well in relation to the piece as to its author. He composed, besides these, many discourses on literary subjects, for the exercise of his own thoughts, and for the better discovery of truth, which went no farther than his own closet, and, from a principle of modesty, were not communicated even to his most intimate friends.

In his private character he was as amiable as he was respectable in the public. There were certain circumstances that determined him to part with an estate, that was left him by his father; and it being foreseen that he would employ the produce of it, and the money he had acquired by his profession, in a new purchase, there were many decayed families who solicited him to take their land upon his own terms, relying entirely on that equity which they conceived to be the rule of his actions. It appeared that their opinion of him was perfectly well grounded; for, being at length prevailed upon to lay out his money on the estate of an unfortunate family, who had a debt upon it of more than it was worth, he first put their affairs into order, and by classing the different demands, and compromising a variety of claims, secured some thousand pounds to the heirs, without prejudice to any, and of which they had never been possessed but from his interposition and vigilance in their behalf; so far was he either from making any advantage to himself of their necessities, or of his own skill in his profession; a circumstance justly mentioned to his honour, and which is an equal proof of his candor, generosity, and compassion. His piety was sincere and unaffected, and his love for the Church of Scotland was shewn, in his recommending moderation and charity to the clergy as well as laity, and engaging the former to insist upon moral duties as the clearest and most convincing proofs of men's acting upon religious principles; and his practice, through his whole life, was the strongest argument of his being thoroughly persuaded of those truths, which, from his love to mankind, he laboured to inculcate. He was charitable without ostentation, disinterested in his friendships,

ships, and beneficent to all who had any thing to do with him. He was not only strictly just, but so free from any species of avarice, that his lady, who was a woman of great prudence, finding him more intent on the business committed to him by others than on his own, took the care, of placing out his money, upon herself; and, to prevent his postponing, as he was apt to do, such kind of affairs, when securities offered, she caused the circumstances of them to be stated in the form of cases, and so procured his opinion upon his own concerns, as if they had been those of a client. These little circumstances are mentioned as more expressive of his temper than actions of another kind could be; because, in matters of importance, men either act from habit, or from motives that the world cannot penetrate; but, in things of a trivial nature, are less upon their guard, shew their true disposition, and stand confessed for what they are. He passed a long life in ease and honour. His sincerity and steady attachment to his principles recommended him to all parties, even to those who differed from him most; and his charity and moderation converted this respect into affection, so that not many of his rank had more friends, and perhaps none could boast of having fewer enemies. He left behind him three sons and five daughters; his eldest son Archibald Grant, esq. served in his father's life-time for the shire of Aberdeen; and becoming by his demise Sir Archibald Grant, bart. served again for the same county in 1717. His second son, William, followed his father's profession, was several years lord-advocate for Scotland; and, in 1757, one of the lords of session, by the title of lord Prestongrange. Francis, the third son, was a merchant; three of the daughters were married to gentlemen of fortune; and the two youngest were unmarried in 1761. The arms of the family are, Gules, three antique Crowns, Or, [as descended from Grant of That-llk] within a border Ermine, in quality of a judge, supported with two angels proper; Crest, a book expanded; Motto, on a scroll above, " Suum Cuique;" and on a compartment, " Jehovah," Greek; as appears by a special warrant under his majesty's hand, dated May 17, 1720.

GRANT (PATRICK, esq.). He was born at Edinburgh, 1698, and studied the law first in the university of Glasgow, and afterwards at Paris, and Leyden. In 1724, he was called to the bar in the Court of Session, and became a most eminent pleader. He was several times a member of the House of Commons; and, in 1746, was promoted to be lord advocate of Scotland. In 1754, he left the bar, and took his seat on the bench under the title of lord Prestongrange. He wrote several ingenious pieces against the Rebellion 1745, and decisions of the Court of Session. He died at Edinburgh 1762, aged 64.

GRANVILLE (GEORGE), viscount Lansdowne, an english poet, was descended of a family distinguished for their loyalty; being second son of Barnard Granville, Esq. brother to the first earl of Bath of this name, who had a principal share in bringing about the restoration of Charles II. and son of the loyal Sir Bevil Greenvile, who lost his life fighting for Charles I. at Lansdowne in 1643; and whose spirit was in some measure revived by the birth of his grandson George, which happened about 1667. In his infancy he was sent to France, under the tuition of Sir William Ellys, a gentleman bred up under Dr. Busby, and who was afterwards eminent in many public stations. From this excellent tutor he not only imbibed a taste for classical learning, but was also instructed in all other accomplishments suitable to his birth. Nature, indeed, had been very liberal to him, and endowed him with a genius worthy of all the advantages that could be given it by education; wherein he made so quick a proficiency, that after he had distinguished himself above all the youths of France in martial exercises, he was sent to Trinity-college in Cambridge, at eleven years of age; and before he was twelve, spoke a fine copy of verses of his own composing to the duchess of York, afterwards queen-consort to James II. who made a visit to that university in 1679 [A]. On account of his extraordinary merit, he was created M. A. at the age of thirteen.

In the first stage of his life, he seems rather to have made his Muse subservient to his ambition and thirst after military glory, wherein there appeared such a force of genius as raised the admiration of Mr. Waller. But his ambition shewed itself entirely on the duke of Monmouth's rebellion; an opportunity he could by no means let slip. He applied earnestly to his father to let him arm in defence of his sovereign; but he received a check which did not a little mortify him. He had not yet left the academy, and, being then only eighteen years of age, was thought too young for such an enterprize. It was not without extreme reluctance that he submitted to the tenderness of paternal restraint; which was brooked the worse, as his uncle the earl of Bath had on this occasion raised a regiment of foot for the king's service; with the behaviour and discipline of which his majesty was so well pleased, that, on reviewing them at Hounslow, as a public mark of his approbation he conferred the honour of knighthood upon our author's elder brother Bevil, who was a captain therein, at the head of the regiment. Thus, forbidden to handle his pike

[A] They are inserted in his works, near the beginning of Vol. I.

in assisting to crush that rebellion, he took up his pen after it was crushed, and addressed some congratulatory lines to the king.

When the prince of Orange declared his intended expedition to England, our young hero made a fresh application, in the most importunate terms, to let him approve his loyalty. But the danger was now increased in a greater proportion than his age. The king's affairs were become desperate, he was therefore kept from engaging at a juncture, when the attempt could evidently serve no purpose so surely as that of involving him in his royal master's ruin. Broken with this last denial, he sat down a quiet spectator of the revolution; in which most of his family acquiesced.

But he was far from being pleased with the change; he saw no prospect of receiving any favours from the new administration; and resolving to lay aside all thoughts of pushing his fortune either in the court or the camp, he diverted that chagrin and melancholy (which naturally attends disappointed ambition) in the company and conversation of the softer sex. The design was natural at his age, and with his accomplishments easy to execute, and might have been pursued too with safety enough by one that carried a breast less sensible than his was to the impressions of beauty. But in his compositions the tender had at least an equal share with the terrible; and as the present situation of his mind, in regard to the latter quality, disposed him to give a full indulgence to the former, it could be no surprise to any body, that he presently became a conquest of the countess of Newbourg.

Poetry is the handmaid of love He exerted all the powers of verse in singing the force of his enchantress's charms, and the sweets of his own captivity. But he sang in vain, hapless like Waller in his passion, while his poetry raised Myra to the same immortality as had been conferred by that rival poet on Sacharissa. ' In the mean time, some of his friends were much grieved at this conduct in retiring from business, as unbecoming himself and disgraceful to his family. One of these in particular, a female relation, whose name was Higgins, took the liberty to send an expostulatory ode upon it in 1690, in hopes of shaming him out of his enchantment, but he stood impregnable; the address only served him with an opportunity of asserting the unalterableness of his resolution, not to tread the public stage as a courtier, together with the happiness of his condition as a lover.

In this temper he passed the course of king William's reign in private life, enjoying the company of his Muse, which he employed in celebrating the reigning beauties of that age, as Waller, whom he strove to imitate, had done those of the
preceding.

preceding We have also several dramatic pieces written in this early part of life, of which the "British Enchanters," he tells us himself, was the first essay of a very infant Muse; being written at his first entrance into his teens, and attempted rather as a task in hours free from other exercises, than any way meant for public entertainment. But Betterton, the famous actor, having had a casual sight of it many years after it was written, begged it for the stage, where it found so favourable a reception, as to have an uninterrupted run of at least forty days. His other pieces for the stage were all well received; and we are assured they owed that reception to their own merit, as much as to the general esteem and respect that all the polite world professed for their author. Wit and learning know no party; and Addison joined with Dryden in sounding out Granville's praises [B].

Thus debarred, as we have seen, from those passages to fame in which the martial disposition of his family would have inclined him to tread, he struck out a road untrodden by any of his ancestors, by which he reached the temple of honour, and that too much sooner than most of his contemporaries. So that, upon the accession of queen Anne, he stood as fair in the general esteem as any man of his years, which were about thirty-five. He had always entertained the greatest veneration for the queen, and he made his court to her in the politest manner [c]. He entered heartily into the measures for carrying on the war against France; and, in the view of exerting a proper spirit in the nation, he translated the second "Olynthian" of Demosthenes, in 1702. This new specimen of his learning gained him many friends, at the same time that it added highly to his reputation; and, when the design upon Cadiz was projected the same year; he presented to Mr. Harley, afterwards earl of Oxford, an authentic journal of Mr. Wimbledon's expedition thither, in 1625; with a view that, by avoiding the errors committed in a former attempt upon the same place, a more successful plan might be formed. But, little attention being given to it, the very same mistakes again happened, and the very same disappointment was the consequence; with this difference only, that my lord of Ormond had an opportunity to take his revenge at Vigo, and to return with glory, which was not the lord Wimbledon's good fortune.

[B] The former, in the "Epilogue to the British Enchanter;" and the latter, in a copy of verses addressed to him upon his tragedy of "Heroic Love."

[c] This was in Urganda's prophecy,

spoken by way of epilogue at the first representation of the "British Enchanters," where he introduced a scene representing the queen, and the several triumphs of her reign.

Our

Our patriot stood now upon a better footing as to his finances. His father, who was just dead, had made some provision for him; which was increased by a small annuity left him by his uncle the earl of Bath, who died not long after. These advantages, added to the favours which his cousin John Grenville had received from her majesty in being raised to the peerage by the title of lord Grenville of Potheridge, and his brother being made governor of Barbadoes, with a fixed salary of 2000l. the same year, engaged him to come into parliament; and he was accordingly chosen for Fowey in Cornwall, in the first parliament of the queen, with John Hicks, Esq. In 1706, his fortune was improved farther by a very unwelcome accident in the loss of his eldest brother, Sir Bevil, who died that year, in his passage from Barbadoes, in the flower of his age, unmarried, and universally lamented. Hence our younger brother stood now as the head-branch of his family, and he still held his seat in the house of commons, both in the second and third parliaments of the queen. But the administration being taken out of the hands of his friends, with whom he remained steadily connected in the same principles, he was cut off from any prospect of being preferred at court.

In this situation he diverted himself among his brother poets; and in that humour we find him at this time introducing Wycherley and Pope to the acquaintance of Henry St. John, Esq; afterwards lord viscount Bolingbroke. This friend, then displaced, having formed a design of celebrating such of the poets of that age as he thought deserved any notice, had applied for a character of the former to our author, who, in reply, having done justice to Mr. Wycherley's merit, concludes his letter thus: " In short, Sir, I'll have you judge for yourself. I am not satisfied with this imperfect sketch; name your day, and I will bring you together; I shall have both your thanks, let it be at my lodging. I can give you no Falernian that has out-lived twenty consulships, but I can promise you a bottle of good claret, that has seen two reigns. Horatian wit will not be wanting when you meet. He shall bring with him, if you will, a young poet newly inspired in the neighbourhood of Cooper's-hill, whom he and Walsh have taken under their wing. His name is Pope, he is not above seventeen or eighteen years of age, and promises miracles. If he goes on as he has begun in the pastoral way, as Virgil first tried his strength, we may hope to see English poetry vie with the Roman, and this Swan of Windsor sing as sweetly as the Mantuan. I expect your answer."

Sacheverell's trial, which happened not long after, brought on that remarkable change in the ministry in 1710, when Mr. Granville's friends came again into power. He was elected for the borough of Helston, but being returned too for the county of Cornwall, he chose to represent the latter; and, September 29, he was declared secretary at War, in the room of the late earl of Orford, then Robert Walpole, Esq. He continued in this office for some time, and discharged it with reputation; and, towards the close of the next year, 1711, he espoused the lady Mary, daughter of Edward Villiers, earl of Jersey, at that time possessed of a considerable jointure, as widow of Thomas Thynne, Esq. by whom she was mother of the late lord Weymouth. He had just before succeeded to the estate of the elder branch of his family, at Stow; and December 31, he was created a peer of Great Britain, by the title of lord Lansdowne, baron of Bideford, in the county of Devon. It is true, he was one of the twelve peers who were all created at the same time; a step taken to serve the purpose of this party. So numerous a creation, being unprecedented, made a great noise, but none gave less offence than his. His lordship was now the next male-issue in that noble family, wherein two peerages had been extinguished almost together: his personal merit was universally allowed; and with regard to his political sentiments, those who thought him most mistaken, allowed him to be open, candid, and uniform. He stood always high in the favour of queen Anne; and with great reason, having upon every occasion testified the greatest zeal for her government, and the most profound respect for her person. It is no wonder, therefore, that in the succeeding year, 1712, we find him sworn of her majesty's privy-council, made controller of her household, and about that time twelve-month advanced to the post of treasurer in the same office. His lordship continued in this post till the decease of his beloved mistress, when he kept company with his friends in falling a sacrifice to party-violence, being removed from his treasurer's place, by George I. Oct. 11, 1714.

His lordship still continued steady to his former connections, and in that spirit entered his protest with them against the bills for attainting lord Bolingbroke and the duke of Ormond, in 1715. He even entered deeply into the scheme for raising an insurrection in the West of England, and was at the head of it, if we may believe lord Bolingbroke, who represents him possessed now with the same political fire and frenzy for the pretender as he had shewn in his youth for the father.

Accordingly, we find lord Lansdowne was seized as a suspected person, September 26, 1715, and committed prisoner to the Tower of London, where he continued a long time. He was, however, at length set free from his imprisonment, February 8, 1717, when all dangers were over. However sensible he might be at this time of the mistake in his conduct, which had deprived him of his liberty, yet he was far from running

running into the other extreme. He seems, indeed, to be one of those tories, who are said to have been driven by the violent persecutions against that party in Jacobitism, and who returned to their former principles as soon as that violence ceased. Hence we find him, in 1719, as warm as ever in defence of those principles, the first time of his speaking in the house of lords, in the debates about repealing the act against occasional conformity.

His lordship continued steady in the same sentiments, which were so opposite to those of the court, and inconsistent with the measures taken by the administration, that he must needs be sensible a watchful eye was kept ever upon him. Accordingly, when the flame broke out against his friends, on account of what is sometimes called Atterbury's plot, in 1722, his lordship, apparently to avoid a second imprisonment in the Tower, withdrew to France. He had been at Paris but a little while, when the first volume of Burnet's "History of his own Times" was published. Great expectations had been raised of this work, so that he perused it with attention; and finding the characters of the duke of Albemarle and the earl of Bath treated in a manner he thought they did not deserve, he formed the design of doing them justice. This led him to consider what had been said by other historians concerning his family; and, as Clarendon and Echard had treated his uncle Sir Richard Granville more roughly, his lordship, being possessed of memoirs from which his conduct might be set in a fairer light, resolved to follow the dictates of duty and inclination, by publishing his sentiments upon these heads [D].

He continued abroad at Paris almost the space of ten years; and, being sensible that many juvenilities had escaped his pen in his poetical pieces, made use of the opportunity furnished by this retirement, to revise and correct them, in order to republication. Accordingly, at his return to England in 1732, he published these, together with a vindication of his kinsman just mentioned, in two volumes, 4to. The late queen Caroline having honoured him with her protection, the last verses he wrote were to inscribe two copies of his poems, one of which was presented to her majesty, and the other to the princess royal Anne, late princess dowager of Orange [E]. The

---

[D] These pieces are printed in his works, under the title of "A Vindication of General Monk, &c." and "A Vindication of Sir Richard Greenville, General of the West to King Charles I. &c." They were answered by Oldmixon, in a piece, intituled, "Reflections historical and politic, &c. 1732," 4to. and by judge Burnet, in "Remarks, &c." a pamphlet. His lordship replied, in "A Letter to the author of the Reflections, &c. 1732. 4to." and the spring following, there came out an answer in defence of Echard, by Dr. Colbatch, intituled, "An Examination of Echard's Account of the Marriage Treaty, &c."

remaining

remaining years of his life were passed in privacy and retirement, to the day of his death, which happened January 30, 1735, in his 68th year; having lost his lady a few days before, by whom having no male issue, the title of Lansdowne became in him extinct.

GRAPALDUS (FRANCIS MARIUS), a learned man, who lived in the sixteenth century. He was of Parma, distinguished himself on an embassy to the Pope so much, that Julius the second crowned him with his own hand. The work for which he is most eminent is that in which he describes all the parts of a house, and which really discovers much taste, improved by learning. His book has been often printed.

GRAS (ANTONY LE), a Parisian and a priest. After some time spent in retirement from the world, he appeared in the Theatre of Letters, and published the lives of great men, being a translation of Cornelius Nepos. He also wrote an account of the fathers who lived in the times of the apostles. He is not to be confounded with James le Gras, who was a native of Rome, and published a translation of Hesiod.

GRASWINCKEL (THEODORE), a native of Delft, was a very learned civilian in the seventeeth century, and published several works. He was not only well versed in matters of law; but also in the Belles Lettres and latin poetry. He dedicated his book "De jure Majestatis" to the queen of Sweden; and the Republic of Venice made him a knight of St. Mark, in return for his having published a tract in vindication of the Venetians against the duke of Savoy. He also wrote many books in Dutch. He died at Mechlin, and was buried at the Hague, where a monument, with an inscription highly to his honour, was erected to his memory.

GRATAROLUS (WILLIAM), a learned physician of the sixteenth century. He was born at Bergamo in Italy, and, quitting his country, went into Germany, that he might live undisturbed in the protestant religion. After some stay at Bazil, he was invited to Marpurg to be physic-professor. After a little stay in this town, he returned to Bazil and died there in 1562, at fifty two-years of age. He wrote a great many books, as, "De Memoria Reparanda, Augenda, Conservanda, ac Reminiscentia. De Prædictione Morum, Naturarumque Hominum facili, & Inspectione partium corporis. Prognostica Naturalia de Temporum mutatione perpetua, ordine Literarum. De Literatorum & eorum qui Magistratibus funguntur, conservanda, preservandaque valetudine. De Vini Natura, artificio & usu; Deque omni Re Potabili. De Regi-

[1] See his works, Vol. III. p. 263, 264.

mine iter Agentium, vel Equitum, vel Peditum, vel Navi, vel Curru viatoribus quibusque Utilissimi Libri duo." He likewise made a collection of several tracts touching the sweating-sickness in England. Lindenius Renovatus, p. 376, 377. Paulus Freherus in Theatro. Bayle Diction. Histor.

GRATIAN, son of Valentinian, by the empress Severa. He succeeded to the empire in 367. His character is thus given by Gibbon: "The fame of Gratian, before he had accomplished the twentieth year of his age, was equal to that of the most celebrated princes. His gentle and amiable disposition endeared him to his private friends; the graceful affability of his manner engaged the affection of the people. The men of letters, who enjoyed the liberality, acknowledged the taste and eloquence, of their sovereign. His valour and dexterity in arms were equally applauded by the soldiers, and the clergy considered the piety of Gratian as the first and most useful of his virtues. This sneer of Gibbon in the concluding paragraph is unworthy of his pen. Gratian, however, was the first roman emperor who refused the title of Pontifex Maximus. He was assassinated by Andragathus, in the twenty-fourth year of his age.

GRATIAN, a famous Benedictine monk, in the twelfth century, who employed twenty-four years in a work, whose object it was to reconcile the contradictory canons to each other. To this monk's. decretals the popes are principally indebted for the authority which they enjoyed in the thirteenth and subsequent centuries.

GRATIANI (JEROME), an italian writer of the last century. His poetry was rather sweet than animated, and his prose compositions were rather elegant than profound. He wrote the "Conquest of Grenada," and a tragedy, called, "Cromwell," which was highly esteemed. He published also some agreeable miscellanies in prose.

GRATIUS (FALISCUS), an eminent latin poet, is supposed to have been contemporary with Ovid, and pointed out by him in the last elegy of the fourth book "De Ponto:" "Aptaque venanti Gratius arma dedit." We have a poem of his, intitutled, "Cynegeticon, &, The Art of Hunting with Dogs:" but it is imperfect towards the end, so that in strictness it can only be called a fragment. The style of this poem is reckoned pure, but without elevation; the poet having been more solicitous to instruct than to please his reader. He is also censured by the critics as dwelling too long on fables; and as he is counted much superior to Nemesianus, who has treated the same subject, so he is reckoned in all points inferior to the greek poet, Oppian, who wrote his Cynegetics and Halieutics under Severus and Caracalla, to whom he presented them, and who is said to have rewarded the poet very magnificently. The "Cynegetica"

negetica" were publifhed at Leyden, 1645, in 12mo. with the learned notes of Janus Ulitius; and afterwards with Nemefianus, at London 1699, in 8vo. " cum Notis perpetuis Thomæ Jonfon, M. A." The lateft edition is that of Leyden 1728, in 4to, in which Nemefianus, and the other writers " rei venaticæ," are publifhed with him.

GRATIUS (ORTUINUS), born at Helvick, in the diocefe of Munfter. He was a very learned man, and wrote feveral books. He was the inftructor of the wits, who joined in writing the Epiftolæ obfcurorum virorum, which being condemned by the Pope, as too much favouring the growth of Lutheranifm, Gratius publifhed the Lamentationes obfcurorum virorum non prohibitæ per fedem apoftolicam. His real name was Graes: He died in 1542.

GRAVELOT (HENRY FRANCIS BOURGUIGNON), born at Paris in 1699, an eminent engraver. He fpent fome time of his early life at St. Domingo, where he affifted in drawing a chart of the ifland. On his return to France, he applied ferioufly to his profeffion; but, conceiving that he fhould have a fairer fcope for his abilities in England, he came to London, where he refided for thirteen years. The fineft editions of the beft french poets have been adorned by his pencil. Gravelot was alfo a man of wit and talents, and was admired for his manners as much as for his fkill in his art. He died in 1773.

GRAVEROL (FRANCIS), a french advocate, born at Nimes, in 1635. He was the author of many works, and in particular of the Sorberiana. He had the reputation, when living, of being an excellent fcholar, and perfectly verfed in the knowledge of antiquity. He died in 1694. He had a brother, John Graverol, who wrote feveral theological works, and in particular one againft bifhop Burnet, which he called " Archeologia Philofophica."

GRAVESANDE (WILLIAM JAMES), was born 1688, at Delft, in Holland, of an ancient and honourable family. He was educated with the greateft care, and very early difcovered an extraordinary genius for mathematical learning. He was fent to the univerfity of Leyden, in 1704, with an intention to ftudy the civil law; but at the fame time he cultivated with the greateft affiduity his favourite fcience. Before he was nineteen, he compofed his treatife on perfpective, which gained him great credit among the moft eminent mathematicians of his time. When he had taken his doctor's degree in 1707, he quitted the college, and fettled at the Hague, where he practifed at the bar. In this fituation he contracted and cultivated an acquaintance with learned men; and made one of the principal members of the society that composed a periodical review, intituled, "Le Journal Littéraire." This journal began in May 1713, and was continued without interruption till 1722. The parts of it written or extracted by Gravesande were principally those relating to physics and geometry. But he enriched it also with several original pieces entirely of his composition, viz. "Remarks on the Construction of Pneumatical Engines;" "A moral Essay on Lying;" and a celebrated "Essay on the Collision of Bodies;" which, as it opposed the Newtonian philosophy, was attacked by Dr. Clarke and many other learned men.

In 1715, when the states sent to congratulate George I. on his accession to the throne, Gravesande was appointed secretary to the embassy. During his stay in England, he was admitted a member of the Royal Society, and became intimately acquainted with Sir Isaac Newton. On his return to Holland, when the business of the embassy was over, he was chosen professor of the mathematics and astronomy, at Leyden: and he had the honour of first teaching the Newtonian philosophy there, which was then in its infancy. The most considerable of his publications is, "An Introduction to the Newtonian Philosophy, or, a Treatise on the Elements of Physics, confirmed by Experiments." This performance, being only a more perfect copy of his public lectures, was first printed in 1720; and hath since gone through many editions, with considerable improvements. He published also "A small Treatise on the Elements of Algebra, for the Use of young Students." After he was promoted to the chair of philosophy in 1734, he published "A Course of Logic and Metaphysics." He had a design too of presenting the public with "A System of Morality," but his death, which happened in 1742, prevented his putting it in execution. Besides his own works, he published several correct editions of the valuable works of others.

He was amiable in his private and respectable in his public character; for, few men of letters have done more eminent services to their country. The ministers of the republic consulted him on all occasions in which his talents were requisite to assist them, which his skill in calculation often enabled him to do in money-affairs. He was of great service also in detecting the secret correspondence of their enemies, as a decipherer. And, as a professor, none ever applied the powers of nature with more success, or to more useful purposes.

GRAVINA (PETER), an italian poet, wrote a quarto volume of poems, which have been admired for the harmony ⁓⁓ versification and the delicacy of the sentiment. He was favourite with Sannazarius, who preferred him to all

the

the poets of his time. Paul Jovius has also commended the tenderness of his elegies,

GRAVINA (JOHN VINCENT), an eminent scholar, and illustrious lawyer of Italy, was born of genteel parents at Roggiano, February 18, 1664; and educated under Gregory Caloprese, a famous philosopher of that time, and withal his cousin-german. He went to Naples at sixteen, and there applied himself to latin eloquence, to the greek language, and to civil law: which application, however, did not make him neglect to cultivate, with the utmost exactness, his own native tongue. He was so fond of study, that he pursued it ten or twelve hours a day, to the very last years of his life; and, when his friends remonstrated against this unnecessary labour, he used to tell them, that he knew of nothing which could afford him more pleasure. He went to Rome in 1696, and some years after was made professor of canon law, in the college of Sapienzi, by Innocent XI. who esteemed him much; which employment he held as long as he lived. He does not seem to have been of an amiable cast: at least, he had not the art of making himself beloved. The free manner in which he spoke of all mankind, and the contempt with which he treated the greatest part of the learned, raised him up many enemies: and among others the famous Settano, who has made him the subject of some of his satires. Many universities of Germany would have drawn Gravina to them, and made proposals to him for that purpose; but nothing was able to seduce him from Rome. That of Turin offered him the first professorship of law, at the very time that he was attacked by the distemper of which he died, and which seems to have been a mortification in his bowels. He was troubled with pains in those parts for many years before; but they did not prove fatal to him till Jan. 6, 1718. He had made his will in April 1715, in which he ordered his body to be opened and embalmed.

We shall now proceed to give an account of his works: His first publication was a piece, intituled, 1. "Prisci Censorini Photistici Hydra Mystica; sive, de corrupta morali Doctrina Dialogus, Coloniæ, 1691," 4to; but really printed at Naples. This was without a name, and is very scarce; the author having printed only fifty copies, which he distributed among his friends. 2. "L'Endimione di Erilo Cleoneo, Pastore Arcade, con un Discorso di Bione Crateo. In Roma, 1692," 12mo. The Endymion is Alexander Guidi's, who, in the academy of the Arcadians, went under the name of Erilo Cleoneo; and the discourse annexed, which illustrates the beauties of this pastoral, is Gravina's, who conceals himself under that of Bione Crateo. 3. "Delle Antiche Favola, Roma 1696, 12mo. 4. A Collection of pieces under the name of "Opufcula," at Rome in 1696, 12mo; containing, firft, "An Effay upon an ancient Law;" fecondly, "A Dialogue concerning the Excellence of the Latin Tongue:" thirdly, "A Difcourfe of the Change which has happened in the Sciences, particularly in Italy;" fourthly, "A Treatife upon the Contempt of Death;" fifthly, upon Moderation in Mourning;" fixthly, "The Laws of the Arcadians."

But the greateft of all his works, and for which he will be ever memorable, is, 5. His three books, "De Ortu & Progreffu Juris Civilis;" the firft of which was printed at Naples, in 1701, 8vo. and at Leipfic, in 1704, 8vo. Gravina, afterwards fent the two other books of this work to John Burchard Mencken, librarian at Leipfic, who had publifhed the firft there, and who publifhed thefe alfo in 1708, together with it, in one volume, 4to. They were publifhed alfo again at Naples in 1713, in two volumes, 4to. with the addition of a book, "De Romano Imperio;" and dedicated to pope Clement XI. who was much the author's friend. This is reckoned the beft edition of this famous work; for, when it was reprinted at Leipfic with the "Opufcula" above-mentioned, in 1717, it was thought expedient to call it in the title-page, "Editio noviffima ad nuperam Neapolitanam emendata & aucta." Gravina's view, in this "Hiftory of Ancient Law," was to induce the Roman youth to ftudy it in its original records; in the Pandects, the Inftitutes, and the Code; and not to content themfelves, as he often complained they did, with learning it from modern abridgements, drawn up with great confufion, and in very barbarous latin. Such knowledge and fuch language, he faid, might do well enough for the bar, where a facility of fpeaking often fupplied the place of learning and good fenfe, before judges who had no extraordinary fhare of either; but were what a real lawyer fhould be greatly above. As to the piece "De Romano Imperio," Le Clerc pronounces it to be a work in which Gravina has fhewn the greateft judgement and knowledge of Roman antiquity.

The next performance we find in the lift of his works is, 6. "Acta Confiftorialia creationis Emin. & Rev. Cardinalium inftitutæ à S. D. N. Clemente XI. P.M. diebus 17 Maii & 7 Junii anno falutis 1706. Acceffit eorundem Cardinalium brevis delineatio. Coloniæ, 1707," 4to. 7. "Della Ragione Poetica Libri duo. In Roma, 1708," 4to. 8. "Tragedie cinque. In Napoli, 1712, 8vo. Thefe five tragedies are, "Il Papiniano," "Il Palamede," "L'Andromeda," "L'Appio Claudio," "Il Servio Tullio." Gravina faid, that he compofed thefe tragedies in three months, without interrupting his lectures; yet declares in his preface, that he fhould look upon all thofe as either ignorant or envious, who fhould fcruple to prefer them to what Taſſo, Bonarelli, Triffino, and others, had compoſed of the ſame kind. Not having the volume before us, we take this upon Niceron's authority; and, if it be true, it ſhews, that Gravina, great as his talents were, had yet too high an opinion of them. 9. "Orationes. Neap. 1712," 12mo. Theſe have been reprinted more than once, and are to be found with his "Opuſcula" in the edition of "Origines Juris Civilis," printed at Leipſic, in 1717. 10. "Della Tragedia Libro uno. Napoli, 1715," 4to. This work, his two books "Della Ragione Poetica," his diſcourſe upon the "Endymion" of Alexander Guidi, and ſome other pieces, were printed together at Venice in 1731, 4to.

GRAUNT (EDWARD), was head-maſter of Weſtminſter-ſchool, and died in 1601. He publiſhed "Græcæ linguæ Spicelegium & Inſtitutio Græcæ Grammaticæ," which obtained the eſteem of the age in which he lived.

GRAUNT (JOHN), the celebrated author of the "Obſervations on the Bills of Mortality," was the ſon of Henry Graunt of Hampſhire, who being afterwards ſettled in Birchin-lane, London, had this child born there, April 24, 1620. Being a rigid puritan, he bred him up in all the ſtrictneſs of thoſe principles; and deſigning him for trade, gave him no more education than was barely neceſſary for that purpoſe: ſo that, with the ordinary qualifications of reading, writing, and arithmetic, without any grammar-learning, he was put apprentice to a haberdaſher in the city, which trade he afterwards followed; but he was free of the drapers company. He came early into buſineſs, and in a ſhort time grew ſo much into the eſteem of his fellow-citizens, that he was frequently choſen arbitrator for compoſing differences between neighbours, and preventing law-ſuits. With this reputation he paſſed through all the offices of his ward, as far as that of a common-council-man, which he held two years, and was firſt captain and then major of the train bands. Theſe diſtinctions were the effects of a great ſhare of good ſenſe and probity, rendered amiable by a mild and friendly diſpoſition; and this was all that could be expected from a tradeſman of no great birth, and of ſmall breeding. But Graunt's genius was far from being confined within thoſe limits: it broke through all the diſadvantages of his ſlender education, and enabled him to form a new and noble deſign, and to execute it with as much ſpirit as there appeared ſagacity in forming it.

We do not know the exact time when he firſt began to collect and conſider the Bills of Mortality; but he tells us himſelf, that he had turned his thoughts that way ſeveral years, before he had any deſign of publiſhing the diſcoveries he had made. As his character muſt have been eminently diſtinguiſhed in 1650, when, though not above thirty years of age, his
intereſt

interest was so extensive, as to procure the music professor's chair at Gresham, for his friend doctor (afterwards Sir William) Petty; so it is more than probable, that his acquaintance and friendship with that extraordinary virtuoso was the consequence of a similarity of genius; and that our author had then communicated some of his thoughts upon this subject to that friend, who, on his part, is likewise said to have repaid the generous confidence with some useful hints towards composing his book. This piece, which contained a new and accurate thesis of policy, built upon a more certain reasoning than was before that time known, was first presented to the public in 1661, 4to. and met with such an extraordinary reception as made way for another edition the next year.

In short, our author's fame spread, together with the admirable usefulness of his book, both at home and abroad. Immediately after the publication of it, Lewis XIV. of France, or his ministers, provided, by a law, for the most exact register of births and burials, that is any where in Europe; and in England Charles II. conceived such a high esteem for his abilities, that, soon after the institution of the Royal Society, his majesty recommended him to their choice for a member; with this charge, that if they found any more such tradesmen, they should be sure to admit them all. He had dedicated the work to Sir Robert Moray, president of the Royal Society, and had sent fifty copies to be dispersed among their members, when he was proposed, (though a shopkeeper) and admitted into the society, February 26, 1661-2 [F]; and an order of council passed, June 20, 1765, for publishing the third edition, which was executed by the society's printer [G], and came out that same year. After receiving this honour, he did not long continue a shopkeeper, but left off his business; and September 25, 1666, became a trustee for the management of the New-river. He was so for one of the shares belonging to Sir William Backhouse, who dying in 1669, his relict, afterwards countess of Clarendon, appointed him one of her trustees in the said company.

This account of the time of our author's admission into the government of the New-river is taken from the minute books, or register, of the general court of that company, and sufficiently clears him from an imputation thrown upon his memory by bishop Burnet; who, having observed that the New-river was brought to a head at Islington, where there is a great room full of pipes that conveys it through the streets of London, and that the constant order was to set all the pipes run-

[F] Birch's "History of the Royal Society," Vol. I.

[G] The order is prefixed to this edition, which contained large additions.

ning on Saturday night, that so the cisterns might be all full on Sunday morning, there being a more than ordinary consumption of water on that day, relates the following story, which he says was told him by Dr. Lloyd (afterwards bishop of Worcester) and the countess of Clarendon. "There was," says he, "one Graunt, a papist, who under Sir William Petty published his Observations on the Bills of Mortality. He had some time before applied himself to Lloyd, who had great credit with the countess of Clarendon, and said he could raise that estate considerably, if she would make me a trustee for her. His schemes were probable; and he was made one of the board that governed that matter, and by that he had a right to come as often as he pleased to view their works at Islington. He went thither the Saturday before the fire broke out, and called for the key where the heads of the pipes were, and turned all the cocks of the pipes that were then open, stopt the water, and went away and carried the keys with him; so, when the fire broke out next morning, they opened the pipes in the streets to find water, but there was none. Some hours were lost in sending to Islington, where the door was broke open and the cocks turned, and it was long before the water got to London. Graunt, indeed, denied that he had turned the cocks; but the officer of the works affirmed, that he had, according to order, set them all running, and that no person had got the keys from him besides Graunt, who confessed he had carried away the keys, but said he did it without design [H]." This, indeed, as the right reverend story-teller observes, is but a presumption; and, if he had the same thirst after searching out the truth as he had for extraordinary story-telling, he would have added that it is a groundless calumny; since it is evident, from the above account, that Graunt was not admitted into the government of the New-river company till twenty-three days after the breaking out of the fire of London. To which may be added, that the parliament met September 18, 1666, and, on the very day that he was admitted a member of the New-river company, they appointed a committee to enquire into the causes of the fire.

The report made by Sir Robert Brooke, chairman of that committee, contains abundance of extraordinary relations; but not one word of the cocks being stopped, or any suspicions of Graunt [I]. It is true, indeed, that he changed his religion, and was reconciled to the church of Rome some time before his death; but it is more than probable he was no

[H] Burnet's "History of his own Times," Vol. I. p. 23.

[I] See a true and faithful account of the several informations exhibited to the honourable committee, appointed by the parliament to enquire into the late dreadful burning of the city of London, printed in 1667.

papist at this juncture, since the additions to his book in 1665 speak him then otherwise, being in the title-page styled captain, and Wood informs us, that he had been two or three years a major when he made this change; whence it follows, that this change in his religion could not happen before 1667 or 1668 at soonest. However, the circumstances of the countess of Clarendon's saying he was her trustee makes it plain that the story was not invented till some years after the fire, when Graunt was known to be a papist [K].

Happy it was, for the good of the public, that it never reached his ears, and so could not disturb him in the prosecution of his studies, which he carried on after this change with the same assiduity as before, and made some considerable observations within two years of his death, which happened April 18, 1674, in the vigour of his age, having not quite completed his 54th year. He was interred on the 22d of the same month in St. Dunstan's church, in Fleet-street, the corpse being attended by many of the most ingenious and learned persons of the time, and particularly by Sir William Petty, who paid his last tribute with tears to his memory. He left his papers to this friend, who took care to adjust and insert them in a fifth edition of his work, which he published in 1676, 8vo. and that with so much care, and so much improved, that he frequently cites it as his own: which probably gave occasion to bishop Burnet's mistake, who, as we have seen, called it Sir William's book, published under Graunt's name. It is evident, however, that his observations were the elements of that useful science, which was afterwards happily styled "Political Arithmetic," and greatly advanced under that title by this friend. In a word, Graunt must have the honour of being the first founder of this science; and whatever merit may be ascribed to Sir William Petty, Mr. Daniel King, Dr. Davenant, and others [L], upon the subject, it is all originally derived from the first author of the "Observations on the Bills of Mortality."

[K] It was apparently not coined till after his death. The first time of its appearance in public seems to have been in Echard's "History of England." And according to bishop Burnet's account, the story could not be told to him till after the year 1667, when Graunt was appointed trustee for the countess of Clarendon.

[L] Among the rest, our author's reasoning in defence of a particular providence, from the constant proportion that is kept up between the number of males and females, is pushed to the utmost by the late Dr. John Arbuthnet; who, by an excellent skill in calculation, has demonstrated, that it is forty-eight millions of millions of millions of millions to one, that the proportion should not constantly come so near the same as experience shews it to be, if it depended on chance, Phil. Transf. No. 328. But the most extraordinary, as well as the most extensively useful improvement that has hitherto appeared of our author's remarks, was made by Dr. Halley, for which we must refer to his article.

GRAY

GRAY (THOMAS), eminent for a few excellent poems he has left us, and of whom it is as truly said, as it was of Persius by Quintilian, "multum & veræ gloriæ, quamvis uno libro, meruit," was the son of a reputable citizen; and born in Cornhill, December 26, 1716. He was educated at Eton-school, and thence removed to St. Peter's college, Cambridge, in 1734. In April 1738, he removed to town, intending to apply himself to the study of the law, for which purpose his father had procured him a set of chambers in the Temple; but on an invitation which Mr. Horace Walpole, his intimate friend, gave him to be his companion in his travels, his intention was laid aside for the present. He left England, March 29, 1739; made the tour of France and Italy; and arrived in London again about September 1741.

About two months after his return, his father died; when, finding his patrimony too small to enable him to prosecute the study of the law, he changed the line of that study; and, at the latter end of 1742, went to Cambridge to take the degree of LL. B. His principal residence, henceforwards, was at this place; and he was seldom absent from college any considerable time, except between the years 1759 and 1762; when, on the opening of the British Museum, he took lodgings in Southampton-row, in order to have recourse to the Harleian and other MSS. there deposited; from which he made several curious extracts. In 1747, he became acquainted with Mr. Mason, who has shewn himself so faithful to his memory, and so just to his reputation; and this acquaintance presently ripened into the closest friendship. In 1768, he was appointed professor of modern history; but, his health being now upon the decline, he never was able to execute the duties of it. He died of the gout, July 30, 1771.

In an anonymous character of him [M], which seems to be drawn by a very impartial hand he is represented to have been "perhaps the most learned man in Europe; equally acquainted with the elegant and profound parts of science, and that not superficially but thoroughly; knowing in every branch of history, both natural and civil, as having read all the original historians of England, France, and Italy; a great antiquarian; who made criticism, metaphysics, morals, politics, a principal part of his plan of study who was uncommonly fond of voyages and travels of all sorts; and who had a fine taste in painting, prints, architecture, and gardening."

Upon the whole, there is good reason to allow, that he was indeed a very extraordinary person. We have only to

[M] This well-written character, adopted both by Mr. Mason and Dr. Johnson, was drawn by the Rev. Mr. Temple, rector of St. Gluvius, in Cornwall.

lament,

lament, that he has left us no other proofs of it, but a very small collection; highly finished indeed, and excellent in their kind, but shewing him only under one single attitude of greatness, while, in the mean time, he was capable of appearing under many. These "Poems" were collected and published together by his friend Mr. Mason, 1775, in 4to. who hath also prefixed "Memoirs of his Life and Writings." In these memoirs is interwoven a large collection of letters of Mr. Gray and his intimate friends, which abound with curious and interesting anecdotes; and which, like all such collections, may be read with more edification, to private persons at least, that even some histories of large and pompous stature.

GRAZZINI (ANTONY FRANCIS), one of the principal founders of the academy of La Crusca. He was also a poet, and a writer of comedies. The work by which he obtained his highest reputation was a "Collection of Novels," printed at Paris, in 1756. He had the appellation of Lasca assigned him, and, among his countrymen of Italy, was thought almost upon a par with Boccace. His works are recommended by a considerable portion of elegance and purity.

GREATRAKES (VALENTINE), an irish gentleman, had a strong impulse upon his mind to attempt the cure of diseases by touching or stroking the parts affected. He first practised in his own family and neighbourhood, and several persons, to all appearance, were cured by him of different disorders. He afterwards came into England, where his reputation soon rose to a prodigious height; but it declined almost as fast, when the expectation of the multitudes that resorted to him were not answered. Mr. Glanville imputed his cures to a sanative quality inherent in his constitution; some to fiction, and others to the force of imagination in his patients; of this there were many instances, one of which, if a fact, is related by Monsieur St. Evremond, in a peculiar strain of pleasantry. It is certain that the great Mr. Boyle believed him to be an extraordinary person, and that he has attested several of his cures. His manner of treating some women was said be very different from his usual mode of operation.

GREAVES (JOHN), an eminent mathematician and antiquary, was eldest son of John Greaves, rector of Colmore, near Alresford in Hampshire, where his son was born to him in 1602, and probably instructed in grammar-learning by himself, as being the most celebrated school-master in that country. At fifteen years of age our author was sent to Baliol-college, in Oxford, where he proceeded B. A. July 6, 1621. Three years after which, his superiority in classical learning procured him the first place of five in an election to a fellowship of Merton-college. June 25, 1628, he commenced M. A.

and,

and, being made complete fellow, was more at liberty to pursue the bent of his inclination, which leading him chiefly to oriental learning, and the mathematics, he quickly distinguished himself in each of these studies; and his eminent skill in the latter procured him the geometry-lecture in Gresham, into which he was chosen, February 22, 1630.

At this time he had not only read the writings of Copernicus, Regiomontanus, Purbach, Tycho Brahe, and Kepler, with other celebrated astronomers of that and the preceding age, but had made the antient greek, arabian, and persian authors familiar to him, having before gained an accurate skill in the oriental languages: but he was far from being satisfied; the acquisitions he had already made serving to create a thirst for more. This ambition prompted him to travel. In which spirit he crossed the sea to Holland, in 1635; and having attended for some time the lectures of Golius, the famous professor of arabic at Leyden, he proceeded to Paris, where he conversed with the learned Claudius Hardy, about the persian language; but finding little or no assistance there, he continued his journey to Rome, in order to view the antiquities of that city. He also visited other parts of Italy: and before his departure, meeting with the earl of Arundel, was offered 200l. a year to live with his lordship, and attend him as a companion in his travels to Greece; the earl also promised all other acts of friendship that should lie in his power. This was a very advantageous proposal, and would have been eagerly accepted by Mr. Greaves, as being highly agreeable to his inclination in general; but he had now formed another and greater design, which soon brought him back to England, in order to furnish himself with every thing proper to complete the execution of it. This was a voyage to Egypt.

Immediately after his return, he acquainted his patron, archbishop Laud, with his intentions, and, being encouraged by his grace, set about making preparations for it. His primary view was, to measure the pyramids with all proper exactness; and, withal, to make astronomical and geographical observations, as opportunities offered, for the improvement of those sciences. A large apparatus of proper mathematical instruments was consequently to be provided; and, as the expence of purchasing these would be considerable, he applied for assistance to the city of London, but met with an absolute denial. This he resented to that degree, that, in relating the generosity of his brothers upon his own money falling short, he observes, "That they had strained their own occasions, to enable him, in despite of the city, to go on with his designs." He had been greatly disappointed in his hopes of meeting with curious books in Italy; he therefore
proposed

proposed to make that another principal part of his business; and, to compass it in the easiest manner, he bought several books before his departure, in order to exchange them with others in the East. Besides his brothers, he had probably some help from Laud, from whom he received a general discretionary commission to purchase for him arabic and other MSS. and likewise such coins and medals as he could procure. Laud also gave him a letter of recommendation to Sir Peter Wyche, the English ambassador at Constantinople.

Thus furnished, he embarked in the river Thames for Leghorn, June 1637, in company with his particular friend, Mr. Pocoke, whom he had earnestly solicited to that voyage [N]. After a short stay in Italy, he arrived at Constantinople before Michaelmas. Here he met with a kind reception from Sir Peter Wyche, and became acquainted with the venerable Cyril Lucaris, the greek patriarch, by whom he was much assisted in purchasing greek MSS. He promised Mr. Greaves to recommend him to the monks of Mount Athos, where he would have had the liberty of entering into all the libraries, and of collecting a catalogue of such books as either were not printed, or else, by the help of some there, might have been more correctly set out. These, by dispensing with the anathemas which former patriarchs had laid upon all greek libraries, to preserve the books from the latins, Cyril proposed to present to archbishop Laud, for the better prosecution of his designs in the edition of greek authors; but this likewise was frustrated by the cruel death of that patriarch, who was barbarously strangled June 1638, by express command of the Grand Seignior, on pretence of holding a correspondence with the emperor of Muscovy.

[N] Our author's generosity on this occasion deserves a particular mention. In a letter to this friend, Dec. 23, 1636, he writes thus: " I shall desire your favour in sending up to me, by my brother Thomas, Ulug Beig's astronomical tables, of which I purpose to make this use. The next week I will shew them to my lord's grace [Laud] and highly commend your care in procuring those tables, being the most accurate that ever were extant; then will I discover my intention of having them printed and dedicated to his grace; but because I presume that there are many things which in these parts cannot perfectly be understood, I shall acquaint my lord with my desire of taking a journey into those countries, for the more emendate edition of them; afterwards, by degrees, fall down upon the business of the consulship, and how honourable a thing it would be if you were sent out a second time, as Golius, in the Low Countries, was by the States, after he had been once there before. If my lord shall be pleased to resolve and compass the business, I shall like it well; if not, I shall procure 300l. for you and myself, besides getting a dispensation for the allowances of our places in our absence, and, by God's blessing, in three years dispatch the whole journey. It shall go hard, but I will too get some citizen in, as a benefactor to the design; if not, 300l. of mine, whereof I give you the half, together with the return of our stipends, will, in a plentiful manner, if I be not deceived, in Turkey maintain us." Biog. Brit. vol. IV. p. 2268.

Nor

Nor was this the only lofs which our traveller fuftained by Cyril's death; for having procured, out of a blind and ignorant monaftery, which depended on the patriarch, fourteen good MSS. of the fathers, he was forced privately to reftore the books and lofe the money, to avoid a worfe inconvenience. Thus Conftantinople was no longer agreeable to him, and the lefs fo, becaufe he had not been able to perfect himfelf in the arabic tongue for want of fufficient mafters, which he had made no doubt of finding there. In thefe circumftances, parting with his fellow-traveller, Pococke, he embraced the opportunity then offered of paffing in company with the annual Turkifh fleet to Alexandria, where, having in his way touched at Rhodes, he arrived before the end of September 1638. This was the boundary of his intended progrefs. The country afforded a large field for the exercife of his curious and inquifitive genius; and he omitted no opportunity of remarking whatever the heavens, earth, or fubterraneous parts, offered, that feemed any way ufeful and worthy of notice; but, in his aftronomical obfervations, he was too often interrupted by the rains, which, contrary to the received opinion, he found to be frequent and violent, especially in the middle of winter. He was alfo much difappointed here in his expectations of purchafing books, finding very few of thefe, and for learned men none at all. But the grand purpofe of his coming here being to take an accurate furvey of the pyramids, he went twice to the defarts near Grand Cairo, where they ftand; and, having executed his undertaking entirely to his fatisfaction, embarked at Alexandria, in April 1639. Arriving in two months at Leghorn, he made the tour of Italy a fecond time, in order to examine more accurately into the true ftate of the Roman weights and meafures, now that he was furnifhed with proper inftruments for that purpofe, made by the beft hands.

From Leghorn he proceeded to Florence, where he was received with particular marks of efteem by the great duke of Tufcany, Ferdinand II. to whom he had infcribed a latin poem from Alexandria, in which he exhorted that prince to clear thofe feas of pirates, with whom they were extremely infefted. He obtained, likewife, admittance into the Medicean library, which had been denied to him as a ftranger when he was here before in his former tour. From Florence he went to Rome, and took moft exact meafurements of all the antique curiofities in that city and neighbourhood; after which he returned to Leghorn, where taking his paffage in a veffel called the Golden Fleece, at the end of March, he arrived at London before Midfummer 1640, with a rich cargo, confifting of a curious collection of arabic, perfic, and greek MSS. together with a great number of gems, coins, and other valuable antiquities, having spent full three years in this agreeable tour.

But upon his return, he met with a different scene at home from what he had left at his departure; and the ensuing national troubles proved greatly detrimental to his private affairs, in which he suffered much by his loyalty to the king and his gratitude to Laud. After a short stay at Gresham-college, which was no longer agreeable to him, he went to Oxford, and set about digesting his papers, and preparing such of them as might be most useful for the press. In this business he was assisted by archbishop Usher, to whom he had been long known; and now he drew a map of the Less Asia at his grace's request, who was writing his dissertation of that country, printed in 1641.

All this while he gave himself no concern about his Gresham-lecture, whereupon he was removed from it November 15, 1643. But this loss had been more than abundantly compensated by the Savilian professorship of astronomy, to which he was chosen the day before, in the room of Dr. Bainbridge, lately deceased; and he had a dispensation from the king, to hold his fellowship at Merton-college, because the stipend was much impaired by the means of the civil wars. The lectures being also impracticable on the same account, he was at full leisure to continue his attention to his papers; and accordingly we find, that he had made considerable progress in it by September the following year; some particulars whereof may be seen in a letter of that date to archbishop Usher. Among other things it appears, that he had made several extracts from them concerning the true length of the year; and happening, in 1645, to fall into discourse with some persons of figure at the court then at Oxford, with whom he was much in company, about amending the Kalendar, he proposed a method of doing it by omitting the intercalary day in the leap-year for forty years, and to render it conformable to the Gregorian [o]. He drew up a scheme for that purpose, which was approved by the king and council; but the state of the times would not permit the execution of it. The publication of his " Pyramidographia," and the " Description of the Roman Foot and Denarius," employed him the two subsequent years: he determined to begin with these, as they contained the fruit of his labours in

[o] The same method had been proposed to Pope Gregory, who rejected it, as Mr. Greaves says, that he might have the honour of doing it at once, and thereby of calling that year Annus Gregorianus, which our author did not doubt might justly be called Annus Confusionis, as the ancients called that year in which Julius Cæsar corrected the calendar, by a subtraction of days, after the same manner. But we have lately seen this method of doing it at once put in practice, without any ill consequences at all. This piece of Mr. Greaves is in the Phil. Transf. No. 237.

the primary view of his travels [P], and he was not in a condition to proceed any farther at prefent.

Hitherto he had been able, in a good meafure, to weather his difficulties, there being ftill left fome members in the Houfe of Commons who had a good regard for learning, among whom Selden made the greateft figure. That gentleman was burgefs for the univerfity of Oxford; and, being well known to our author before his travels, he dedicated his "Roman Foot" to him, under the character of his noble and learned friend: and his friendfhip was very ferviceable to Greaves, in a profecution in the parliament, in 1647, occafioned by his executorfhip to Dr. Bainbridge. This truft had involved him in law-fuits fo much as entirely to fruftrate his defign of going to Leyden to confult fome perfian MSS. neceffary for publifhing fome treatifes in that language. Upon the coming of the parliament's commiffioners to Oxford, feveral complaints were made to them againft him on the fame account; which being fent by them to the committee of the Houfe of Commons, our author, probably by the intereft of Selden (who was a member of that committee), was there cleared. After which he applied to the court of aldermen and the committee of Camden-houfe for reftitution. But though he evaded this farther difficulty, by the affiftance of fome powerful friends, yet this refpite was but fhort; however, he made ufe of that time in publifhing a piece begun by Dr. Bainbridge, and completed by himfelf. This was printed at Oxford, in 1648, under the title of "Johannis Bainbriggii Canicularia, &c." He dedicated this piece to doctor (afterwards Sir George) Ent, with whom he had commenced an acquaintance at Padua, in Italy; and that gentleman gave many proofs of his fincere friendfhip to our author, as well as to Dr. Pococke, in thefe times.

But the violence of the parliamentary vifitors was now grown above all reftraint, and a frefh charge was drawn up againft Greaves. Dr. Walter Pope informs us, that, confidering the violence of the vifitors, Greaves faw it would be of no fervice to him to make any defence; and, finding it impoffible to keep his profefforfhip, he made it his bufinefs to procure an able and worthy perfon to fucceed him. By the advice of Dr. Charles Scarborough, the phyfician, having pitched upon Mr. Seth Ward, he opened the matter to that gentleman, whom he foon met with there; and at the fame time propofed a method

[P] Thefe are the moft generally-ufeful part of his works. The latter is ranked among the claffics, and is nearly allied to the former: the exactnefs of which is put beyond all doubt in a piece of Sir Ifaac Newton, publifhed along with the moft correct editions of it in 1737, 8vo. Mr. Greaves took care to preferve, to the lateft times, the prefent ftandard of the meafures ufed in all nations, by taking the dimenfions of the infide of the largeft pyramid with the Englifh foot.

of compassing it, by which Ward did not only obtain the place, but the full arrears of the stipend, amounting to 500l. due to Greaves, and designed him a considerable part of his salary. The king's death, which happened soon after, was a shock to Greaves, and lamented by him in the most mournful terms, in a letter to Dr. Pococke: " O my good friend," says he, "my good friend, never was sorrow like our sorrow; excuse me now, if I am not able to write to you, and to answer your questions. O Lord God, avert this great sin, and thy judgements from this nation." However, he bore up against his own injuries with admirable fortitude; and, fixing his residence in London, he married, and, living upon his patrimonial estate, went on as before, and produced some other curious arabic and persic treatises, translated by him with notes every year. Besides which, he had prepared several others for the public view, and was meditating more when he was seized by a fatal disorder, which put a period to his life, October 8, 1652, before he was full fifty years of age. He was interred in the church of St. Bennet Sherehog, in London. His loss was much lamented by his friends, to whom he was particularly endeared, by joining the gentleman to the scholar. He had the happiness to be endowed with great firmness of mind, zeal in the interest which he espoused, and steadiness in his friendship; though, as he declares himself, not at all inclined to contention. He was highly esteemed by the learned in foreign parts, with many of whom he corresponded. Nor was he less valued at home by all who were judges of his great worth and abilities. He had no issue by his wife, to whom he bequeathed his estate for her life; and having left his cabinet of coins to his friend Sir John Marsham, author of the "CanonChronicus," he appointed the eldest of his three younger brothers (Dr. Nicolas Greaves) his executor, who by will bestowed our author's astronomical instruments to the Savilian library at Oxford, where they are reposited, together with several of his papers; but a great many of these were sold by his widow to a bookseller, and lost or dispersed.

GREEN (ROBERT), an author in queen Elizabeth's reign, was first of St. John's college, Cambridge, where he took the degree of B. A. in 1578; afterwards removed to Clare-hall, and, in 1583, became M. A. It is said, he was likewise incorporated at Oxford. He was a man of great wit and humour, but prostituted his talents to the purposes of vice and obscenity; and, upon the whole, both in theory and practice, seems to have been a most perfect libertine. Unable to support his extravagances, he was forced to recur to his pen for maintenance; and is believed to be the first english poet who wrote for bread. After a course of years, spent in dissipation

sipation, riot, and debauchery, we find him fallen into a state of the most wretched penury, disease, and self-condemnation; as appears from a letter written to a much-injured wife, and inserted in Cibber's "Lives of the Poets." His letter, we hope, was truly penitential and sincere; yet from the titles of some of his later works, such as Green's "Never too Late," Green's "Farewell to Folly," Green's "Groatsworth of Wit," &c. it should seem as if he was more solicitous about appearances than realities. Wood says, that he died in 1592 of a surfeit, gotten by eating too great a quantity of pickled-herrings, and drinking Rhenish wine with them; so that he died as he lived, and was consistent throughout. His works of different kinds are very numerous; but, as to his dramatic ones, there are many difficulties in coming, with any degree of certainty, at a knowledge of them. What are undoubtedly his, amounting to four or five pieces, may be seen in the "Biographia Dramatica."

GREEN (JOHN), born about 1706, at or near Hull, in Yorkshire, received the rudiments of his education at a private school, and was sent to St. John's college, Cambridge; after taking his degrees in arts, and being chosen fellow, he engaged himself as usher to a school at Lichfield, before Dr. Johnson and Mr. Garrick had left that city to launch into the world, with both of whom he was of course acquainted. In 1744, Charles duke of Somerset, chancellor of the university, appointed Mr. Green (then B D.) his domestic chaplain. In January, 1747, Green was presented by his noble patron to the rectory of Borough-green, near New-market, which he held with his fellowship. In December 1748, on the death of Dr. Whalley, he was elected regius professor of divinity; and soon after was appointed one of his majesty's chaplains. In June 1750, on the death of dean Castle, master of Corpus Christi or Benet-college, a majority of the fellows (after the headship had been declined by their president, Mr. Scottowe) agreed to apply to archbishop Herring for his recommendation; and his grace, at the particular request of the duke of Newcastle, recommended professor Green, who was immediately elected.

Among the writers on the subject of the new regulations proposed by the chancellor, and established by the senate, Dr. Green took an active but anonymous part, in a pamphlet published in the following winter, intituled, "The Academic, or a Disputation on the State of the University of Cambridge." March 22, 1751, on the advancement of his friend Dr. Keene, master of St. Peter's college, to the bishopric of Chester, Dr. Green preached the consecration-sermon in Ely-house-chapel, which, by order of the archbishop of York, was soon after published. In October 1756, on the death of Dr. George, he was preferred to the deanery of Lincoln, and resigned his professorship. Being then eligible to the office of vice-chancellor,

cellor, he was chosen in November following. In June, 1761, the dean moſt ably exerted his polemical talents in two letters (publiſhed without his name) "on the Principles and Practices of the Methodiſts," 1. addreſſed to Mr. Berridge, 2. to Mr. Whitfield. On the tranſlation of biſhop Thomas to the biſhopric of Saliſbury: Green was promoted to the ſee of Lincoln, the laſt mark of favour which the duke of Newcaſtle had it in his power to ſhew him. In 1762, archbiſhop Secker (who had always a juſt eſteem of his talents and abilities) being indiſpoſed, the biſhop of Lincoln viſited as his proxy the dioceſe of Canterbury. In 1763, he preached the 30th of January ſermon before the Houſe of Lords, which was printed.

The biſhop reſigned the maſterſhip of Benet-college, viz. in July 1764. After the death of lord Willoughby of Parham, in 1765, the literary *converſatione* of the Royal Society, &c. which uſed to be held weekly at his lordſhip's houſe, was transferred to the biſhop of Lincoln, in Scotland-yard, as one of their moſt accompliſhed members. In July 1771, on a repreſentation to his majeſty, that, with diſtinguiſhed learning and abilities, and a moſt entenſive dioceſe, biſhop Green (having no commendam) had a very inadequate income, he was preſented to the reſidentiaryſhip of St. Paul's, which biſhop Egerton vacated on his tranſlation to the ſee of Durham. He now removed to his reſidentiary-houſe in Amen-corner, and took a ſmall country-houſe at Tottenham. It ſhould ever be remembered, to our prelate's honour, that, in May 1772, when the Bill for relief of Proteſtant Diſſenters, &c. after having paſſed the Houſe of Commons, was rejected, on the ſecond reading, by the Houſe of Lords, (102 to 27,) he nobly diſſented from his brethren, and was the only biſhop who voted in its favour. Without any particular previous indiſpoſition, his lordſhip died ſuddenly in his chair at Bath, on Sunday, April 25. 1779.

GREEN (EDWARD BURNABY), was the author of various poetical works. He was educated at Benet college, Cambridge. He tranſlated Anacreon and Apollonius Rhodius, He publiſhed a paraphraſe of Perſius, and a tranſlation of parts of Pindar; but he had more taſte then animation, and more accuracy than harmony. His talents were of the reſpectable kind, indeed the moſt reſpectable; but he cannot be placed in the firſt rank of our authors.

GREEN (MATTHEW), a reſpectable poet, was born of a reputable family among the diſſenters. He was a man of great integrity of mind and ſweetneſs of manner. His converſation was full of wit, which neverthelefs he ſo tempered as never to give offence. He had an appointment in the Cuſtom-houſe, the duty of which he diſcharged with great diligence and ability. He died at the age of forty-one. He wrote many elegant poems;
but

but the one, which more particularly entitled him to a place among the English poets, is called the "Spleen," and which is full of witty and original thoughts. Mr. Green's fame has received much honour from a publication of his more distinguished pieces by Dr. Aikin, with critical and explanatory notes.

GREENE (Dr. MAURICE), an eminent musician, was the son of a London clergyman, and nephew of John Greene, serjeant at law. He was brought up in St. Paul's choir, and apprenticed to the organist of that cathedral. He soon distinguished himself in his profession; and, about 1716, when he was not yet twenty, was chosen organist of St. Dunstan in the West. In 1717, he became organist of St. Andrew's, Holborn; and the year after of St. Paul's; upon which last preferment he quitted the two former. In 1727, upon the decease of Croft, he was appointed organist and composer to the Royal Chapel, and thus placed at the head of his profession in England. In 1730, he took the degree of doctor in music at Cambridge: his exercise for it was Pope's "Ode for St Cecilia's Day," which he set very finely to music. It was performed with great applause; and he was honoured with the title of professor of music in that university. Greene was a man of understanding, was patronized by many great personages, and, about 1735, appointed master of the royal band. About 1750, he had a considerable estate left to him by a natural son of his uncle, the serjeant; and this state of affluence inspired him with a project of reforming our church-music, which was greatly corrupted by a multiplication of copies, and the ignorance and carelessness of transcribers. To correct, and also secure it against such injuries for the future, he began with collating a great number of copies of services and anthems, and reducing them into score. He had made a considerable progress in the work; but, his health failing him, he made his will, and transmitted the farther prosecution of it to his friend Dr. William Boyce, who completed and published it. Dr. Greene died Sept. 1, 1755. An account of his performances may be seen in Sir John Hawkins.

GREENHILL (JOHN), a very ingenious english painter, was descended from a good family in Salisbury, where he was born. He was the most excellent of all the disciples of Sir Peter Lely, who is said to have considered him so much as a rival, that he never suffered him to see him paint. Greenhill, however, prevailed with Sir Peter to draw his wife's picture, and took the opportunity of observing how he managed his pencil; which was the great point aimed at. This gentleman was finely qualified by nature for both the sister-arts of painting and poetry; but his loose and unguarded manner of living was probably the occasion of his early death; and only suffered him just to leave enough of his hand, to make

make us wifh he had been more careful of a life fo likely to do honour to his country. This painter won fo much on the celebrated Mrs. Behn, that fhe endeavoured to perpetuate his memory by an elegy, to be found among her works. He painted a portrait of bifhop Ward, which is now in the town-hall of Salifbury. He died May 19, 1676.

GREENVILE (SIR RICHARD), grandfather to the famous Sir Bevil Greenvile, was vice-admiral under lord Thomas Howard, fon to the duke of Norfolk, who was fent with a fquadron of feven fail to America, to intercept the fpanifh galeons; but, Sir Richard happening to be feparated from the reft of the fquadron, unfortunately fell in with the enemy, whofe fleet confifted of fifty-two fail, which he engaged and continued fighting till he was covered with blood and wounds, and nothing remained of his fhip but a battered hulk. He died on-board the fpanifh fleet three days after, expreffing the higheft courage in the article of death, and his having acted an englifh part, 1591.

GREGORY, furnamed the GREAT, was born of a patrician family, equally confpicuous for its virtue and nobility at Rome, where his father Gordian was a fenator, and extremely rich; and, marrying a lady of diftinction, called Sylvia, had by her this fon, about 544. From his earlieft years he difcovered genius and judgement; and, applying himfelf particularly to the apophthegms of the ancients, he fixed every thing worth notice in his memory, where it was faithfully preferved as in a ftore-houfe; he alfo improved himfelf by the converfation of old men, in which he took great delight. By thefe methods he made a great progrefs in the fciences, and there was not a man in Rome, who furpaffed him in grammar, logic, and rhetoric; nor can it be doubted but he had early inftructions in the civil law, in which his letters prove him to have been well verfed: he was neverthelefs entirely ignorant of the greek language. Thefe accomplifhments in a young nobleman procured him fenatorial dignities, which he filled with great reputation; and he was afterwards appointed præfect of the city by the emperor Juftin the Younger; but, being much inclined to a monaftic life, he quitted that poft, and retired to the monaftery of St. Andrew, which he himfelf had founded at Rome in his father's houfe, and put it under the government of an abbot, called Valentius. Befides this, he founded fix other convents in Sicily; and, felling all the reft of his poffeffions, he gave the purchafe-money to the poor.

However, he had not enjoyed his folitude in St. Andrew's long, when he was removed from it by pope Pelagius II. who made him his feventh deacon, and fent him as his nuncio to the emperor Tiberius at Constantinople, to demand succours against the Lombards. The Pope could not have chosen a man better qualified than Gregory for so delicate a negociation; of which, however, the particulars are unknown. Meanwhile, he was not wanting in exerting his zeal for religion. While he was in this metropolis, he opposed Eutychius the patriarch, who had advanced an opinion bordering on Origenism, and maintained, that after the resurrection the body is not palpable, but more subtile than air. In executing the business of his embassy, he contracted a friendship with some great men, and gained the esteem of the whole court, by the sweetness of his behaviour; insomuch, that the emperor Maurice chose him for a godfather to one of his sons, born in 583. Soon after this he was recalled to Rome, and made secretary to the Pope; but, after some time, obtained leave to retire again into his monastery, of which he had been chosen abbot.

Here he had indulged himself with the hopes of gratifying his wish, in the enjoyment of a solitary and unruffled life, when Pelagius II. dying Feb. 8, 590, he was elected Pope by the clergy, the senate, and the people of Rome; to whom he had become dear by his charity to the poor, whom the overflowing of the Tiber, and a violent plague, had left perishing with hunger. This promotion was so disagreeable to him, that he employed all possible methods to avoid it; he wrote a pressing letter to the emperor, conjuring him not to confirm his election, and to give orders for the choice of a person who had greater capacity, more vigour, and better health than he could boast; and hearing his letter was intercepted by the governor of Rome, and that his election would be confirmed by the imperial court, he fled, and hid himself in the most solitary part of a forest, in a cave; firmly resolved to spend his days there, till another Pope should be elected: and, the people despairing to find him, a new election ensued. In such cases, the ecclesiastics of that church never slip the opportunity of introducing miracles; accordingly, we are told, that Gregory would never accept the papal chair, till he had manifestly found, by some celestial signs, that God called him to it. It is pretended, that a dove flying before those who sought for him, shewed them the way they were to go; or that a miraculous light, appearing on a pillar of fire over his cavern, pointed out to them the place of his retreat [Q].

However that be, it is almost as certain that his reluctance was sincere [R] as it is that he at length accepted the dignity,

and

[Q] St. Gregory, ond and credulous as he was of miracles, says nothing of these.

[R] His famous pastoral is alleged on the side of his sincerity. Gregory wrote it in answer to John, bishop of Ravenna, and was enthroned Pope, Sept. 3, 590. And it appeared by his conduct, that they could not have elected a person more worthy of this exalted station; for, besides his great learning, the pains he took to instruct the church, both by preaching and writing, he had a very happy talent to win over princes, in favour of the temporal as well as spiritual interests of religion. It would be tedious to run over all the particulars of his conduct on these occasions; and his converting the English to Christianity, a remarkable fact in our history is on that account vulgarly known [s]; but there is one circumstance in it worth noting. It is observable, that Gregory owed his success to the assistance of a woman. The queen [Ethelburga] had a great share in these conversions, since she not only prompted the king [Ethelbert] her consort, to treat the Pope's missionaries kindly, but also to become himself a convert.

The new Pope, according to custom, held a synod at Rome the same year, 591; whence he sent letters to the four patriarchs of the East, with a confession of his faith, declaring his reverence to the four general councils, and the fifth too, as well as the four gospels. In this modesty he was not followed by his successors; and he even exceeded some of his predecessors in that and other virtues, which for many ages past have not approached the pretended chair of St. Peter. As he had governed his monastery with a severity unparalleled in those times; so now he was particularly careful to regulate his house and person according to St. Paul's directions to Timothy, 1 Ep. iii. 5. Even in performing divine worship, he used ornaments of but a moderate price, and his common garments were still more simple. Nothing was more decent than the furniture of his house, and he retained none but clerks and religious in his service. By this means his palace became a kind of monastery, in which there were no useless people; every thing in his house had the appearance of an angelic life, and his charity surpassed all description. He employed the revenues of the church entirely for the relief

Ravenna, who had given him a friendly reproof for hiding himself, in order to avoid the pontificate. This conduct is ascribed, and not undeservedly, to his humility; and, after his promotion, he gave another evidence of his sincerity, in constantly declaring his dislike of the appellation, " Your Beatitude, &c." which had been given to his predecessors. Bayle, in viewing his subsequent conduct in this post, observes, that those who forced him into the papal chair knew him better than he knew himself; that they saw in him a fund of all the cunning and suppleness that is requisite to acquire great protectors, and bring upon the church the blessings of the earth. Dict. under this Pope's art.

[S] He first set out on his mission himself, while he was a monk only, and was advanced three days journey, when Pelagius, then Pope, recalled him to Rome at the instigation of the people, who even clamorously pressed him to it.

of

of the poor; he was a constant and indefatigable preacher, and devoted all his talents for the instruction of his flock.

In the mean time, he extended his care to the other churches under his pontifical jurisdiction, and especially those of Sicily, for whom he had a particular respect; he put an end to the schism in the church of Iberia the same year; this was effected by the gentle methods of persuasion, to which, however, he had not recourse till after he had been hindered from using violence. Upon this account he is censured as an intolerant; and it is certain his maxims on that head were a little inconsistent. He did not, for instance, approve of forcing the Jews to receive baptism, and yet he approved of compelling heretics to return to the church. In some of his letters too he exclaims against violence in the method of making converts, yet at the same time was for laying heavier taxes on such as would not be converted by persuasive means; and, 593, he sent a nuncio to Constantinople, and wrote a letter the same year to the emperor Maurice, declaring his humility and submission to that sovereign; he also shewed the same respect to the kings of Italy, even though they were heretics.

The same year he composed his " Dialogues," a work filled with false miracles and incredible stories; the style is also low, and the narration coarse; however, they were received with astonishing applause; and Theodilinda, queen of the Lombards, having converted her spouse to the catholic faith, the Pope was exceedingly rejoiced at it, and sent his " Dialogues," composed the following year, to that princess. She is thought to have made use of his book at this time for the conversion of that people, who were the fittest in the world to be wrought upon by such pious fooleries. For, the same Pope Zachary, about 150 years after, translated it into greek for the use of those people, who were so delighted with it, that they gave St. Gregory the surname of Dialogitt. In 594, he excommunicated and suspended the bishop of Salona, the metropolis of Dalmatia, who, however, paid no regard to the exercise of his power in these censures. The same year he laboured to convert the infidels in Sardinia by gentle methods, according to his system: which was, to punish heretics, especially at their first rise, as rebels and traitors, but to compel infidels only indirectly; that is, treating the obstinate with some rigour, and persuading them as much by promises, threats, and gentle severities, as by argument and reason. This was the distinction he made in treating with the Manichees and Pagans.

In 595, he refused to send the empress Constantia any relics of St. Paul, which she had requested, desiring to look at the body of that apostle: he thereupon relates several miraculous

culous punishments for such a rash attempt, all as simply devised as those in his "Dialogues." The same year he warmly opposed John patriarch of Constantinople, for assuming the title œcumenical or universal, which he himself disclaimed, as having no right to reduce the other bishops to be his substitutes; and afterwards forbad his nuncio there to communicate with that patriarch, till he should renounce the title. His humility, however, did not keep him from resenting an affront put upon his understanding, as he thought, by the emperor, for proposing terms of peace to the Lombards, who besieged Rome this year: the same year he executed the famous mission into England; and as Brunehaut, queen of France, had been very serviceable therein, he wrote a letter of thanks to her on the occasion. The princess is represented as a profligate woman, but very liberal to the ecclesiastics; founding churches and convents, and even sueing to the Pope for relics. This was a kind of piety which particularly pleased Gregory; and accordingly, he wrote to the queen several letters, highly commending her conduct in that respect, and carried his complaisance so far as to declare the French happy above all other nations in having such a sovereign. In 598, at the request of the christian people at Caprita, a small island at the bottom of the gulph of Venice, he ordered another bishop to be ordained for that place, in the room of the present prelate, who adhered to the Istrian schism. This was done contrary to the orders of the emperor Maurice against taking any violent measures with schismatics.

In 599, he wrote a letter to Serenus bishop of Marseilles, commending his zeal in breaking some images which the people had been observed to worship, and throwing them out of the church; and the same year a circular letter to the principal bishops of Gaul, condemning simoniacal ordinations, and the promotions of laymen to bishoprics: he likewise forbad clerks in holy orders to live with women, except such as are allowed by the canons; and recommended the frequent holding assemblies to regulate the affairs of the church. The same year he refused, on account of some foreseen opposition, to take cognizance of a crime alleged against the primate of Byzacena, a province in Africa. About the same time he wrote an important letter to the bishop of Syracuse, concerning ceremonies, in which he says, "That the church of Rome followed that of Constantinople, in the use of ceremonies; and declares that see to be undoubtedly subject to Rome, as was constantly testified by the emperor and the bishop of that city." He had already this year reformed the office of the church, which is one of the most remarkable actions of his pontificate. In this reform, as it is called,

he

he introduced several new customs and superstitions; amongst the rest, Purgatory. He ordered pagan temples to be consecrated by sprinkling holy water, and an annual feast to be kept, since called wakes in England, on that day; with the view of gaining the pagans in England to the church-service. Besides other less important ceremonies, added to the public forms of prayer, he made it his chief care to reform the psalmody, of which he was excessively fond. Of this kind he composed the "Antiphone [T]," and such tunes as best suited the psalms, the hymns, the prayers, the verses, the canticles, the lessons, the epistles, and gospels, the prefaces, and the Lord's prayer. He likewise instituted an academy of chanters for all the clerks, as far as the deacons exclusively: he gave them lessons himself, and the bed, in which he continued to chant amidst his last illness, was preserved with great veneration in the palace of St. John Lateran for a long time, together with the whip, with which he used to threaten the young clerks and singing boys, when they sang out of tune. He was so rigid in regard to the chastity of ecclesiastic, that he was unwilling to admit a man into the priesthood who was not strictly free from defilement by any commerce with women. The candidates for orders were according to his commands questioned particularly on that subject. Widowers were excepted, if they had observed a state of continency for some considerable time.

At this time, as well as the next year 600, he was confined to his bed by the gout in his feet, which lasted for three years; yet he celebrated mass on holidays, with much pain all the time. This brought on a painful burning heat all over his body, which tormented him in 161. His behaviour in this sickness was very exemplary. It made him feel for others,

[T] It is to this Pope that we owe the invention, used to this day, of expressing musical sounds by the seven first letters of the alphabet. Indeed the Greeks made use of the letters of their alphabet to the like purpose: but in their scale they wanted more signs, or marks, than there were letters, which were supplied out of the same alphabet, by making the same letter express different notes, as it was placed upright, or reversed, or otherwise put out of the common position; also making them imperfect by cutting of something, or by doubling some strokes. For example, the letter Pi expresses different notes in all these positions and forms, Π Π ⊏ ⊐ Π ΙΙ &c. They who are skilled in music, need not be told what a loss the scholar had in this method to learn. In Boethius's time the Romans eased themselves of this difficulty as unnecessary, by making use only of the first 15 letters of their alphabet. But afterwards, this Pope, considering that the octave was the same in effect with the first note, and that the order of degrees was the same in the upper and lower octave of the diagram, introduced the use of seven letters, which were repeated in a different character. Malcolm on Music, chap. xiv. § 4.—N. B. Platina says, that Gregory was the inventor of the whole church-office; and it is certain he introduced many new ceremonies, calculated to strike the beholders with their pomp and magnificence, and thereby make them converts.

whom

whom he compassionated, exhorting them to make the right use of their infirmities, both by advancing in virtue and forsaking vice. He was always extremely watchful over his flock, and careful to preserve discipline; and while he allowed that the misfortunes of the times obliged the bishops to interfere in worldly matters, as he himself did, he constantly exhorted them not to be too intent on them. This year he held a council at Rome, which made the monks quite independent by the dangerous privileges which he granted them. Gregory forbad the bishops to diminish in any shape the goods, lands, and revenues, or titles of monasteries, and took from them the jurisdiction they ought naturally to have over the converts in their dioceses. But many of his letters shew, that though he favoured the monks in some respects, he nevertheless knew how to subject them to all the severity of their rules. The same year he executed a second mission into England, and, in answer to the bishop of Iberia, declared the validity of baptism by the Nestorians, as being performed in the name of the trinity.

The dispute about the title of Universal Bishop and the equality of the two sons of Rome and Constantinople still subsisting, and the emperor Maurice having declared for the latter, our Pope saw the murder of him and his family without any concern by Phocas. This usurper having sent his picture to Rome in 603, Gregory received it with great respect, and placed it with that of the empress his consort [Leontia] in the oratory of St. Cæsarius in the palace; and soon after congratulated Phocas's accession to the throne. There are still extant, written upon this occasion, by the holy pontiff, three letters, wherein he expresses his joy, and returns thanks to God, for that execrable parricide's accession to the crown, as the greatest blessing that could befal the empire; and he praises God, that, after suffering under a heavy galling yoke, his subjects begin once more to enjoy the sweets of liberty under his empire: flatteries unworthy a man of honour, and especially a pope [υ]; but Gregory thought himself in conscience obliged to assert the superiority of his see above that of Constantinople, and he exerted himself much to secure it. In general he had the pre-eminence of the holy see much at heart; accordingly this same year, one Stephen, a Spanish bishop, having complained to him of an unjust deprivation from his bishopric, the pope sent a delegate to judge the matter upon the spot, giving him a memorial of his instructions, wherein among other particulars he orders thus: "If it be said, that bishop Stephen had neither metropolitan nor patriarch, you must answer, that he ought to

[υ] His historian Maimbourg, though a jesuit, condemns him on this occasion.

be tried, as he requested, by the holy fee, which is the chief of all churches. It was in the same spirit of preserving the dignity of his pontificate, that he resolved to repair the celebrated churches of St. Peter and St. Paul; in which view, he gave orders this year to the subdeacon Sabinian (afterwards his successor in the popedom), to have felled all the timber necessary for that purpose in the country of the Brutii, and shipt for Rome: he wrote several other letters on this occasion, which are so many proofs of his zeal for carrying on the work [w].

But while he was thus intent in repairing the mischiefs of the late war, he saw it break out again in Italy, and still to the disadvantage of the empire, the affairs of which were in a very bad situation, not only in the provinces of the West, but every where else. Gregory was much afflicted with the calamities of this last war, and at the same time his illness intolerable. The Lombards made a truce in November 603, which was to continue in force till April 605. Some time after, the pope received letters from queen Theodilinda, with the news of the birth and baptism of her son Adoaldus. She sent him also some writings of the abbot Secundinus upon the fifth council, and desired him to answer them. Gregory "congratulates her on having caused the young prince, destined to reign over the Lombards, to be baptised in the catholic church." And as to Secundinus, he excuses himself on account of his illness: " I am afflicted with the gout," says he, "to such a degree, that I am not able even to speak, as your envoys know; they found me ill when they arrived here, and left me in great danger when they departed. If God restores my health, I will return an exact answer to all that the abbot Secundinus has written to me. In the mean time, I send you the council held under the emperor Justinian, that by reading it he may see the falsity of all that he has heard against the holy see and the catholic church. God forbid that we should receive the opinions of any heretic, or depart in any respect from the letter of St. Leo, and the four councils:" he adds, " I send to the prince Adoaldus, your son, a cross, and a book of the gospel in a persian box ; and to your daughter three rings, desiring you to give them these things with your own hand, to enhance the value of the present. I likewise beg of you, to return my thanks to the king, your consort, for the peace he

---

[w] Lib. x. epist. 24, 25, 26, 27. It is observable, that this pope built no new churches, but took care of the old ones. For instance, he made a silver ciborium in the church of St. Peter, that is, a canopy to hang over the altar, and another in the church of St. Paul. He also appropriated several adjacent lands to supply this church with lights. Greg. Epist. book xii. epist. 9.

made

made for us, and engage him to maintain it, as you have already done."

This letter, written in January 604, is the laſt of Gregory's that has any date to it; he died the 12th of March following, worn out with violent and almoſt inceſſant illneſs. His remains were interred in a private manner, near the old ſacriſty of St. Peter's church, at the end of the great portico, in the ſame place with thoſe of ſome preceding popes. It is thought he was not above ſixty years of age. We ſhall only add one particular relating to our own country. Auguſtin the miſſionary having followed the rule approved by former popes of dividing the revenues of all the Engliſh churches into four parts, the firſt for the biſhop, the ſecond for the clergy, the third for the poor, and the fourth for repairing the church; this diviſion was confirmed by Gregory, who directed farther, that the biſhop's ſhare ſhould be not only for himſelf, but likewiſe for all his neceſſary attendants, and to keep up hoſpitality.

We muſt not conclude without obſerving, in juſtice to this pope, that the charge of his cauſing the noble monuments of the ancient ſplendor of the Romans to be deſtroyed, in order to prevent thoſe who went to Rome from paying more attention to the triumphal arches, &c. than to things ſacred, is rejected by Platina as a calumny. Nor is the ſtory, though credited by ſeveral learned authors, of his reducing to aſhes the Palatine library founded by Auguſtus, and the burning an infinite number of pagan books, particularly Livy, abſolutely certain. However, it is undeniable, he had a prodigious averſion to all ſuch books, which he carried to that exceſs, that he flew in a violent paſſion with Didier, archbiſhop of Venice, for no other reaſon than becauſe he ſuffered grammar to be taught in his dioceſe. In this he followed the apoſtolical conſtitutions: the compiler whereof ſeems alſo to have copied from Gregory Nazianzen, who thought reading pagan books would turn the minds of youth in favour of their idolatry; and we have ſeen in our days the ſame practice zealouſly defended, and upon the ſame principle too, by Mr. Tillemont. Notwithſtanding, Julian the apoſt..te is charged with uſing the ſame prohibition, as a good device to effect the ruin of chriſtianity, by rendering the profeſſors contemptible on account of their ignorance. Upon the whole, Bayle ſcruples not, all things conſidered, to pronounce this pope to have juſtly merited the title of Great.

We have more of his writings left than of any other pope; and they were held in ſuch eſteem in his life-time, as occaſioned ſome miſapplication of them, that troubled him: they have gone through no leſs than ſeventeen editions, the laſt of which was printed at Paris in 1675. Du Pin ſays, that his genius was well ſuited to morality, and he had acquired an inexhauſtible

ble fund of fpiritual ideas, which he expreffed nobly enough generally in periods, rather than fentences: his compofition was laboured, and his language inaccurate, but eafy, well connected, and always equally fupported.

GREGORY (JAMES), an eminent mathematician in Scotland, was born in 1639, at Aberdeen; and, being educated at that univerfity, made a good progrefs in claffical learning, but was more delighted with philofophical refearches, into which a new door had been lately opened by the key of the mathematics. Kepler and Des Cartes were the great mafters of this new method: their works, therefore, Gregory made his principal ftudy, and began early to make improvements upon their difcoveries in optics. The firft of thefe improvements was the invention of the reflecting telefcope, which ftill bears his name; and which was fo happy a thought, that it has given occafion to the moft confiderable improvements made in optics, fince the invention of the telefcope. He publifhed the conftruction of this inftrument in 1663, at the age of twenty-four; and coming next year, or the year after that, to London, he became acquainted with Mr. John Collins, who recommended him to the beft optic glafs-grinders there, in order to have it executed. But as this could not be done, for want of fkill in the artifts to grind a plate of metal for the object fpeculum into a true parabolic concave, which the defign required, he was much difcouraged; and after a few imperfect trials made with an ill-polifhed fpherical one, which did not fucceed to his wifh, he dropt the purfuit, and refolved to make the tour of Italy, then the mart of mathematical learning, in the view of profecuting his favourite ftudy with greater advantage.

He had not been long abroad, when the fame inventive genius, which had before fhewn itfelf in practical mathematics, carried him to fome new improvements in the fpeculative part. The fublime geometry on the doctrine of curves was then hardly paffed its infant ftate, and the famed problem of fquaring the circle ftill continued a reproach to it; when our author difcovered a new analytical method of fumming up an infinite converging feries, by which the area of the hyperbola, as well as the circle, may be computed to any degree of exactnefs. He was then at Padua; and getting a few copies of his invention printed there in 1667, he fent one to his friend Mr. Collins, who communicated it to the Royal Society, where it met with the commendation of lord Brounker and Dr. Wallis. He reprinted it at Venice, and publifhed it the following year 1668, together with another piece, wherein he firft of any one entertained the public with a method for the transformation of curves. An account of this piece was alfo read by Mr. Collins

lins before the Royal Society, of which Gregory, being returned from his travels, was chosen a member, admitted the 14th of January this year, and communicated to them an account of the controversy in Italy about the motion of the earth, which was denied by Riccioli and his followers..

The same year, his quadrature of the circle being attacked by Mr. Huygens, a controversy arose between those two eminent mathematicians, in which our author produced some improvements of his series. But in this dispute it happened, as it generally does in most others, that the antagonists, though setting out with temper enough, yet grow too much heated in the combat. This was the case here, especially on the side of Gregory, whose defence was, at his own request, inserted in the "Philosophical Transactions." He received from Mr. Collins, about this time, an account of the series invented by Sir Isaac Newton; who therein had actually effected what our author was stiffly contending against Huygens to be utterly impossible: that is, the ratio of the diameter of a circumference, expressed in a series of simple terms, independent of each other, and entirely freed from the magic vinculum of surds, in which they had till then been indissolubly held. It must be confessed, that our author had not the better in this dispute.

However, he was in so great esteem with the Royal Academy at Paris, that, in the beginning of 1671, it was resolved by that academy to recommend him to their grand monarch for a pension; and the design was approved even by Mr. Huygens, though he said, he had reason to think himself disobliged by Mr. Gregory, on account of the controversy between them. Accordingly, several members of that academy wrote to Mr. Oldenburg, desiring him to acquaint the council of the Royal Society with their proposal; informing him likewise, that the king of France was willing to allow pensions to one or two learned Englishmen, whom they should recommend. But no answer was ever made to that proposal; and our author, with respect to this particular, looked upon it as nothing more than a compliment.

In 1672, Sir Isaac Newton, on his wonderful discoveries in the nature of light, having contrived a new reflecting telescope, and made several objections to Mr. Gregory's, this gave birth to a dispute between those two philosophers, which was continued during that and the following year, in the most amicable manner on each side; Mr. Gregory defending his own construction, so far, as to give his antagonist the whole honour of having made the catoptric telescopes preferable to the dioptric; and shewing, that the imperfections in these instruments were not so much owing to a defect in the object-
speculum

speculum as to the different refrangibility of the rays of light. In the course of this dispute, our author described a burning concave mirrour, which was approved by Sir Isaac, and is still in good esteem. All this while he attended the proper business of his professorship with great diligence, which taking up the greatest part of his time, especially in the winter season, interrupted him in the pursuit of his proper studies. These, however, led him to farther improvements in the invention of infinite series, which he occasionally communicated to his intimate friend and correspondent Mr. Collins, who might have had the pleasure of receiving many more, had not our professor's life been cut short by a fever, December 1675, at the age of thirty-six years.

The most shining part of Gregory's character is that of his mathematical genius as an inventor. In this view, particularly, he merits a place in these memoirs; and therefore we shall conclude this article with a list of the most remarkable of his inventions. His reflecting telescope; burning concave mirrour; his quadrature of the circle, by an infinite converging series; and his method for transformation of curves have been already mentioned. Besides these, he first of any one gave a geometrical demonstration of lord Brounker's series for squaring the hyperbola, as it had been explained by Mercator, in his "Logarithmotechnia." He was likewise the first who demonstrated the Meridian Line to be analogous to a scale of Logarithmic Tangents, of the half compliment of latitude [x]. He also invented and demonstrated geometrically, by the help of the hyperbola, a very swift converging series for making the logarithms, and therefore recommended by Dr. Halley as very proper for practice. He also sent to Mr. Collins the solution of the famous Keplerian problem by an infinite series. He found out a method of drawing tangents to curves geometrically, without any previous calculations. He gave a rule for the direct and inverse method of tangents, which stands upon the same principal [of exhaustions] with that of fluxions, and differs not much from it in the manner of applications. He likewise gave a series for the length of the arc of a circle from the tangent, and *vice versa*; as also for the secant and logarithmic tangent and secant, and *vice*

[x] This invention is of great use in navigation; and his just merit as the inventor of the demonstration of it was afterwards asserted by Dr. Halley, who, however, at the same time observes, that it was performed, not without a long train of consequences, and complications of proportions, whereby the evidence of the demonstration was in a great measure lost, and the reader wearied before he attains it. Miscel. Curios. Vol. II. 1727. The truth is, complication, tediousness, and intricacy, were faults complained of in all his series, before he had learned to improve them by a sight of those of Sir Isaac Newton. Commerc. Epistol. No. 53.

versa.

verfa. Thefe, with others, for certifying, or meafuring the length of the elliptic and hyperbolic curves, were fent to Mr. Collins, in return for fome received from him of Sir Ifaac Newton's; and their elegance being admirable, and above whatever he had produced before, and after the manner of Sir Ifaac, gave room to think he had improved himfelf greatly by that mafter, whofe example he followed, in delivering his feries in fimple terms, independent on each other [y]

We are affured, that at his death he was in purfuit of a general method of quadrature, by infinite feries, like that of Sir Ifaac. This appeared by his papers, which came into the hands of his nephew, Dr. David Gregory, who publifhed feveral of them; and he himfelf affured Mr. Collins, he had found out the method of making Sir Ifaac's feries; who thereupon concluded he muft have written, a treatife upon it. This encouraged Mr. Stewart, profeffor of mathematics in Aberdeen, to take the trouble of examining his papers, then in the hands of Dr. David Gregory, the late dean of Chriftchurch, Oxford; but no fuch treatife could be found, nor any traces of it, and the fame had been declared before by Dr. David Gregory; whence it happens, that it is ftill unknown what his method was of making thofe feriefes. However, Mr. Stewart affirms, that, in turning over his papers, he faw feveral curious ones upon particular fubjects, not yet printed. On the contrary, fome letters which he faw confirmed Dr. David Gregory's remark, and made it evident, that our author had never compiled any treatife, containing the foundations of this general method, a very fhort time before his death; fo that all that can be known about his method can only be collected from his letters, publifhed in the fhort hiftory of his "Mathematical Difcoveries," compiled by Mr. Collins, and his letters to that gentleman in the "Commercium Epiftolicum." From thefe it appears, that, in the beginning of 1670, when Mr Collins fent him Sir Ifaac Newton's feries for fquaring the circular zone, it was then fo much above every thing he comprehended in this way, that after having endeavoured in vain, by comparing it with feveral of his own, and combining them together, to difcover the

---

[y] We fhall here give a lift of his works, which contain thefe feveral inventions. 1. "Optica Promota, &c. 1663," 4to, contains the conftruction of his telefcope. 2. Vera Circuli & Hyperbolæ Quadratura, Padua, 1667." It was firft publifhed in fuch hafte, that he found it neceffary for his reputation, to quicken as much as poffible the publication, with a preface, of his third piece, "Geometriæ pars Univerfalis, &c. 1667," 4to, containing his method of transforming curves. The reft of his inventions make the fubject of feveral letters and papers, printed either in the Philof. Tranf. the Commerc. Epiftol. Joh. Collins, & alior. 1715, 8vo, and in the Appendix to the englifh edition of Dr. David Gregory's "Elements of Optics, 1735," 8vo, by Dr. Defaguliers.

method of it, he concluded it to be no legitimate series; till, being assured of his mistake by his friend, he went again to work, and after almost a whole year's indefatigable pains, as he acknowledges, spent therein, he discovered, at last, that it might be deduced from one of his own, upon the subject of the logarithms, wherein he had given a method for finding the power to any given logarithm, or of turning the root of any pure power into an infinite series; and in the same manner, viz. by comparing and combining his own series together, or else by deduction therefrom, he fell upon several more of Sir Isaac's, as well as others like them, in which he must needs become daily more ready by continual practice; and this seems to have been the utmost he ever actually attained to, in the progress towards the discovering any universal method for those series. For, to speak ingenuously, he was not of a temper to conceal those discoveries: as is evident from the hurry he was in to print his treatise, "De vera Circuli & Hyperbolæ Quadratura," even before he had well revised it.

GREGORY (DAVID), nephew of the preceding, was born June 24, 1661, at the same place, Aberdeen; where he also received the first grounds of his learning, but was afterwards removed to Edinburgh, and took his degree of M. A. in that university. The great advantage of his uncle's papers induced his friends to recommend the mathematics to him; and he had a natural subtilty of genius particularly fitted for that study, to which he applied with indefatigable industry, and succeeded so well that he was advanced to the mathematical chair, at Edinburgh, at the age of twenty-three. The same year he published a treatise, intituled, "Exercitatio Geometrica de dimensione figurarum, Edinb. 1684." 4to. wherein, assuming the doctrine of indivisibility, and the arithmetic of infinites, as already known, he explained a method which not only suited his uncle's examples, left by him without any way of finding them, but discovered others, whereby an infinite number of curve-lines, and the areas contained between them and right lines (such as no other method then known extended to) might be measured. He had already seen some hints in his uncle's paper's concerning Sir Isaac Newton's method, of which he made the best use he could [z];

---

[z] In his latin "Treatise of Practical Geometry," there is a series of his uncle's, which he recommends for squaring the circle, though it converges so slow, as to be utterly of no use in practice, without some farther artifice. This is observed by Mr. Maclaurin, who published an English translation of it in 1745, 8vo. with additions, and the second edition was printed at Edinburgh 1751, 8vo. However Mr. Maclaurin's remark shews our author's skill in infinite series to be very imperfect, at the time of reading those lectures, from which the tract was compiled after his death; and Mr. Cotes, of Cambridge, spoke slightly of his abilities in that doctrine. Gen. Dict. Vol. IV. p. 144.

and the advantage he found thereby raised an ardent desire in him to see that method published. Under this impatient expectation, the "Principia" was no sooner out in 1687, but our author took it in hand, and presently made himself so much master of it [A] as to be able to read his professorial lectures upon the philosophy contained in it, and, causing his scholars to perform their exercises for their degrees upon several branches of it, became its first introducer into the schools.

He continued at Edinburgh till 1691, when, hearing of Dr. Bernard's intention to resign the Savilian professorship of astronomy at Oxford, he left Scotland, and, coming to London, was admitted a member of the Royal Society: and made his addresses to Sir Isaac Newton, who took the first opportunity of recommending him to Mr. Flamstead [master of the mathematical school in Christ's-Hospital, London,] with a letter, recommending his mathematical merit above all exception in these terms: "Sir, it is almost a fortnight since I intended, with Mr. Paget and another friend or two, to have given you a visit at Greenwich; but sending to the Temple Coffee-house, I understood you had not been in London for two or three weeks before, which made me think you were retired to your living for a time. The bearer hereof, Mr. Gregory, mathematic professor of Edinburgh college, in Scotland, intended to have given you a visit with us. You will find him a very ingenious person, and a good mathematician, worth your acquaintance." In proceeding, he mentions our author as a fit person, in case of Mr. Flamstead's death, to carry on his astronomical views [B]. Thus recommended, the royal astronomer used his best interest to procure him success at Oxford, where he was elected astronomy-professor this year, having been first admitted of Baliol college, and incorporated M. A. February 8, and he was created M. D. on the 18th of the same month. He had no relish for the technical part of his profession, and was seldom seen in the observatory. His genius lay more to geometry, and in that way he succeeded very well, both in his elements of optics [C], and of physical and geometrical astronomy. This last is reckoned

[A] Among his papers there was found a commentary upon it; and we learn from Mr. Flamstead, that his countryman gave out he had found a great many errors therein.

[B] The whole letter is under our author's article. Ibid.

[C] It was published in 1695, in latin, intituled, "Catoptricæ & Dioptricæ Sphericæ Elementa, Oxon." 8vo. and was compiled from his lectures, read at Edinburgh in 1684. In it he gives the preference to Sir Isaac Newton's reflecting telescope, above that of his uncle James Gregory. It was much esteemed for the neatness and easiness of the demonstrations, and a second edition in English came out in 1705, by Dr. Browne; and a third in 1735, by Dr. Desaguliers, who added an appendix, containing the history of the two reflecting telescopes, with their several improvements at that time.

his master-piece; and, having finished it in 1702 [D], he immediately engaged in carrying on the noble design of his predecessor, Dr. Bernard, to print all the works of the ancient mathematicians, the first-fruits of which appeared in an edition of Euclid's works in greek and latin, folio, the following year. In the same design, he afterwards joined with his colleague, Dr. Halley, in preparing an edition of "Apollonius's Conics:" Dr. Bernard had left materials for the four first books, which our author undertook to complete, but was prevented by his death, which happened October 16, 1710. He died at a country retirement at Maidenhead, in Berkshire; and there is a handsome marble monument erected to his memory in St. Mary's church at Oxford [E], by his wife, whom he left a widow with several children. His eldest son, David Gregory, was bred at Christ-church in Oxford, and appointed regius professor of modern history in that university, at the institution thereof by George I. He afterwards commenced D. D. and succeeded to a canonry, and afterwards became dean of that church.

Our professor's genius lay chiefly in inventing new and elegant demonstrations of the discoveries made by others. For instance, he gave the first demonstration of that curve, which is well known since by the name of catenaria, or the curve that is formed by a chain fastened at each end; and first discovered, that this curve inverted gave the form of a true and legitimate arch, all the parts supporting each other [F]. There are several other papers of his in the "Philosophical Transactions," a list of which, with some account of the most considerable, may be seen in "Biographia Britannica," under his article. His explication of Sir Isaac Newton's method, to construct the orbit of a comet by three accurate observations, is commended by Dr. Halley.

GREGORY (JOHN), a learned divine, was born November 10, 1607, at Agmondesham, in Buckinghamshire. There appeared in his infancy such a strong inclination to learning as recommended him to the notice of some persons of the best rank in the town; and, his parents being well respected for their piety and honesty, it was resolved to give him a liberal education at the university, the expence of which they were not able to support. To this purpose, he was chosen at the age of fifteen, by Dr. Crooke, to go with Sir William Drake

---

[D] It was published that year in folio; it was afterwards reprinted in 4to. at Geneva, and lastly in English by Mr. Stone, 1726, at Lond. 8vo.

[E] The inscription may be seen in Biog. Brit.

[F] This is printed in Phil. Transf. No. 231. He observes, that arches of all other forms, in stone, brick, and the like, are only supported by including some catenary curve, within the breadth of their forming stones.

to Chrift-church, in Oxford, whom he attended in the ftation of a fervitor, and he was foon after retained by Sir Robert Crook in the fame capacity ; Dr. George Morley, afterwards bifhop of Winchefter, was their tutor. Mr. Gregory made the beft ufe of this favour, and applied fo clofely to his ftudies, that he became almoft a prodigy for learning. He took his firft degree in arts in 1621, and commenced mafter in 1631 ; about which time, entering into orders, the dean, Dr. Brian Duppa, gave him a chaplain's place in that cathedral. 'In 1634, he publifhed a fecond edition of Sir Thomas Ridley's "View of the Civil and Ecclefiaftical Law," with notes; which piece was well received, and brought our author's merit into the knowledge of the world: the notes fhewing him well verfed in the hiftorical, ecclefiaftical, ritual, and oriental learning, and a confiderable mafter in the faxon, french, italian, fpanifh, and all the eaftern languages. All thefe acquifitions were the pure fruit of his own induftry ; for he had no affiftance, only for the hebrew tongue, wherein Mr. John Dod, the decalogift [G], gave him fome directions. His merit engaged the farther kindnefs of Dr. Duppa; and, when that prelate was promoted to the bifhopric of Chichefter in 1638, he made Mr. Gregory his domeftic chaplain, and fome time after gave him a prebend in that church. His patron alfo continued his favours after his tranflation to the fee of Salifbury in 1641, when he feated him in a ftall of that cathedral.

But he did not enjoy the benefit of thefe preferments long; being a firm loyalift, as well as his patron, he was deprived of both by the iniquity of the times, whence he was reduced fome years before his death to great diftrefs. In thefe circumftances, he was taken into the houfe of one Sutton, to whofe fon he had been tutor; this was an obfcure ale-houfe on Kiddington-green, near Oxford, where he lived till his death, which happened March 13, 1646; occafioned by an hereditary gout, with which he had been troubled for above twenty years, and which at laft feized his ftomach. His corpfe was carried to Oxford, and interred, at the expence of fome friends, in that cathedral. He was honoured with the acquaintance and favour of the greateft men of the age, and held a correfpondence with feveral eminent perfons abroad, as well Jews and Jefuits, as others. His works are, 1. "Notes and Obfervations on fome Paffages of Scripture," publifhed a little before his death in 1646, 4to. and tranflated into latin, and inferted in the "Critici Sacri." 2. "Gregorii Pofthuma; or certian learned Tracts, written by John Gregory, &c. Lond. 1650:" and again in 1664, 1671, 1683. 4to.

[G] So called from an expofition written by him, together with Robert Cleaver, another puritan minifter, on the Ten Commandments.

GREGORY (EDMUND), the author of the "Historical Anatomy of Christian Melancholy," and a "Meditation on Job ix. 4." printed in 1 vol. 8vo. to which is prefixed his head; was some time a student at Trinity-college, in Oxford; but left that university after he had taken one degree in arts. Mr. Granger says, it is uncertain whether he ever received episcopal ordination. He died after 1650.

GREGORY (NAZIANZEN), was born A. D. 324, at Azianzum, an obscure village belonging to Nazianzum, a town of the second Cappadocia, situated in a poor, barren, and unhealthy country. His parents were persons of rank, and no less eminent for their virtues: his father, whose name was also Gregory, had been educated in an odd sort of religion, called Hypsistarianism [u], to which, being the religion of his ancestors, he was a bigot in his younger years; and the deserting it not only lost him the kindness of his friends, but estranged him from his mother, and deprived him of his estate. This, however, he bore with great chearfulness for the sake of christianity, to which he was converted by his wife, though not without the help of an emphatical dream; he was afterwards made bishop of Nazianzum, being the second who sat in that chair, where he behaved with great prudence and diligence. Nor was our author's mother less eminent: descended of a pious family, she was herself, for piety, so much the wonder of her age, that this son was said to have been the pure effect of her prayers, and of a vow to devote him to God, after the example of Hannah: and, as in that case, the Deity here also not only gratified her importunity, but was pleased in a vision to communicate to her both the shape of the child she should bear, and the name by which he was to be called; and, upon his birth, she was careful to perform her vow.

Thus advantageously born, he proved a child of pregnant parts; by which, and the advantage of a domestic institution under his parents, he soon outstript his contemporaries in learning. Nature had formed him of a grave and serious temper, so that his studies were not obstructed by the little sports and pleasures of youth. After some time, he travelled abroad for his farther improvement: in which rout, the first step he took was to Cæsarea; and, having rifled the learning of that university, he travelled to Cæsarea Philippi in Palestine, where

---

[u] This was a kind of Samaritan mixture, made of Judaism and Paganism, or rather some select rites of each. With the Gentiles, they did honour to fire and burning lights, but rejected idols and sacrifices; with the Jews, they observed the sabbath, and a strict abstinence from some kind of meats, but disowned circumcision. They pretended to worship no other deity but the almighty, supreme, and most high God; whence they assumed their characteristic above mentioned, ὕψιςος, signifying The Most High.

some

some of the most celebrated masters of that age resided, and where Eusebius then sat bishop. Here he studied under the famous orator Thespasias, and had, among other fellow pupils, Euzoïus, afterwards the Arian bishop of that place. He applied himself particularly to rhetoric, minding the elegance, not the vanity and affectation, which then too much affected that profession. Hence he removed to Alexandria, whose schools were famous next to those of Athens, which he designed for his last stage; and, in order thereto, went aboard a ship belonging to Ægina, an island not far from Athens, the mariners of which were his familiar acquaintance; but it being about the middle of November, a season for rough weather, they were taken with a storm in the road near Cyprus; and the case was become desperate, when suddenly the tempest, it was affirmed, ceased by the prayers of our author. Thus miraculously preserved, he arrived safe at Athens, where he was joyfully entertained, his great abilities rendering him the admiration both of the scholars and professors. Here he commenced a friendship with St. Basil, the great companion of his life: here too he fell into the acquaintance of Julian, afterwards emperor and apostate, an event which, it is pretended, he now remarkably foretold: here also he was visited in a vision, or a dream, by two ladies, who called themselves Wisdom and Chastity, and in a familliar embrace told him, they were sent by God to take up their residence in his soul, where he had prepared them so neat and pleasant an habitation.

After the departure of his friend, Nazianzen was prevailed upon by the students, to undertake the professor's place of rhetoric, and he sat in that chair with great applause for a little while; but being now thirty years of age, and much solicited by his parents to return home, he complied, taking his journey by land to Constantinople. Here he met his brother Cæsarius, just then arrived from Alexandria, so accomplished in all the polite learning of that age, and especially in physic, which he had made his particular study, that he had not been there long before he had public honours decreed him, matches proposed from noble families, the dignity of a senator offered him, and a committee appointed to wait upon the emperor, to intreat him, that though the city at that time wanted no learned men in any faculty, yet this might be added to all its other glory, to have Cæsarius for its physician and inhabitant. But Nazianzen's influence prevailed against all these temptations; and the two brothers returned home together, to the great joy of their aged parents.

Nazianzen now thought it time to fulfil a vow which he had made to consecrate himself to God by baptism. Soon afterwards he was ordained a presbyter by his father, to make

him

him more useful to himself, and there soon happened an occasion for that help. Gregory, the father, among several of the eastern bishops, had received a creed composed by a convention at Constantinople, anno 395, in which the word consubstantial being laid aside, that article was expressed thus: "That the Son was in all things like the Father, according to the Scriptures." In consequence, the monks of Cappadocia in denying him communion were followed by a great part of the people. Nazianzen, therefore, bestirred himself to make up this breach. He first convinced his father of the error, which he found him as ready to recant, and give public satisfaction to the people; then he dealt with the other party, whom he soon prevailed with to be reconciled: and, to bind all with a lasting cement, he made on this occasion his first oration, "Concerning Peace."

Julian had now ascended the throne; and, in order to suppress and stifle christianity, published a law, prohibiting christians not only to teach, but to be taught the books and learning of the Gentiles. The defeat of this design, next to the two Apollinarii in Syria, was chiefly owing to Nazianzen, who upon this occasion composed a considerable part of his poems, comprehending all sorts of divine, grave, and serious subjects, in all kinds of poetry; by which means the christian youth of those times were completely furnished, and found no want of those heathen authors that were taken from them. Julian afterwards coming to Cæsarea, in the road to his persian expedition, one part of the army was quartered at Nazianzum, where the commander peremptorily required the church (which the elder Gregory had not long since built) to be delivered to him. But the old man stoutly opposed him, daily assembling the people to public prayers, who were so affected with the common cause, that the officer was forced to retire for his own safety. Julian being slain not long after, Nazianzen published two invective orations against him, which are at once remarkable proofs of his wit and eloquence, and no less so of the abuse of these talents by too much virulence and acrimony.

Having by Julian's death obtained some respite from public concerns, he made a visit to his friend Basil, who was then in monastic solitude upon a mountain in Pontus, whither he had often solicited Nazianzen's company. The latter was naturally inclined to such a course of life, and always looked upon his entering into orders as a kind of force and tyranny put upon him, which he could hardly digest; yet he knew not how to desert his parents. But his brother Cæsarius being now returned from court, where he had been for some years, with a purpose to fix in his possession at home, gave him an opportunity to indulge his inclination. He accordingly ly retired to his old companion, with whom in his solitary recess he remained several years, passing the time in watching, weeping, fasting, and all the several acts of mortification. He was thus employed when the necessity of affairs at home forcibly ravished him from his retirement. His father stooped under the infirmities of age, and, being no longer able to attend his charge, prevailed with him to come home; he returned about Easter, and published a large apologetic in excuse of his flight, which had been much censured. He had not been long entered upon his charge of assistant to his father, when the family had the misfortune to lose his brother Cæsarius, who departed this life soon after the terrible earthquake that happened in Bithynia, October 11, 358. Some time after died, of a malignant fever, his sister Gorgonia, whose funeral-sermon he preached; as he did also that of his father, the aged bishop of Nazianzum, who died not long after, being then near one hundred years old, having been forty-five years bishop of that place. In the conclusion of this latter oration, he addressed himself to his mother Norma, to support her mind under so great a loss. And the consolations were proper and seasonable: for she, being thus deprived of the main staff of her life, and nearly of equal years to her husband, expired, as may probably be conjectured, soon after.

By these breaches in the family, Nazianzen was sufficiently weaned from the place of his nativity; and, though he was not able to procure a successor to his father, he resolved to throw up his charge, and accordingly retired to Selucia, famous for the temple of St. Thecla, the virgin-martyr; where, in a monastery of devout virgins dedicated to that saint, he continued a long time, and did not return till the death of St. Basil; whom, to his great trouble, he could not attend to his last hours, being himself confined by sickness. About this time, he was summoned to a council at Antioch, holden anno 378, to consider how to make the best use of the emperor's late edict for tolerating the catholics, in order to suppress Arianism; and, being ordered by the council to fix himself for that purpose at Constantinople, he presently repaired thither. Here he found the catholic interest at the lowest ebb: the Arians, favoured by Valens, had possessed themselves of all the churches, and proceeded in such extremities that scarcely any of the orthodox durst avow their faith. He first preached in his lodgings to those that repaired thither, and the congregation soon growing numerous, the house was immediately consecrated by Nazianzen, under the name of the church of Anastasia, or the Resurrection; because the catholic faith, which in that city had been hitherto oppressed, here seemed to have its resurrection. The opposition to his measures but increased his

fame,

fame, together with the number of his auditors, and even drew admirers and followers from foreign parts; among whom St. Jerom, lately ordained prefbyter, came on purpofe to put himfelf under his tutelage and difcipline; an honour in which Jerome glories on every occafion. As the catholics grew more confiderable, they chofe him for their bifhop, and the choice was confirmed by Meletus of Antioch, and Peter who fucceeded Athanafius at Alexandria; but he was oppofed by the Arians, who confecrating Maximus, a famous cynic philofopher and chriftian, gave him a great deal of trouble. The Arian bifhop, however, was at length forced to retire, and his fucceffor Demophilus was depofed by the emperor Theodofius, who directed an edict to the people of Conftantinople, February 27, 380, re-eftablifhing the orthodox faith; and afterward coming thither in perfon, he treated Nazianzen with all poffible kindnefs and refpect, and appointed a day for his inftalment in the fee.

But this ceremony was deferred for the prefent at his own requeft; and falling fick foon after, he was vifited by crowds of his friends, who all departed when they had made their compliments, except a young man with a pale look, long hair, in fqualid and tattered cloaths, who, ftanding at the bed's feet, made all the dumb figns of the bittereft forrow and lametation. Nazianzen, ftarting, afked him, "Who he was, whence he came, and what he wanted?" To which he returned no anfwer, but expreffed fo much the more paffion and refentment, howling, wringing his hands, and beating his breaft in fuch a manner that the bifhop himfelf was moved to tears. Being at length forced afide by one who ftood by, he told the bifhop, "This, Sir, is the affaffin, whom fome had fuborned to murder you; but his confcience has molefted him, and he is here come ingenuoufly to confefs his fault, and to beg your pardon." The bifhop replied, "Friend, God Almighty be propitious to you, his gracious prefervation of me obliges me freely to forgive you; the defperate attempt you defigned has made you mine, nor do I require any other reparation, than that henceforth you defert your party, and fincerely give up yourfelf to God."

Theodofius being highly folicitous about the peace of the church, fummoned a council to meet at Conftantinople in May, anno 382. This is called the fecond General Council, in which the Nicene Creed was ratified; and, becaufe the article concerning the Holy Ghoft was but barely mentioned, which was become one of the prime controverfies of the age, and for the determination of which the council had been principally fummoned, the fathers now drew up an explanatory creed, compofed, as it is faid, by Gregory of Niffen: it is the creed, which in our liturgy takes place under the name of of the NICENE CREED. The see of Constantinople was also now placed next in precedence to that of Rome. Our author carried a great sway in that council, where all things went on smoothly, till at last they fell into disturbances on the following occasion.

There had been a schism for some time in the church of Antioch, occasioned by the ordination of two bishops to that see; and one of those named Melitus, happening to die before the end of the council, Nazianzen proposed to continue the other, named Paulinus, then grown old, for his life. But a strong party being made for one Flavianus, presbyter of the church, these last carried it; and, not content with that, resolved to deprive their grand opposer of his seat at Constantinople. To prevent this he made a formal resignation to the emperor, and went to his paternal estate at Nazianzum, resolving never to episcopize any more; insomuch, that though, at his return, he found the see of Nazianzum still vacant, and over-run with the heresy of Apollinarius, yet he pertinaciously resisted all intreaties that were made to take that charge upon him. And, when he was summoned to the re-assembling of the council the following year, he refused to give his attendance, and even did not stick to censure all such meetings as factious, and governed by pride and ambition. Mean while, in defence of his conduct, he wrote letters to the Roman Prætorian Præfect, and the Consul; assuring them, that, though he had withdrawn himself from public affairs, it was not, as some imagined, from any discontent for the loss of the great place he had quitted; and that he would not abandon the common interests of religion; that his retirement was a matter of choice more than necessity, in which he took as great pleasure as a man that has been tossed in a long storm at sea does in a safe and quiet harbour. And, indeed, being now freed from all external cares, he entirely gave himself up to solitude and contemplation, and the exercise of a strict and devout life. At vacant hours, he refreshed the weariness of his old age with poetry, which he generally employed upon divine subjects, and serious reflections upon the former passages of his life; an account of which he drew up in Iambics, whence no inconsiderable part of his memoir is derived. Thus he passed the remainder of his days till death put a period to them, anno 389, in his 66th year. He made a will, by which, except a few legacies to some relations, he bequeathed his whole estate to the poor of the diocese of Nazianzum. In this spirit, during the three years that he enjoyed the rich bishopric of Constantinople, he never touched any part of the revenues, but gave it all to the poor, to whom he was extremely liberal.

He

He was one of the ableſt champions of the orthodox faith concerning the trinity, whence he had the title given him of ὁ Θεόλογος, "THE DIVINE," by unanimous conſent. His moral and religious qualities were attended with the natural graces of a ſublime wit, ſubtle apprehenſion, clear judgement, and eaſy and ready elocution, which were all ſet off with as great a ſtock of human learning as the ſchools of the Eaſt, as Alexandria, or Athens itſelf, was able to afford. All theſe excellences are ſeen in his works, of which we have the following character by Eraſmus; who, after having enriched the Weſtern church with many editions of the antient fathers, confeſſes, that he was altogether diſcouraged from attempting the tranſlation of Nazianzen, by the acumen and ſmartneſs of his ſtyle, the grandeur and ſublimity of his matter, and thoſe ſomewhat obſcure alluſions that are frequently interſperſed among his writings. Upon the whole, Eraſmus doubts not to affirm, that, as he lived in the moſt learned age of the church, ſo he was the beſt ſcholar of that age.

GREGORY (NYSSEN), was the younger brother of St. Baſil, and had an equal care taken of his education, being brought up in all the polite and faſhionable modes of learning; but, applying himſelf particularly to rhetoric, he valued himſelf more upon being accounted an orator than a chriſtian. On the admonition of his friend Gregory Nazianzen, he quitted thoſe ſtudies; and, betaking himſelf to ſolitude and a monaſtic diſcipline, he turned his attention wholly to the Holy Scriptures, and the controverſies of the age; ſo that he became as eminent in the knowledge of theſe as he had before been in the courſe of more pleaſant ſtudies. Thus qualified for the higheſt dignity in the church, he was placed in the ſee of Nyſſa, a city on the borders of Cappadocia. The exact time of his promotion is not known, though it is certain he was biſhop in 371. He proved in this ſtation a ſtout champion for the Nicene faith, and ſo vigorouſly oppoſed the Arian party, that he was ſoon after baniſhed by the emperor Valens; and, in a ſynod held at Nyſſa by the biſhop of Pontus and Galatia, was depoſed, and met with very hard uſage. He was hurried from place to place, heavily fined, and expoſed to the rage and petulancy of the populace, which fell heavier upon him, as he was both unuſed to trouble and unapt to bear it. In this condition he remained for ſeven or eight years, during which, however, he went about, countermining the ſtratagems of the Arians, and ſtrengthening thoſe in the orthodox faith; and in the council of Antioch 378, he was among others delegated to viſit the eaſtern churches lately harraſſed by the Arian perſecution.

He

He went not long after to Arabia; and, having difpatched the affairs of the Arabian churches, he proceeded to Jerufalem, having engaged to confer with the bifhops of thofe parts, and to affift in their reformation. Upon his arrival, finding the place overrun with vice, fchifm, and faction, fome fhunning his communion, and others fetting up altars in oppofition to him, he foon grew weary of it, and returned with a heavy heart to Antioch: and being on this occafion confulted afterwards, whether it was an effential part of religion to make pilgrimages to Jerufalem (which, it feems, was the opinion of the monaftic difciplinarians at that time), he declared himfelf freely in the negative. After this, he was fummoned to the great council at Conftantinople, where he made no inconfiderable figure, his advice being chiefly relied on in the moft important cafes; and particularly the compofition of the creed, called by us the Nicene creed, was committed to his care. He compofed a great many other pieces, a lift of which may be feen in Cave. He lived to a great age, and was alive when St. Jerom wrote his "Catalogue of Ecclefiaftical Writers" in 392; and two years after was prefent at the fynod of Conftantinople, on adjufting the controverfy between Agapius and Bagadius, as appears by the acts of that council. No notices are extant concerning his death, more than that the memory of it is celebrated in the Weftern Martyrologies, March ix. in the Greek, on Jan. x.

He was a married man, and lived with his wife Theofebia, even after he was bifhop: Gregory Nazianzen, in a confolatory letter to his fifter on her death, gives her extraordinary commendations.

GREGORY (THEODORUS), furnamed Thaumaturgus, was defcended of parents eminent for their birth and fortune, at Neo-Cefarea the metropolis of Cappadocia, where he was born. He was educated very carefully in the learning and religion of the Gentiles by his father, who was a warm zealot, but, lofing his father at fourteen years of age, he, enlarging his enquiries, began by degrees to perceive the vanity of that religion in which he had been bred, and turned his inclinations to chriftianity. Having laid the neceffary ground-work of his education at home, he refolved to accomplifh himfelf by foreign travels, to which purpofe he went firft to Alexandria, then become famous by the platonic fchool lately erected there Departing from Alexandria, he came back probably through Greece, and ftaid a while at Athens; whence returning home, he applied himfelf to his old ftudy of the law: but quickly growing weary of it, he turned to the more agreeable fpeculations of philofophy.

The

The fame of Origen, who at that time had opened a school at Cæsarea in Paleſtine, and whoſe renown no doubt was great at Alexandria, ſoon reached his ears. To that city therefore he betook himſelf, where meeting with Fermilian a Cappadocian gentleman, and afterwards biſhop of Cæſarea in that country, he commenced a friendſhip with him, there being an extraordinary ſympathy and agreement in their tempers and ſtudies; and they jointly put themſelves, together with his brother Athenodorus, under the tutorage of that celebrated maſter. Origen endeavoured to ſettle him in the full belief of chriſtianity, of which he had ſome inſight before, and to ground him in the knowledge of the Holy Scriptures, as the beſt ſyſtem of true wiſdom and philoſophy.

Neo-Cæſarea was a large and populous place, but miſerably overgrown with ſuperſtition and idolatry; chriſtianity had as yet ſcarce made its entrance there. However, our young philoſopher was appointed to be a guide of ſouls in the place of his nativity. Phædinius, biſhop of Amaſia, a neighbouring city in that province, caſt his eye upon him for that purpoſe; and it was thought his relation to the place would more endear the employment to him. But, upon receiving the firſt intimation of the deſign, he ſhifted his quarters, and, as oft as ſought for, fled from one deſert to another; ſo that the biſhop by all his arts and induſtry cou'd not obtain intelligence of him; he therefore conſtituted him biſhop of the place in his abſence, and how averſe ſoever he ſeemed to be before, he now accepted the charge, when perhaps he had a more formal and ſolemn conſecration. The province he entered upon was difficult; the city and neighbourhood being wholly addicted to the worſhip of demons, and there not being above ſeventeen chriſtians in thoſe parts, ſo that he muſt find a church before he could govern it. The country was overrun with hereſies; and himſelf, though accompliſhed ſufficiently in human learning, was altogether unexerciſed in theological ſtudies and the myſteries of religion. But here again he had immediate aſſiſtance from heaven; for, one night, as it is related, while he was muſing upon theſe things, and diſcuſſing matters of faith in his own mind, he had the following viſion wherein St. John the Evangeliſt and the bleſſed Virgin appeared in the chamber where he was, and diſcourſed before him concerning thoſe points. In conſequence, after their departure, he immediately penned that canon and rule of faith which they had declared. To this creed he always kept himſelf, and bequeathed it as an ineſtimable depoſit to his ſucceſſors. The original, written with his own hand, we are informed, was preserved in that church in his name.

Thus furnished, he began to apply himself more directly to the charge committed to him. In the happy success of which he was infinitely advantaged by a power of working miracles bestowed upon him: and hence the title of Thaumaturgus, or wonder-worker, is constantly ascribed to him in the writings of the church. St. Basil assures us, that, upon this account the Gentiles used to call him a second Moses. In this faithful and successful government of his flock he continued quietly till about anno 250, when he fled from the Decian persecution; but, as soon as the storm was over, he returned to his charge, and in a general visitation of his diocese, established in every place anniversary festivals and solemnities in honour of the martyrs who had suffered in the late persecution. In the reign of Galienus, the year about 260, upon the irruption of the northern nations into the Roman empire; the Goths breaking into Pontus, Asia, and some parts of Greece, created such confusion, that a neighbouring bishop of those parts wrote to Gregory for advice what to do: our author's answer, sent by Euphrasymus, is called his " Canonical Epistle," still extant among his works. Not long afterwards was convened that synod at Antioch, wherein Paul of Samosata bishop of the place, which he did not care to lose, made a feigned recantation of his heretical opinions. Our St. Gregory was among the chief persons in this synod which met in 264, but did not long survive it, dying either this or most probably the following year.

GREGORIUS (GEORGIUS FLORENTIUS, or GREGORY OF TOURS). He was one of the most illustrious bishops, and distinguished writers of the sixth century. In 573 he was chosen bishop of Tours. He went to Rome to visit the tomb of the Apostles, and was a great friend of Gregory the Great. He wrote the history of France, the lives of the Saints, with other works. His style, says Mr. Gibbon, is devoid of elegance and simplicity; nevertheless, his performances, considering the period at which he lived, must be considered as of some importance to literature.

GREGORY (PETER), a native of Toulouse. He flourished in the sixteenth century, was a learned man, and wrote many books full of erudition. He had, however, more learning than judgement. He died in 1527.

GRENAN (BENIGNUS), a latin poet, and professor of rhetoric at Harcourt. He died at Paris in 1723. His compositions in latin verses are remarkable for much purity and elegance, and for very noble and delicate sentiments.

GRENEE, a french painter of diftinguifhed merit. His St. Ambrofe, and the apotheofis of St. Lewis, are correctly defigned, finely touched, and the folds of the drapery in the moft perfect ftyle of Guido himfelf. His Clemency appeafing Juftice is a very fine piece: the character of the heads, the delicacy of the pencil, and the frefhnefs of the colours deferve great praife. His Sacrifice of Jephtha is elegant and delicate. His Magdalen finely coloured. His Roman Charity of admirable expreffion, particularly in the countenance of the daughter. His Return of Abraham is well defigned—likewife his Diana and Endymion, claims great praife; the body of the latter is finely defigned, and very well coloured. His Sufannah, furprized in the bath by the two old men, has great expreffion, particularly in the head of Sufannah, and the defign of her whole figure is very happy; the old men are finely contrafted to her. His Aurora quitting Tithonius is yet more brilliant, and of a finer expreffion than the preceding, and the colours are wonderfully happy. His Soft Captivity, in which is reprefented the buft of a young woman careffing a pigeon, which fhe holds between her hands, is delicate and pleafing. His fmall piece of a Virgin careffing an infant Jefus: and another of a Virgin preparing food for the Divine Infant, are exquifite in defign, colouring, and compofition.

GRESHAM (Sir Thomas), defcended of an ancient family diftinguifhed by many honourable perfons, which took its name from a town fo called in Norfolk, was born in 1519 at London, and bound apprentice to a mercer there while he was young: but, to enlarge his mind by an education fuitable to his birth and fortune, was fent to Caius-college, then Gonvil-hall, in Cambridge; where he ftayed a confiderable time, and made fuch improvements in learning, that Caius the founder of the college ftyles him " doctiffimus mercator," the very learned merchant. However, the profits of trade were then fo great, and fuch large eftates had been raifed by it in his own family, that he afterwards engaged in it, and was admitted a member of the mercers company in 1543. About this time he married: and not long after fucceeded his father in the office of agent to king Edward for taking up money of the merchants at Antwerp, and removed to that city with his family in 1551.

The bufinefs of his employ gave him a great deal of trouble and much uneafinefs. The money he had taken up for his majefty not being paid at the time ftipulated, he found himfelf obliged to get it prolonged, which was not to be done without the confideration of the king's purchafing jewels or fome other commodities to a large amount. This way of proceeding, he neither thought for his majesty's honour nor his own credit as his agent, and therefore projected a scheme to bring the king wholly out of debt in two years, as follows. —Provided the king and council would assign him 1200, or 1300l. to be secretly received at one man's hands, that so it might be kept secret, he would so use that matter in Antwerp, that every day he would be seen to take up in his own name 200l. sterling by exchange, which would amount in one year to 72000l. and so doing it should not be perceived nor give occasion to make the exchange fall. He proposed farther, that the king should take all the lead into his own hands, and making a staple of it should put out a proclamation or shut up the Custom-house, that no lead should be conveyed out of the kingdom for five years; by which the king might cause it to rise, and feed them at Antwerp from time to time, as they should have need. By which means he might keep his money within the realm, and bring himself out of the debts which his father and the duke of Somerset had brought upon him. This scheme being put into execution, had the proposed effect in discharging his majesty's debts, which were very considerable: and, by the advantageous turn which by this means was given to the exchange in favour of England, not only the price of all foreign commodities was greatly sunk and abated; but likewise gold and silver, which before had been exported in large quantities, were most plentifully brought back again.

However, upon the accession of queen Mary, Gresham was removed from his agency. He accordingly drew up a memorial of his services to the late king, and sent it to a minister of state to be laid before her majesty. The services represented in it as done, not only to the king, but to the nation in general, by the increase both of money and trade, and the advancement of the public credit, being observed to be fact, he was taken soon after into the queen's service, and reinstated in his former employ, as appears by the commissions given him at different times during that reign. He was not much above 30, when he first entered upon the employ under king Edward, and his prudence and dexterity in the conduct of that important trust discovered an uncommon genius in mercantile affairs. After the decease of queen Mary, he was taken immediately into the service of queen Elizabeth, who employed him on her accession to provide and buy up arms; and, in 1559, she conferred on him the honour of knighthood, and appointed him her agent in foreign parts. In this eclat of credit and reputation, he thought proper to provide himself with a mansion-house in the city, suitable to his station and dignity; and with this spirit built

that

GRESHAM.

that large and fumptuous houfe for his own dwelling, on the weft-fide of Bifhopfgate-ftreet, London, called Grefham-college, where he maintained a port becoming his character and ftation. But this flow of profperity received a heavy check by the lofs of his only fon, aged 16 years, who died in 1564, and was buried in St. Helen's church oppofite to his manfion-houfe.

At this time the merchants of London met in Lombard-ftreet, expofed to the open air and all the injuries of the weather. To remedy which inconvenience, Sir Thomas's father during his fhrievalty wrote a letter to Sir Thomas Audeley then lord privy feal, acquainting him that there were certain houfes in that ftreet belonging to Sir George Monoux, which if purchafed and pulled down, a handfome exchange might be built on the ground; he therefore defired his lordfhip to move his majefty, that a letter might be fent to Sir George, requiring him to fell thofe houfes to the mayor and commonalty of the city of London for that purpofe. The building he fuppofes would coft upwards of 2000l. 1000l. of which he doubts not to raife before he was out of his office; but nothing effectual was done it. Sir Thomas therefore took up his father's defign, and improving upon his fpirit, propofed, that if the citizens would give him a piece of ground in a proper place large enough for the purpofe, he would build an exchange at his own expence with large and covered walks, where the merchants and traders of all forts might daily affemble, and tranfact bufinefs, at all feafons, without interruption from the weather or impediments of any kind. This generous offer was gratefully accepted, and in 1566 feveral houfes upon Cornhill and the back of it, with three alleys, called Swan-alley, New-alley, and St. Chriftopher's alley, containing in all 80 houfes, were purchafed by the citizens for more than 3532l. and fold for 478l. on condition of pulling them down, and carrying off the ftuff. This done, the ground plot was made plain at the charges of the city, and poffeffion given to Sir Thomas, therein ftyled " Agent to the queen's highnefs;" who, on the 7th of June, laid the firft ftone of the foundation; and the work was forthwith followed with fuch diligence, that, by Nov. 1567, the fame was covered with flate, and the fhell fhortly after fully finifhed.

The plan of this edifice was formed from the exchange at Antwerp, being like that of an oblong fquare, with a portico fupported with pillars of marble, ten on the north and fouth fides, and feven on the eaft and weft: under which ftood the fhops each feven feet and a half long, and five feet broad; in all 120, twenty-five on each fide eaft and weft, and thirty-

thirty-four and an half north, and thirty-five and an half south, each of which paid Sir Thomas 4l. 10s. a year upon an average. There were likewise other shops fitted up at first in the vaults below, but the dampness and darkness rendered these so inconvenient, that the vaults were soon let out to other uses; upon the roof stood at each corner, upon a pedestal, a grasshopper, which was the crest of Sir Thomas's arms. This edifice was fully completed, and the shops opened in 1569: and Jan. 29, 1570, queen Elizabeth, attended by her nobility, came from Somerset-house thither, and caused it by a trumpet and a herald to be proclaimed "The Royal Exchange."

Though Sir Thomas had purchased very large estates in several counties of England, yet he thought a country-seat near London, to which he might retire from business, and the hurry of the city as often as he pleased, would be very convenient. With this view he bought Osterley-park near Brentford in Middlesex, where he built a large magnificent seat within the park, which he impaled, being well wooded, and furnished with many ponds stocked with fish and fowl, and of great use for mills, as paper-mills, oil-mills, and corn-mills.

Before this seat was completed, he projected and executed that noble design of converting his mansion-house in Bishopsgate-street into a seat for the Muses, and endowing it with the revenues arising from the Royal Exchange after his decease. While he was meditating this design, the university of Cambridge wrote him an elegant latin letter, reminding him of a promise, as they had been informed, to give them 500l. either towards building a new college there, or repairing one already built. This letter was dated March 14, 1574-5; and it was followed by another of the 25th, to acquaint him with a report they had heard, that he had promised lady Burghley both to found and endow a college for the profession of the seven liberal sciences. They observe, that the only place proper for such a design was either London, Oxford, or Cambridge: they endeavour to dissuade him from London, lest it should prove prejudicial to the two universities; and they hope he will not make choice of Oxford, since he was himself bred at Cambridge, which might presume upon a superior regard from him on that account. At the same time, they wrote another letter to the lady Burghley, in which they earnestly request, that she will please to use her interest with him, to fix upon Cambridge for the place of his intended college [1].

[1] See these Letters in Ward's Lives of the Gresham Professors, Appen. No. 3.

But these letters had not the desired effect: he persisted in his resolution to settle it in his house at London; and accordingly, by an indenture dated May 20, 1575, he made a disposition of his several manors, lands, tenements, and hereditaments; with such limitations and restrictions, particularly as to the Royal Exchange and his mansion-house, as might best secure his views with regard to the uses for which he designed them. This indenture was soon followed by two wills, one of his goods, and the other of his real estates: the former of these bears date July 4th ensuing, whereby he bequeaths to his wife, whom he makes his sole executrix, all his goods, as ready money, plate, jewels, chains of gold, with all his stock of sheep and other cattle if within the realm of England, and likewise gives several legacies to his relations and friends and to all his servants, amounting in the whole to upwards of 2000l. besides some small annuities. The other will is dated July the 5th, wherein he gives one moiety of the Royal Exchange to the mayor and commonalty of London, and the other to the mercers company, for the salaries of seven lecturers in divinity, law, physic, astronomy, geometry, music, and rhetoric, at 50l. per annum for each, with his house in Bishopsgate-street for the lecturer's residence, where the lectures were to be read. He likewise leaves 53l. 6s. 8d. yearly for the provision of eight alms-folks residing in the almshouses behind his house, and 10l. yearly to each of the prisons in Newgate, Ludgate, King's bench, the Marshalsea, and Compter in Wood-street, and the like sum to each of the hospitals of Christ-church, St. Bartholomew, Bedlam, Southwark, and the Poultry-compter; and 100l. yearly to provide a dinner for the whole mercers company in their hall on every of their quarter-days, at 25l. each dinner. By this disposition, sufficient care was taken, that the two corporations, to whom the affair was trusted, should receive no damage by the execution of it; for, the stated annual payments amount to no more than 603l. 6s. 8d. and the yearly rents of the Exchange received by Sir Thomas were 740l. besides the additional profits that must arise from time to time by fines, which were very considerable. But the lady Anne his wife was to enjoy both the mansion-house and the Exchange during her life if she survived Sir Thomas, and then they were both vested in the two corporations for the uses declared in the will for the term of 50 years; which limitation was made on account of the statutes of mortmain, that prohibited the alienation of lands or tenements to any corporation, without licence first had from the crown. And that space of time the testator thought sufficient for procuring such licence, the doing of which he earnestly recommends to them

them without delay; in default whereof, at the expiration of 50 years, these estates were to go to his heirs at law.

Having thus settled his affairs so much to his own honour, the interest of the public, and the regards due to his family, he was at leisure to reap the fruits of his industry and success. But he did not long enjoy this felicity; for, Nov. 21, 1579, coming from the Exchange to his house in Bishopsgate-street, he suddenly fell down in his kitchen, became speechless, and presently died. He was buried in his own parish-church of St. Helen's. His obsequies were performed in a very solemn manner, the corpse being attended by 100 poor men, and the like number of poor women, whom he had ordered to be cloathed in black gowns of 5s. 8d. per yard at his own expence. The charges of the funeral amounted to 800l. His corpse was deposited in a vault at the north-east corner of the church, which he had before provided for himself and family, with a curious marble tomb over it: on the south and west sides of which are his own arms, and on the north and east the same impaled with those of his lady. The arms of Sir Thomas, together with the city of London and mercers company, are likewise painted in the glass of the east window of the church above the tomb, which stood as he left it without any inscription till 1736, when the following words taken from the parish-register were cut on the stone that covers it by order of the church-wardens; " Sir Thomas Gresham knight, was buried December 15, 1579. By his death many large estates in several counties of England, amounting at that time to the clear yearly value of 2300l. and upwards, came to his lady, who survived him many years, and continued to reside after his decease in the mansion-house at London in the winter, and at Osterley-park in the summer season, at which last place she died Nov. 23, 1596, very aged. Her corpse was brought to London, and buried in the same vault with her husband.

Mr. Ward has drawn Sir Thomas's character, and observes, that he had the happiness of a mind every way suited to his fortune, generous and benign; ready to perform any good actions and encourage them in others. He was a great friend and patron of our celebrated martyrologist John Fox. He was well acquainted with the ancient and several modern languages; he had a very comprehensive knowledge of all affairs relating to commerce, whether foreign or domestic; and his success was not less, being in his time esteemed the highest commoner in England. He transacted queen Elizabeth's mercantile affairs so constantly, that he was called " The Royal Merchant," and his house was sometimes appointed for the reception of foreign princes upon their first arrival at London. As no one could be more ready to perform any generous actions which might contribute to the honour of this country;

country; fo he very well knew how to make the beſt uſe of them for the moſt laudable purpoſes. Nor was he leſs ſerviceable both to the queen and her miniſtry on other occaſions, who often conſulted him, and ſought his advice in matters of the greateſt importance relating to the welfare of the government. But the moſt ſhining part of his character appears in his public benefactions. The Royal Exchange was not only a ſingular ornament to the city of London, and a great convenience to the merchants who wanted ſuch a place to meet and tranſact their affairs in, but likewiſe contributed very much to the promotion of trade, both by the number of ſhops erected there, and the much greater number of the poor, who were employed in working for them. And the donation of his own manſion-houſe for a ſeat of learning and the liberal arts, with the handſome proviſion made for the endowment and ſupport of it, was ſuch an inſtance of a generous and public ſpirit as has been equalled by few, and muſt perpetuate his memory with the higheſt eſteem and gratitude ſo long as any regard to learning and virtue is preſerved among us Nor ought his charities to the poor, his alms-houſes, and the liberal contributions to the ten priſons and hoſpitals in London and Southwark, to be omitted.

His public benefactions, the Royal Exchange, and his manſion-houſe, on the deceaſe of his lady, immediately came into the hands of the two corporations, the city of London and the mercers company, who, according to their truſt, obtained a patent from the crown, dated Feb. 3, 1614, 12 Jacobi I. to hold them for ever upon the terms expreſſed in the will of the donor.

GRESSET (JOHN BAP. LOUIS), one of the moſt lively and agreeable poets of France. His Ver-vert is lively and elegant, and the beſt of his pieces. They are collected in an edition under the title of Oeuvres diverſes, 12mo. 1748. His letter to the duke de Choiſeul, on the publication of the negociation for peace in 1762, is worth reading. Born at Amiens in 1709, and died there June 16, 1777.

GRETSER (JAMES), a learned German, was born at Maredorf about 1561, and entered among the ſociety of Jeſuits at 17. When he had finiſhed his ſtudies, he was appointed a profeſſor at Ingolſtad. He ſpent 24 years there; teaching philoſophy, morality. and ſchool-divinity. Theſe employments did not hinder him from being conſtant at prayers, and compoſing a prodigious number of books. The catalogue of them, as given by Niceron, conſiſts of near 153 articles; which, he tells us, were copied by him from the propoſals, publiſhed in 1753, for printing an edition of all Gretſer's works at Ratiſbon in 17 vols. folio. His great erudition was attended with a ſurpriſing modeſty; he could not bear to be commended. The inhabitants of Maredorf were

deſirous

defirous of having his picture, to hang it up in their houfe; but, when informed of the earneft application they had made to his fuperiors for that purpofe, he was heartily vexed; and told them, that if they wanted his picture, they need but draw that of an afs. To make themfelves amends, they purchafed all his works, and devoted them to the ufe of the public. He died at Ingolftad, in 1635. He fpent his whole life in writing againft proteftants, and in defending the order to which he belonged. Some authors have beftowed very great encomiums upon him.

His works were printed, according to the propofals abovementioned, at Ratifbon 1739, 17 vols folio.

GREVENBROECK, a flemifh painter, excelled in fea-pieces, and was remarkable for the accuracy with which he delineated minute objects. He flourifhed in the feventeeth century.

GREVILLE (FULK or FOULK), lord Brooke, an ingenious writer, was the eldeft fon of Sir Fulk Greville of Beauchamp-court (at Alcafter) in Warwickfhire, and born there in 1554. It is conjectured, that he was educated at the fchool in Shrewfbury; whence he was removed to Cambridge, and admitted a fellow-commoner at Trinity-college; and fome time after, making a vifit to Oxford, he became a member of that univerfity, but of what college is not certain. Having completed his academical ftudies, he travelled abroad to finifh his education; and upon his return, being well accomplifhed, was introduced to the court of queen Elizabeth by his uncle Robert Greville, where he was efteemed a moft ingenious perfon, and particularly favoured by the lovers of arts and fciences. He was foon nominated to fome beneficial employment in the court of marches of Wales by his kinfman Sir Henry Sidney, then lord prefident of that court and principality.

Our author was not then above twenty-two years of age, fo that this poft may be efteemed an honourable atteftation of his merit. But the nature of it did not pleafe him; his ambition prompted him to another courfe of life. He had already made fome advances in the queen's favour, had attained a competent familiarity with the modern languages, and fome expertnefs in the martial exercifes of thofe times: thefe were qualifications for a foreign employment, which was more agreeable to the activity of his temper, and promifed a quicker way of raifing him to fome of the firft pofts in the ftate. In reality he was fo eager to advance his fortune in this line, that, to gratify his defire, he ventured to incur his royal miftrefs's difpleafure, and made feveral attempts in it, not only with but even without her majefty's confent. Out of many of thefe we have an account of the few following from his own pen. Firft, when the two mighty armies of Don John and the duke Cafimire were to meet in the Low-countries, he applied

plied and obtained her majesty's leave under her own hand to go thither; but, after his horses with all other preparations were shipped at Dover, the queen (who always discouraged these excursions) sent her messenger, Sir Edward Dyer, with her mandate to stop him. He was so much vexed at this disappointment, that afterwards, when secretary Walsingham was sent ambassador in 1578, to treat with those two princes, an opportunity of seeing an affair, in which so much christian blood and so many christian empires were concerned, was so tempting, that he resolved not to risque a denial, and therefore stole away without leave, and went over with the secretary incog. The consequence was, that, at his return, the queen forbade him her presence for many months. To the same ambition may also be referred his engagement with Sir Philip Sidney to accompany Sir Francis Drake in his last expedition but one, to the West-Indies in 1515, in which they were both frustrated by the same authority.

Again, when the earl of Leicester was sent general of her majesty's forces the same year, and had given Mr. Greville the command of one hundred horse, "Then I," to use his own words, "giving my humour over to good order, yet found that neither the intercession of this grandee, seconded with my own humble suit, and many other honourable friends of mine, could prevail against the constant course of this excellent lady [the queen] with her servants, so as I was forced to tarry behind, and for this importunity of mine to change my course, and seem to press nothing before my service about her; this princess of government as well as kingdoms made me live in her court a spectacle of disfavour too long as I conceived."

During his excursions abroad, his royal mistress granted him the reversion of two of the best offices in the court of the marches of Wales, one of which falling to him in 1580, he met with some difficulties about the profits. In this contest, he experienced the friendship of Sir Philip Sidney, who by a letter written to his father's secretary, Mr. Molyneux, April 10, 1581, prevailed on him not to oppose his cousin Greville's title in any part or construction of his patents; and a letter of Sir Francis Walsingham to the president, the next day, April 11, put an end to the opposition that had been made from another quarter. This office appears to be clerk of the signet to the council of Wales, which is said to have brought him in yearly above 2000l. arising chiefly from the processes which went out of that court, all of which are made out by that officer. He was also constituted secretary for South and North Wales by the queen's letters patent, bearing date April 25, 1683. In the midst of these civil employments, he made a conspicuous figure in the martial line, when the french ambassadors,

bassadors, accompanied by great numbers of their nobility, were in England a second time to treat of the queen's marriage with the duke of Anjou, in 1581. Tilts and tournaments were the courtly entertainments in those days; and they were performed in the most magnificent manner on this occasion by two noblemen, beside Sir Philip Sidney and Fulk Greville, who with the rest behaved so gallantly as to win the reputation of a most gallant knight. In 1586, these two friends were separated by the unfortunate death of the former, who in his death bequeathed to his dear friend one moiety of his books.

In 1558 Mr. Greville attended his kinsman, the earl of Essex to Oxford, and among other persons in that favourite's train was created M. A. April 11, that year. In 1558, he was accused to the lords of the council, by a certificate of several gentlemen borderers upon Farickwood in Warwickshire, of having made waste there to the value of 14,000l. but the prosecution seems to have been dropped, and, October 1597, he received the honour of knighthood. In the beginning of March the same year, he applied for the office of treasurer of the war; and about two years afterwards, in the 41st of Elizabeth, he obtained the place of treasurer of marine causes' for life. In 1599, a commission was ordered to be made out for him as rear-admiral of the fleet, which was intended to be sent forth against another threatened invasion by the spaniards.

During this glorious reign, he frequently represented his county in the House of Commons, together with Sir Thomas Lacy; and it has been observed that a better choice could not have been made, as both of them were learned, wise, and honest. He continued a favourite of queen Elizabeth to the end of her reign. The beginning of the next opened no less in his favour. At the coronation of James I. July 15. 1603, he was made K. B. and his office of secretary to the council of the court of marches of Wales was confirmed to him for life, by a patent bearing date July 24. In the second year of this king, he obtained a grant of Warwick castle. He was greatly pleased with this favour, and, the castle being in a ruinous condition, he laid out at least 20,000l. in repairing it.

He was afterwards possessed of several very beneficial places in the marches court of Wales, and at present he seems to have confined his views within the limits of these offices. He perceived the measures of government quite altered, and the state waning from the lustre in which he had seen it shine: besides, he had little hopes of being preferred to any thing considerable in the ministry, as he met with some discouragements from Sir Robert Cecil, the secretary, and the persons in power. In this position of affairs, he seems to have formed some schemes of retirement, in order to write the history of

queen

queen Elizabeth's life. In which view he drew up a plan, commencing with the union of the two rofes in the marriage of Henry VII. and had made fome progrefs in the execution of it; but the perufal of the records in the council cheft being denied him by the fecretary, as he could not complete his work in that authentic and fubftantial manner as became him, he broke off the defign, and difpofed himfelf to revife the product of his juvenile ftudies and his poetical recreations with Sir Philip Sidney.

During the life of the treafurer Cecil, he obtained no advancement in the court or ftate; but, in 1615, fome time after his death, was made under-treafurer and chancellor of the exchequer; in confequence of which, he was called to the board of privy-council. In 1617, he obtained from the king a fpecial charter, confirming all fuch liberties as had been granted to any of his anceftors in behalf of the town of Alcefter, upon a new referved rent of ten fhillings a year; and, in 1620, was created lord Brooke of Beauchamp-court. He obtained this dignity as well by his merit and fidelity in the difcharge of his offices as by his noble defcent from the Nevils, Willoughbys de Brooks, and Beauchamps. September 1621, he was made one of the lords of the king's bed chamber, whereupon refigning his poft in the exchequer, he was fucceeded therein by Richard Wefton, afterwards earl of Portland. After the demife of king James, he continued in the privy-council of Charles I. in the beginning of whofe reign he founded a hiftory-lecture in the univerfity of Cambridge, and endowed it with a falary of 100l. per annum. He did not long furvive this laft act of generofity; for, though he was a munificent patron of learning and learned men, he at laft fell a facrifice to the extraordinary outrage of a difcontented domeftic. The account we have of this fatal event is, that his lordfhip, neglecting to reward one Ralph Heywood, who had fpent the greateft part of his life in his fervice, this attendant expoftulated thereupon with his lordfhip in his bed-chamber, at Brook-houfe in Holborn; and, being feverely reproved for it, prefently gave his lordfhip a mortal ftab in the back with a knife or fword; after which he withdrew into another room, and, locking the door, murdered himfelf with the fame weapon. He died September 30, 1628, and his corpfe being wrapt in lead was conveyed from Brook-houfe, Holborn, to Warwick; where it was interred on the north fide of the choir of St Mary's church there, in his own vault, which had formerly been a chapter-houfe of the church; and where, upon his monument, there is this infcription: "FULKE GREVILLE, Servant to QUEEN ELIZABETH, Counfellor to KING JAMES, and Friend to SIR PHILIP SIDNEY. Tropheum Peccati." Indeed, he made his dear friend the great exemplar of his life in every thing; and Sidney being often celebrated as the patron of the Muses in general as well as Spenser in particular, so we are told, lord Brooke desired to be known to posterity under no other character than that of Shakspeare's and Ben Jonson's master, lord-chancellor Egerton, and bishop Overal's patron. His lordship also obtained the office of Clarencieux at arms for Mr. Camden, who very gratefully acknowledged it in his lifetime, and at his death left him a piece of plate in his will. He also raised John Speed from a mechanic to be an historiographer.

His lordship had an elegant taste for all kinds of polite learning, but his inclination as well as his genius led him particularly to history and poetry. Hence, with respect to the former, it was that lord Bacon submitted his "Life of Henry VII." to his perusal and animadversions. And his extraordinary kindness to Sir William Davenant must be added to other conspicuous evidences of the latter; that poet he took into his family when very young, and was so much delighted with his promising genius, that, as long as the patron lived, the poet had his residence with him, and probably formed the plan of some of his first plays under his lordship's encouragement, since they were published soon after his death. This noble lord was never married, so that his honour falling by the patent to his kinsman Robert Greville, he directed his estate also by his will to go along with it to the same relation, being next of kin to him.

GREVIN (JAMES), a famous french poet and physician, born at Clermont, in Beauvoisis, in 1538. He began early to write, and practised physic with success. He was long retained in the service of Margaret of France, duchess of Savoy, whom he followed to Piedmont. He died at Turin the 5th of November 1573, aged thirty-two. There are three plays extant of his: "The Treasurer's Wife," a comedy, in 1558; the "Death of Cæsar," a tragedy; and the "Frighted Ones, [Les Estahis]" a comedy, both acted the same day at the college of Beauvais. Grevin, though snatched away by a premature-death, had acquired a great reputation, not only as a poet, but as a physician. Our authors give him this favourable testimony, "that he effaced all who preceded him on the french stage, and that eight or ten such poets as he would have put it on a good footing. His versification is easy and smooth, especially in his comedies, and his plots are well contrived." His poems and plays were printed at Paris, 8vo. 1561. He left also a "Treatise on Poisons," and an "Apology for Antimony," both translated into latin and printed in 4to. He was a calvinist and united with Rochandieu and Florence

Christian

Chriftian in writing their ingenious poem, entituled, "The Temple," which they wrote againft Ronfard, who had abufed the calvinifts in his difcourfe on the "Miferies of Time."

GREVIUS, or GRÆVIUS (JOHN GEORGE), a great latin critic, was born January 29, 1632, at Naumbourg, in Saxony; and, having laid a good foundation of claffical learning in his own country, was fent to finifh his education at Leipfic, under the profeffors Rivinus and Strauchius. This laft was his relation by the mother's fide, and fat opponent in the profeffor's chair, when our author performed his exercife for his degree; on which occafion he maintained a thefis, " De Moribus Germanorum." As his father defigned to breed him to the law, he applied himfelf a while to that ftudy, but not without devoting much of his time to polite literature, which he affected moft, and which he afterwards made the fole object of his application. With this view he removed to Deventer in Holland, attended the lectures of John Francis Gronovius; and, converfing with him, became entirely fixed in his refolution. He was fingularly pleafed with this profeffor, fo that he fpent two years in thefe ftudies under his direction, and profited fo much thereby, that he afterwards frequently afcribed all his knowledge to the affiftance of this mafter. However, refolving to make ufe of all advantages for improving himfelf, he went thence firft to Leyden to hear Daniel Heinfius, and next to Amfterdam; where, attending the lectures of Alexander Morus and David Blondel, this laft perfuaded him to renounce the Lutheran religion, in which he had been bred, and to embrace Calvinifm.

Mean while, his reputation increafed daily, and was now raifed fo high, though but twenty-four years of age, that he was judged qualified for the chair; and, upon the death of Schulting, actually nominated to the profefforfhip of Duifburg by the elector of Brandenburgh: who at the fame time yielded to his defire of vifiting Antwerp, Bruffels, Lorrain, and the neighbouring countries; in order to complete the plan he had laid down for finifhing his ftudies before he entered upon the exercife of his office. Young as he was, he appeared every way equal to the employ; but held the place no longer than two years; when he clofed with an offer of the profefforfhip of Deventer, which, though of lefs value than Duifburg, was more acceptable to him on many accounts. He had a fingular affection for the place, where firft he indulged his inclination for thefe ftudies. He had the pleafure of fucceeding his much-beloved Gronovius, and that too by a particular recommendation on his removal to Leyden. It muft be remembered alfo, that he was a profelyte to Calvin in the eftab ifhed religion at Deventer, not eafily, if at all, tolerated at Duifburg;
nd

and laftly, in Holland, there was a fairer profpect of preferment. Accordingly, in 1661, the States of Utrecht made him proffffor of eloquence in that univerfity in the room of Paulus Æmilius.

Here he fixed his ambition, and refolved to move no more. In this temper he rejected folicitations both from Amfterdam and Leyden. The elector Palatine likewife attempted in vain to draw him to Heydelberg, and the republic of Venice to Padua. He was in a manner naturalized to Holland: and the states of Utrecht, being determined not to part if poffible with him, laid frefh obligations upon him; and, in 1673, added to that of eloquence the profefforfhip of politics and hiftory. In thefe ftations he had the honour to be fought after by perfons of different countries; feveral coming from Germany for the benefit of his inftructions, many from England. He had filled all thefe pofts, with a reputation nothing inferior to any of his time, for more than thirty years, when he was fuddenly carried off with an apoplexy, January 11, 1703, in his 71ft year.

He had eighteen children by his wife, whom he married in 1656, but was furvived only by four daughters. One of his fons, a youth of great hopes, died 1692, in his 23d year, while he was preparing a new edition of Callimachus, which was finifhed afterwards by his father, and printed in 1697.

Grevius did great fervice to the republic of letters, not fo much by original productions of his own, as by procuring many editions of authors, which he enriched with notes and excellent prefaces, as Hefiod, Callimachus, Suetonius, Cicero, Florus, Catullus, Tibullus, Propertius, Juftin, Cæfar, Lucian. He publifhed alfo, of the moderns, Cafaubon's "Letters," feveral pieces of Meurfius, Huet's "Poemata," Junius "De pictura veterum," Eremita "De Vita aulica & civili," and others of lefs note. But his *chef d' œuvre* is his "Thefaurus Antiquitatum Romanarum," in 12 vols. folio; to which he added afterwards "Thefaurus Antiq. & Hiftor. Italiæ," which were printed after his death, 1704, in 3 vols. folio. There alfo came out in 1707, "J. G. Grevili Prælectiones & CXX Epiftolæ collectæ ab Alb. Fabricio;" to which was added "Burmanni Oratio dicta in Grævii funere," to which we are obliged for the particulars of this memoir. In 1717 was printed "J. G. Grævii Orationes quas Ultrajecti habuit," 8vo. A great number of his letters were publifhed by Burman in his "Sylloge Epiftolarum," in 5 vols. 4to. And the late Dr. Mead was poffeffed of a collection of original letters in MS. written to Grevius by the moft eminent perfons in learning, as Bafnage, Bayle, Burman, Le Clerc, Faber, Fabricius, Gronovius, Kufter, Limborch, Puffendorff, Salmafius, Spanheim, Spinofa,

Spinofa, Tollius, Bentley, Dodwell, Locke, Potter, Abbé Boſſuet, Bignon, Harduin, Huet, Menage, Spon, Vaillant, &c. from the year 1670 to 1703, when Grevius died.

GREUZE, one of the fineſt painters of whom France can boaſt. His works are diſtinguiſhed by a thouſand circumſtances, which render them the delight of all ſpectators. His Pere de famille, in which is repreſented the old man giving his daughter, with a portion, to an honeſt lad, whom he inſtructs in his duty, is wonderful, natural, and expreſſive: and contains many moſt inimitable touches: his Filial Piety, or the Effect of the old Man's Inſtructions, is likewiſe worthy of all the praiſe that can be beſtowed on it. The figure of the paralytic old man is deſigned in a moſt ſuperior manner; the airs of all the heads, particularly thoſe of the man and his wife, are finely expreſſive: the different characteriſtical degrees of grief, in the group around him, exquiſitively imagined and executed in the happieſt manner; the attitudes fine, the ages all diſtinctly marked; and, in a word, every point of compoſition united to render the picture worthy of the artiſt. His piece repreſenting a young woman, her head reclined upon her hand, bewailing the loſs of a canary bird, which lies dead in a cage, is a work of moſt inimitable expreſſion: nothing but life itſelf can equal the ſpirit and ſtriking truth of this piece. His portraits have all great merit; and his merely grotefque pieces are full of life and expreſſion.

GREW (OBADIAH), a worthy pariſh prieſt, was born, 1607, at Atherſton in Warwickſhire; and, having been well grounded in grammar-learning under his uncle Mr. John Deniſon, was ſent to Baliol-college in Oxford, in 1624. Here purſuing his ſtudies carefully, he became qualified for academical honours; and, taking both his degrees in arts at the regular times, he entered at twenty-eight years of age into the prieſthood. In the beginning of the civil wars, he ſided with the parliament party, took the covenant, and, at the requeſt of the corporation of Coventry, became miniſter of the great pariſh of St. Michael in that city. He filled this ſtation by a conſcientious performance of all his duties. The ſoundneſs of his doctrine according to his perſuaſion, the prudence and ſanctity of his converſation, the vigilancy and tenderneſs of his care, were of that conſtant tenor, that he ſeemed to do all which the beſt writers upon the paſtoral office tell us ſhould be done. As he ſided with the preſbyterians againſt the hierarchy, ſo he joined with that party alſo againſt the deſign of deſtroying the king. In this, as in other things, he acted both with integrity and courage, of which we have the following remarkable inſtance: In 1648, when Cromwell, then lieutenant-general, was at Coventry upon his march towards London,

don, Mr. Grew took this opportunity to reprefent to him the wickednefs of the defign, then more vifibly on foot, for taking off his majefty, and the fad confequences thereof, fhould it take effect; earneftly preffing him to ufe his endeavours to prevent it, and not ceafing to folicit him, till he obtained his promife for it. Nor was he fatisfied with this; afterwards, when the defign became too apparent, he addreffed a letter to him reminding him of his promife, and took care to have his letter delivered into Cromwell's own hands.

In 1651, he accumulated the degrees of divinity, and completed that of doctor the enfuing act, when he preached the "Concio ad Clerum" with applaufe. In 1654, he was appointed one of the affiftants to the commiffioners of Warwickfhire, for the ejection of fuch as were then called fcandalous, ignorant, and infufficient minifters and fchoolmafters. He continued at St. Michael's, greatly efteemed and beloved among his parifhioners, till his majefty's reftoration; after which he feems to have refigned his benefice in purfuance to the act of conformity in 1661. It does not appear that he engaged among the conventiclers after his deprivation; but it is certain that he preferved the refpect and affection of the citizens of Coventry till his death, which happened October 22, 1698. He publifhed "A Sinner's Juftification by Chrift, &c. delivered in feveral Sermons on Jer. ii. 6. 1670." 8vo. and "Meditations upon our Saviour's Parable of the predigal Son, &c. 1678," 4to. both at the requeft, and for the common benefit, of fome of his quondam parifhioners.

GREW (NEHEMIAH), fon of the preceding, a learned writer and phyfician, who, being apparently bred up in his father's principles of nonconformity, was fent abroad to complete his education in one of the foreign univerfities. There he took the degree of M. D. after which, refolving to fettle in London, he ftood candidate for an honorary fellowfhip in the College of Phyficians there, and was admitted September 30, 1680. He grew into an extenfive practice by his merit, which had recommended him to the Royal Society; where he was chofen fellow fome years before, and, upon the death of Mr. Oldenburg their fecretary, fucceeded him in that poft on St. Andrew's day, 1677. In confequence whereof, he carried on the publication of the "Philofophical Tranfactions" from January enfuing till the end of February 1678. In the mean time, purfuant to an order of council of July 18 that year, he drew up "A Catalogue of the natural and artificial Rarities belonging to the Society." This was publifhed under the title of "Mufeum Regalis Societatis, &c. 1681," folio, and was followed by "A comparative Anatomy of the Stomach and Guts, begun, &c. 1681," folio; and

"The

"The Anatomy of Plants, &c. 1612," folio. After this he continued to employ the prefs for the fervice of the public; and his own reputation at the fame time, fince he printed feveral other treatifes much efteemed by the learned world [K], both at home and abroad, being moftly tranflated into Latin by foreigners. Thus he paffed his time with the reputation of a learned author and an able practitioner in his profeffion till his death, which happened fuddenly on Lady-day, 1711.

GREY (Lady JANE), an illuftrious perfonage of the blood royal of England by both parents: her grandmother on her father's fide, Henry Grey, marquis of Dorfet, being queen-confort to Edward IV; and her grandmother on her mother's, lady Frances Brandon, being daughter to Henry VII. queen dowager of France, and mother of Mary queen of Scots. Lady Jane was born, 1537, at Bradgate, her father's feat in Leicefterfhire, and very early gave aftonifhing proofs of the pregnancy of her parts; infomuch, that upon a comparifon with Edward VI. who was partly of the fame age, and thought a kind of miracle, the fuperiority has been given to her in every refpect. Her genius appeared in the works of her needle, in the beautiful character in which fhe wrote; befides which, fhe played admirably on various inftruments of mufic, and accompanied them with a voice exquifitely fweet in itfelf, and affifted by all the graces that art could beftow. Thefe, however, were only inferior ornaments in her character; and, as fhe was far from priding herfelf upon them, fo, through the rigour of her parents in exacting them, they became her grief more than her pleafure.

Her father had himfelf a tincture of letters, and was a great patron of the learned. He had two chaplains, Harding and Aylmer, both men of diftinguifhed learning, whom he employed as tutors to his daughter; and under whofe inftructions fhe made fuch a proficiency as amazed them both. Her own language fhe fpoke and wrote with peculiar accuracy: the french, italian, latin, and it is faid greek, were as

---

[K] Thefe are, 1. "Obfervations touching the Nature of Snow," in Phil. Tranf. No. 92. 2. "The Defcription and Ufe of the Pores in the Skin of the Hands and Feet." Ibid. No. 159. for May, 1684. 3. "Tractatus de falis cathartici amari in agris Ebathameufibus & hujufmodi aliis contenti natura & ufa, 1695," 12mo. 4. "Cofmologia Sacra: or a Difcourfe of the Univerfe, as it is the Creature and Kingdom of God: chiefly written to demonftrate the Truth and Excellence of the Bible, which contains the Laws of this Kingdom in the lower World, 1701," fol. This is his capital piece, was univerfally read, and among others foon drew the eyes of Mr. Bayle; who, finding fome of his principles in danger thereby, thought proper to attack it: but a defence appeared foon after in the "Bibliotheque Choifie," Tom. V. written by Le Clerc, who had printed an abridgement of the "Cofmologia" in Tom. I. II. and III. of the fame "Bibliotheque."

natural to her as her own. She not only understood them, but spoke and wrote them with the greatest freedom: she was versed likewise in hebrew, chaldee, and arabic, and all this while a mere child. She had also a sedateness of temper, a quickness of apprehension, and a solidity of judgement, that enabled her not only to become the mistress of languages, but of sciences; so that she thought, spoke, and reasoned, upon subjects of the greatest importance, in a manner that surprized all. With these endowments, she had so much mildness, humility, and modesty, that she set no value upon those acquisitions. She was naturally fond of literature, and that fondness was much heightened as well by the severity of her parents in the feminine part of her education, as by the gentleness of her tutor Aylmer in this: when mortified and confounded by the unmerited chiding of the former, she returned with double pleasure to the lessons of the latter, and sought in Demosthenes and Plato, who were her favourite authors, the delight that was denied her in all other scenes of life, in which she mingled but little, and seldom with any satisfaction. It is true, her alliance to the crown, as well as the great favour in which the marquis of Dorset her father stood both with Henry VIII. and Edward VI. unavoidably brought her sometimes to court, and she received many marks of Edward's attention; yet she seems to have continued for the most part in the country at Bradgate.

Here she was with her beloved books in 1550, when the famous Roger Ascham called on a visit to the family in August; and all the rest of each sex being out a-hunting, he went to wait upon lady Jane in her apartment, and found her reading the "Phædon" of Plato in the original greek. Astonished at it, after the first compliments, he asked her, why she lost such pastime as there needs must be in the park; at which smiling, she answered, "I wist all their sport in the park is but a shadow to that pleasure that I find in Plato. Alas, good folk, they never felt what true pleasure meant." This naturally leading him to enquire how a lady of her age had attained to such a depth of pleasure both in the Platonic language and philosophy, she made the following very remarkable reply: "I will tell you, and I will tell you a truth, which perchance you will marvel at. One of the greatest benefits which ever God gave me is, that he sent me so sharp and severe parents, and so gentle a schoolmaster. For when I am in presence either of father or mother, whether I speak, keep silence, sit, stand, or go, eat, drink, be merry or sad, be sewing, playing, dancing, or doing any thing else, I am so sharply taunted, so cruelly threatened, yea presently sometimes with pinches, rips, and bobs, and other ways (which I will

not name, for the honour I bear them) fo without meafure
mifordered, that I think myfelf in hell, till time come that I
muft go to Mr. Aylmer, who teacheth me fo gently, fo plea-
fantly, with fuch fair allurements to learning, that I think all
the time nothing while I am with him; and, when I am called
from him, I fall on weeping, becaufe whatfoever I do elfe
but learning is full of grief, trouble, fear, and wholly mif-
liking unto me. And thus my book hath been fo much my
pleafure, and bringeth daily to me more pleafure and more,
and that in refpect of it all other pleafures in very deed be but
trifles and troubles unto me." What reader is not melted
with this fpeech? What fcholar does not envy Afcham's felicity at this interview? He was indeed very deeply affected
with it, and to that impreffion we owe the difcovery of fome
farther particulars concerning this lovely fcholar.

At this juncture he was going to London in order to attend
Sir Richard Morrifon on his embaffy to the emperor Charles
V, and in a letter wrote the December following to the dearest
of his friends [L], having informed him that he had had the
honour and happinefs of being admitted to converfe familiarly
with this young lady at court, and that fhe had written a very
elegant letter to him, he proceeds to mention this vifit at
Bradgate, and his furprife thereon, not without fome degree
of rapture. Thence he takes occafion to obferve, that fhe
both fpoke and wrote greek to admiration; and that fhe had
promifed to write him a letter in that language, upon condition
that he would fend her one firft from the emperor's court.
But this rapture rofe much higher while he was penning a letter addreffed to herfelf the following month. There, fpeaking
of this interview, he affures her, that among all the agreeable
varieties which he had met with in his travels abroad, nothing
had occurred to raife his admiration like that incident in the
preceding fummer when he found her, a young maiden by
birth fo noble, in the abfence of her tutor, and in the fumptuous houfe of her moft noble father, at a time too when all the
reft of the family, both male and female, were regaling themfelves with the pleafures of the chace; I found, continues he,
ὦ Ζεῦ καὶ Θεοί, O Jupiter and all ye Gods! I found, I fay,
the divine virgin diligently ftudying the divine "Phædo" of
the divine Plato in the original greek. Happier certainly in
this refpect than in being defcended, both on the father and
mother's fide, from kings and queens. He then puts her in
mind of the greek epiftle fhe had promifed; prompted her to
write another alfo to his friend Sturmius, that what he had
faid of her, whenever he came, might be rendered credible by
fuch authentic evidence

[L]. Sturmius. See art. ASCHAM.

If lady Jane received this letter in the country, yet it is probable she did not stay there long after, since some changes happened in the family which must have brought her to town; for, her maternal uncles, Henry and Charles Brandon, both dying at Bugden, the bishop of Lincoln's palace, of the sweating sickness, her father was created duke of Suffolk, October 1551; Dudley earl of Warwick was also created duke of Northumberland the same day, and in November the duke of Somerset was imprisoned for a conspiracy against him as privy-counsellor. During this interval, came the queen-dowager of Scotland from France, who, being magnificently entertained by king Edward, was also, among other ladies of the blood royal, complimented as her grandmother, by lady Jane, who was now at court, and much in the king's favour. In the summer of 1552, the king made a great progress through some parts of England, during which, lady Jane went to pay her duty to his majesty's sister, the lady Mary, at Newhall in Essex; and in this visit her piety and zeal against popery prompted her to reprove the lady Anne Wharton for making a curtesy to the host, which, being carried by some officious person to the ear of the princess, was retained in her heart, so that she never loved lady Jane afterwards; and, indeed, the events of the following year were not likely to work a reconciliation.

The dukes of Suffolk and Northumberland, who were now, upon the fall of Somerset, grown to the height of their wishes in power, upon the decline of the king's health in 1553, began to think how to prevent that reverse of fortune which, as things then stood, they foresaw must happen upon his death. To obtain this end, no other remedy was judged sufficient, but a change in the succession of the crown, and transferring it into their own families. What other steps were taken, preparatory to this bold attempt, may be seen in the general history, and is foreign to the plan of this memoir, which is concerned only in relating the part that was destined for lady Jane to act in the intended revolution: but this was the principal part; in reality the whole centered in her. Those excellent and amiable qualities, which had rendered her dear to all who had the happiness to know her, joined to her near affinity to the king, subjected her to become the chief tool of an ambition, notoriously not her own. Upon this very account she was married to the lord Guilford Dudley, fourth son to the duke of Northumberland, without being acquainted with the real design of the match, which was celebrated with great pomp in the latter end of May, so much to the king's satisfaction, that he contributed bounteously to the expence of it from the royal wardrobe. In the mean time, though the populace

populace were very far from being pleased with the exorbitant greatness of the duke of Northumberland, yet they could not help admiring the beauty and innocence which appeared in lord Guilford and his bride.

But the pomp and splendor attending their nuptials was the last gleam of joy that shone in the palace of Edward, who grew so weak in a few days after, that Northumberland thought it high time to carry his project into execution. Accordingly, in the beginning of June, he broke the matter to the young monarch; and, having first made all such colourable objections as the affair would admit against his majesty's two sisters, Mary and Elizabeth, as well as Mary queen of Scots, he observed, that, "the lady Jane, who stood next upon the royal line, was a person of extraordinary qualities; that her zeal for the reformation was unquestioned; that nothing could be more acceptable to the nation than the prospect of such a princess; that in this case he was bound to set aside all partialities of blood and nearness of relation, which were inferior considerations, and ought to be over-ruled by the public good." To corroborate this discourse, care was taken to place about the king those who should make it their business to touch frequently upon this subject, enlarge upon the accomplishments of lady Jane, and describe her with all imaginable advantages: so that at last, the king's affections standing for this disposition of the crown, he yielded to overlook his sisters and set aside his father's will. Agreeably to which, a deed of settlement being drawn up in form of law by the judges, was signed by his majesty, and all the lords of the council.

This difficult affair once accomplished, and the letters patent having passed the seals before the close of the month, the next step was to concert the properest method for carrying this settlement into execution, and till that was done to keep it as secret as possible. To this end Northumberland formed a project, which, if it had succeeded, would have made all things easy and secure. He directed letters to the lady Mary in her brother's name, requiring her attendance at Greenwich, where the court then was; and she had got within half a day's journey of that place when the king expired, July 6, 1553; but, having timely notice of it, she thereby avoided the snare which had been so artfully laid to entrap her. The two dukes, Suffolk and Northumberland, found it necessary to conceal the king's decease, that they might have time to gain the city of London, and to procure the consent of lady Jane, who was so far from having any hand in this business, that as yet she was unacquainted with the pains that had been taken to procure her the title of queen. At this juncture, Mary sent a letter to the privy council, in which, though she did not take the title of queen,

queen, yet she clearly asserted her right to the crown; took notice of their concealing her brother's death, and of the practice into which they had since entered; intimating, that there was still room for reconciliation, and that, if they complied with their duty in proclaiming her queen, she could forgive and even forget what was past: but in answer to this they insisted upon the indubitable right, and their own unalterable fidelity to queen Jane, to whom they persuaded the lady Mary to submit.

These previous steps being taken, and the tower and city of London secured, the council quitted Greenwich and came to London; and July 10, in the forenoon, the two last mentioned dukes repaired to Durham-house, where the lady Jane resided with her husband, as part of Northumberland's family. There the duke of Suffolk with much solemnity explained to his daughter the disposition the late king had made of his crown by letters patent; the clear sense the privy-council had of her right; the consent of the magistrates and citizens of London; and, in conclusion, himself and Northumberland fell on their knees, and paid their homage to her as queen of England. The poor lady, somewhat astonished at their discourse, but not at all moved by their reasons, or in the least elevated by such unexpected honours, returned them an answer to this effect: "That the laws of the kingdom and natural right standing for the king's sisters, she would beware of burdening her weak conscience with a yoke which did belong to them; that she understood the infamy of those who had permitted the violation of right to gain a scepter; that it were to mock God and deride justice, to scruple at the stealing of a shilling, and not at the usurpation of a crown. Besides," said she, "I am not so young, nor so little read in the guiles of fortune, to suffer myself to be taken by them. If she enrich any, it is but to make them the subject of her spoil; if she raise others, it is but to pleasure herself with their ruins; what she adored but yesterday is to-day her pastime; and, if I now permit her to adorn and crown me, I must to-morrow suffer her to crush and tear me to pieces. Nay, with what crown does she present me! a crown which hath been violently and shamefully wrested from Catharine of Arragon, made more unfortunate by the punishment of Anne Boleyn, and others that wore it after her: and why then would you have me add my blood to theirs, and be the third victim, from whom this fatal crown may be ravished with the head that wears it? But in case it shou'd not prove fatal unto me, and that all its venom were consumed, if fortune should give me warranties of her constancy, should I be well advised to take upon me these thorns, which would dilacerate, though not kill

kill me outright; to burden myself with a yoke, which would not fail to torment me, though I were assured not to be strangled with it? My liberty is better than the chain you proffer me, with what precious stones soever it be adorned, or of what gold soever framed. I will not exchange my peace for honourable and precious jealousies, for magnificent and glorious fetters. And, if you love me sincerely and in good earnest, you will rather wish me a secure and quiet fortune, though mean, than an exalted condition, exposed to the wind, and followed by some dismal fall."

However, she was at length prevailed upon, by the exhortations of her father, the intercession of her mother, the artful persuasions of Northumberland, and, above all, the earnest desires of her husband, whom she tenderly loved, to yield her assent to what had been and was to be done.. And thus, with a heavy heart, she suffered herself to be conveyed by water to the Tower, where she entered with all the state of a queen, attended by the principal nobility, and, which is very extraordinary, her train supported by the duchess of Suffolk, her mother, in whom, if in any of this line, the right of succession remained. About six in the afternoon, she was proclaimed with all due solemnities in the city; the same day she also assumed the regal, and proceeded afterwards to exercise many acts of sovereignty ; but, passing over the transactions of her short reign, which are the subject of the general history, it is more immediately our business to conclude this article with her behaviour on her fall. Queen Mary was no sooner proclaimed, than the duke of Suffolk, who then resided with his daughter in the tower, went to her apartment, and, in the softest terms he could, acquainted her with the situation of their affairs, and that, laying aside the state and dignity of a queen, she must again return to that of a private person : to which, with a settled and serene countenance, she made this answer : " I better brook this message than my former advancement to royalty : out of obedience to you and my mother, I have grievously sinned, and offered violence to myself. Now I do willingly, and as obeying the motions of my soul, relinquish the crown, and endeavour to salve those faults committed by others (if at least so great a fault can be salved) by a willing relinquishment and ingenuous acknowledgement of them.

Thus ended her reign, but not her misfortunes. She saw the father of her husband, with all his family, and many of the nobility and gentry, brought prisoners to the tower for supporting her claim to the crown ; and this grief must have met with some accession from his being soon after brought to the block. Before the end of the month, she had the mortification

tification of seeing her own father, the duke of Suffolk, in the same circumstances with herself; but her mother, the duchess, not only remained exempt from all punishment, but had such an interest with the queen as to procure the duke his liberty on the last day of the month. Lady Jane and her husband, being still in confinement, were November 3, 1553, carried from the Tower to Guildhall with Cranmer and others, arraigned and convicted of high treason before judge Morgan, who pronounced on them sentence of death, the remembrance of which afterwards affected him so far, that he died raving. However, the strictness of their confinement was mitigated in December, by a permission to take the air in the queen's garden, and other little indulgences. This might give some gleams of hope; and there are reasons to believe the queen would have spared her life, if Wiat's rebellion had not happened; but her father's being engaged in that rebellion gave the ministers an opportunity of persuading the queen, that she could not be safe herself, while lady Jane and her husband were alive: yet Mary was not brought without much difficulty to take them off. The news made no great impression upon lady Jane: the bitterness of death was passed; she had expected it long, and was so well prepared to meet her fate, that she was very little discomposed.

But the queen's charity hurt her more than her justice. The day first fixed for her death was Friday February the 9th; and she had, in some measure, taken leave of the world by writing a letter to her unhappy father, who she heard was more disturbed with the thoughts of being the author of her death than with the apprehension of his own [M]. In this

[M] There is something so striking in this letter, and so much above her years, that we cannot debar the reader from it. It is in these terms: "Father, although it pleaseth God to hasten my death by you, by whom my life should rather have been lengthened; yet can I so patiently take it, as I yield God more hearty thanks for shortening my woful days than if all the world had been given into my possession with life lengthened to my will. And albeit I am well assured of your impatient dolors, redoubled many ways, both in bewailing your own wo, and also, as I hear, especially my unfortunate estate; yet, my dear father, if I may without offence rejoice in my mishaps, methinks in this I may account myself blessed; that, washing my hands with the innocency of my fact, my guiltless blood may cry before the Lord, mercy to the innocent; and yet, though I must needs acknowledge, that being constrained, and, as you well know continually assayed in taking the crown upon me, I seemed to consent, and therein grievously offended the queen and her laws; yet do I assuredly trust, that this my offence towards God is so much the less, in that, being in so royal an estate as I was, mine enforced honour never mixed with my innocent heart. And thus, good father, I have opened my state to you, whose death at hand, although to you perhaps it may seem right woful, to me there is nothing that can be more welcome than from this vale of misery to aspire to that heavenly throne of all joys, and pleasure with Christ our Saviour; in whose stedfast faith, if it be lawful for the daughter to write so to her father, the Lord, that hitherto hath strengthened you, so continue you, that at last we may meet in heaven, with the Father, Son, and Holy Ghost." Fox's Acts and Monuments.

serene

serene frame of mind, Dr. Feckenham, abbot of Westminster, came to her from the queen, who was very desirous she should die professing herself a papist, as her father-in-law had done. The abbot was indeed a very fit instrument, if any had been fit for the purpose, having, with an acute wit and a plausible tongue, a great tenderness in his nature. Lady Jane received him with much civility, and behaved towards him with so much calmness and sweetness of temper, that he could not help being overcome with her distress: so that, either mistaking or pretending to mistake her meaning, he procured a respite of her execution till the 12th. When he acquainted her with it, she told him, "that he had entirely misunderstood her sense of her situation; that, far from desiring her death might be delayed, she expected and wished for it as the period of her miseries, and her entrance into eternal happiness. Neither did he gain any thing upon her in regard to popery; she heard him indeed patiently, but answered all his arguments with such strength, clearness, and steadiness of mind, as shewed plainly that religion had been her principal care [N]. On Sunday evening, which was the last she was to spend in this world, she wrote a letter in the greek tongue, as some say, on the blank leaves at the end of a testament in the same language, which she bequeathed as a legacy to her sister the lady Catharine Grey; a piece which, if we had no other left, it is said, were sufficient to render her name immortal. In the morning, the lord Guilford earnestly desired the officers, that he might take his last farewel of her; which though they willingly permitted, yet upon notice she advised the contrary, "assuring him that such a meeting would rather add to his afflictions then increase his quiet, wherewith they had prepared their souls for the stroke of death; that he demanded a lenitive which would put fire into the wound, and that it was to be feared her presence would rather weaken than strengthen him; that he ought to take courage from his reason, and derive constancy from his own heart; that if his soul were not firm and settled, she could not settle it by her eyes, nor conform it by her words; that he shou'd do well to remit this interview to the other world; that there, indeed, friendships were happy, and unions indissoluble, and that theirs would be eternal, if their souls carried nothing with them of terrestrial, which might hinder them from rejoicing." All she could do was, to give him a farewel out of a window, as he passed to the

---

[N] The particulars that passed betwixt her and Feckenham are well worth the reader's perusal in Fox; and an account drawn up by herself of her dispute with him about the real presence is printed in the "Phœnix," Vol. II. p. 28.

place of his diffolution [o], which he fuffered on the fcaffold on Tower-hill with much Chriftian meeknefs. She likewife beheld his dead body wrapped in a linen cloth, as it paffed under her window to the chapel within the Tower.

And, about an hour after, fhe was led to a fcaffold: fhe was attended by Feckenham, but was obferved not to give much heed to his difcourfes, keeping her eyes ftedfaftly fixed on a book of prayers which fhe had in her hand. After fome fhort recollection, fhe faluted thofe who were prefent, with a countenance perfectly compofed: then, taking leave of Dr. Feckenham, fhe faid, "God will abundantly requite you, good Sir, for your humanity to me, though your difcourfes gave me more uneafinefs than all the terrors of my approaching death." She next addreffed herfelf to the fpectators in a plain and fhort fpeech; after which, kneeling down, fhe repeated the Miferere in Englifh. This done, fhe ftood up and gave to her women her gloves and handkerchief, and to the lieutenant of the Tower her Prayer-book. In untying her gown, the executioner offered to affift her; but fhe defired he would let her alone; and turning to her women, they undreffed, and gave her a handkerchief to bind about her eyes. The executioner, kneeling, defired her pardon, to which fhe anfwered, "moft willingly." He defired her to ftand upon the ftraw; which bringing her within fight of the block, fhe faid, "I pray difpatch me quickly;" adding prefently after, "Will you take it off before I lay me down?" The executioner anfwered, "No, Madam." Upon this, the handkerchief being bound clofe over her eyes, fhe began to feel for the block, to which fhe was guided by one of the fpectators. When fhe felt it, fhe ftretched herfelf forward, and faid, "Lord, into thy hands I commend my fpirit;" and immediately her head was feparated at one ftroke.

Her fate was univerfally deplored even by the perfons beft-affected to queen Mary; and, as fhe is allowed to have been a princefs of great piety, it muft certainly have given her much difquiet to begin her reign with fuch an unufual effufion of blood; efpecially in the prefent cafe of her near relation, one formerly honoured with her friendfhip and favour, who had

---

[o] After this fad fight, fhe wrote three fhort fentences in a table-book, in Greek, Latin, and Englifh, to this purport. In Greek: "If his flain body fhall give teftimony againft me before men, his moft bleffed foul fhall render an eternal proof of my innocence in the prefence of God." In Latin to this effect: "The juftice of man took away his body, but the divine mercy has preferved his foul." The Englifh ran thus: "If my fault deferved punifhment, my youth at leaft and my imprudence were worthy of excufe. God and pofterity will fhew me favour."—This book fhe gave to Sir John Bridges, the Lieutenant of the Tower, on the fcaffold, at his intreaty to beftow fome memorial upon him, as an acknowledgement of his civility. Heylin.

indeed

indeed ufurped, but without defiring or enjoying, the royal diadem which fhe affumed, by the conftraint of an ambitious father and an imperious mother, and which at the firft motion fhe chearfully and willingly refigned. This made her exceedingly lamented at home and abroad; the fame of her learning and virtue having reached over Europe, excited many commendations, and fome exprefs panegyrics in different nations and different languages. Immediately after her death, there came out a piece, intyled, "The precious Remains of Lady Jane Grey," in quarto.

GREY (Dr. ZACHARY), an ingenious Englifh fcholar, was of a Yorkfhire family, and born about 1087. He was admitted of Jefus college in 1704, but afterwards removed to Trinity-hall, Cambridge, where he became LL.D. He was rector of Houghton-Conqueft in Bedfordfhire, and vicar of St. Giles's and St. Peter's parifhes in Cambridge; at which laft he ufually paffed the winter. He died November 25, 1766; having been twice married, and leaving two daughters. He was the author of near 30 publications, which any one who is curious about them may fee in the "Anecdotes of Bowyer," by Nichols; but his edition of "Hudibras, 1744," 2 vols. 8vo. is the work which will probably keep his memory alive. Warburton, in his preface to Shakfpeare, "hardly thinks there ever appeared, in any learned language, fo execrable an heap of nonfenfe, under the name of Commentaries, as hath lately been given us on this fatiric poet:" and Henry Fielding, in the preface to his "Voyage to Lifbon," has introduced "the laborious much read Dr. Zachary Grey, of whofe redundant notes on Hudibras he fhall only fay, that it is, he is confident, the fingle book extant, in which above 500 authors are quoted, not one of which could be found in the collection of the late Dr. Mead." This is meant for wit; the former was the effect of a fcurrilous and abufive fpirit: and we think our author has very well obferved, in the language of Mr. Warton upon Shakfpeare, that, "'if Butler is worth reading, he is worth explaining; and the refearches ufed for fo valuable and elegant a purpofe merit the thanks of genius and candor, not the fatire of prejudice and ignorance."

GREY (Dr. RICHARD), an ingenious and learned Englifh divine, was born in 1693, and went through Lincoln-college, Oxford, where he took the degree of M. A. January 16, 1718-19. He obtained early in life the rectory of Kilncote in Leicefterfhire, and afterwards that of Hinton in Northamptonfhire; together with a prebend of St. Paul's. He was alfo, 1746, official and commiffary of the archdeaconry of Leicester. In 1730, he published at Oxford a "Visitation-Sermon;" and, the same year, "Memoria Technica; or a new Method of artificial Memory:" a fourth edition of which came out in 1756. At this time also appeared his "System of English Ecclesiastical Law, extracted from the Codex Juris Ecclesiastici Anglicani" of bishop Gibson, 8vo. This was for the use of young students designed for orders; and for this the university gave him the degree of D. D. May 28, 1731. In 1736, he was the undoubted author of a large anonymous pamphlet, under the title of "The miserable and distracted State of Religion in England, upon the Downfall of the Church established." 8vo; and, the same year, printed another Visitation-Sermon. He had printed an Assize-Sermon, in 1732, called, "The Great Tribunal." He published "A new and easy Method of learning Hebrew without points, 1738;" "Historiæ Josephi," and "Paradigmata Verborum, 1739;" "Liber Jobi, 1742;" "Answer to Warburton's Remarks, 1744;" "The last Words of David, 1749;" "Nova Methodus Hebraicè discendi diligentius recognita, & ad Usum Scholarum accommodata, &c. 1751." and, lastly, an English translation of Mr. Hawkins Browne's poem "De Animi Immortalitate, 1753." He died Feb. 28, 1771, in his 78th year; having been married, and leaving daughters.

GRIBALDUS (MATTHEW), a learned civilian of Padua, who left Italy in the 16th century, in order to make a public profession of the Protestant religion; but who, like some other Italian converts, imbibed the heresy of the Antitrinitarians. After having been professor of civil law at Tubingen for some time, he quitted the employment, in order to escape the punishment he would have incurred, had he been convicted of his errors. He was seized at Bern, where he feigned to renounce his opinions, in order to escape very severe treatment; but as he relapsed again, and openly favoured the Heretics, who had been driven from Geneva, he would, as Beza intimates, certainly have been put to death, if the plague had not snatched him away in September 1564, and so secured him from being prosecuted for heresy. In a journey to Geneva, during the trial of Servetus, he desired to have a conference with Calvin, which Calvin at first refused, but afterwards granted; and then Gribaldus, though he came according to the appointed time and place, refused, because Calvin would not give him his hand, till they should be agreed on the articles of the Trinity. He was afterwards cited to appear before the magistrates, in order to give an account of his faith; but, his answers not being satisfactory, he was commanded to leave the city. He wrote several works, which are

are esteemed by the public; as "Commentarii in legem de rerum mistura, & de jure fisci," printed in Italy. "Commentarii in pandectas juris," at Lyons. "Commentarii in aliquot præcipuos Digesti, Codicis Justiniani, titulos, &c." at Frankfort, 1577. "Historia Francisci Spiræ, cui anno 1548 familiaris aderat, secundum quæ ipse vidit & audivit, Basil, 1550." Sleidan declares, that Gribaldus was a spectator of the sad condition of Spira, and that he wrote and published an account of it. "De methodo ac ratione studendi in jure civili libri tres, Lyons 1544 and 1556." He is said to have written this last book in a week.

GRIBNER (MICHAEL HENRY), a professor of law at Wittenberg. He wrote several dissertations, and many works on jurisprudence in Latin. He was also concerned in the "Leipsic Journal." He died in 1734. He is mentioned by Saxius, in his "Onomasticon," in terms of considerable respect.

GRIERSON (CONSTANTIA), a very extraordinary woman, was born in the county of Kilkenny in Ireland. She died in 1733, at the age of 27; and was allowed to be an excellent scholar, not only in Greek and Roman literature, but in history, divinity, philosophy, and mathematics. She gave a proof of her knowledge in the Latin tongue by her dedication of the Dublin edition of Tacitus to lord Carteret; and by that of Terence to his son, to whom she likewise wrote a Greek epigram. She composed several fine poems in English, several of which are inserted by Mrs. Barber amongst her own. When lord Carteret was lord-lieutenant of Ireland, he obtained a patent for Mr. Grierson, her husband, to be the king's printer; and, to distinguish and reward her uncommon merit, had her life inserted in it. Besides her parts and learning, she was also a woman of great virtue and piety. Mrs. Pilkington has recorded some particulars of her, and tells us, that, " when about eighteen years of age, she was brought to her father, to be instructed in midwifery; that she was mistress of Hebrew, Greek, Latin, and French, and understood the mathematics as well as most men: and what," says Mrs. Pilkington, "made these extraordinary talents yet more surprising was, that her parents were poor illiterate country people; so that her learning appeared like the gift, poured out on the Apostles, of speaking all the languages without the pains of study." Mrs. Pilkington enquired of her, where she had gained this prodigious knowledge: to which Mrs. Grierson said, that "she had received some little instruction from the minister of the parish, when she could spare time from her needle-work, to which she was closely kept by her mother." Mrs. Pilkington adds, that "the wrote

wrote elegantly both in verfe and profe; that her turn was chiefly to philofophical or divine fubjects; that her piety was not inferior to her learning; and that fome of the moft delightful hours fhe herfelf had ever paffed were in the converfation of this female philofopher."

GRIFFET (HENRY), a jefuit, born at Moulins in 1698. He was author of many works of different degrees of eftimation. The principal of which were a new edition of Daniel's "Hiftory of France," with many learned and curious differtations. The hiftory of Louis XIII. in the concluding volumes, are original, and written by Griffet. He publifhed alfo fermons and other pious works. One of his moft popular productions is "Delices des Pays Bas," in 5 vols. 12mo. a new edition, with confiderable augmentations. He died at Bruffels in 1775, and left the character not only of an induftrious but very fuccefsful writer.

GRIFFIER (JOHN), was commonly known by the appellation of old Griffier, and an eminent painter. Though born at Amfterdam, he was on the continent called the gentleman of Utrecht. He was celebrated for his ftyle in painting landfcapes, which he enriched with buildings and figures. He alfo etched feveral prints of birds and beafts after the defigns of Francis Barlow. He died in 1718.

GRIFFIN, prince of Wales. We are induced to introduce the name of this perfonage in our work, becaufe he was the laft fovereign of that country. His fubjects were againft him and delivered him up to Edward the confeffor, who put him to death at London in 1160.

GRIGNON (JAQUES), Florent le Comte calls him John Grignon. He was a native of France, and flourifhed towards the end of the laft century. His beft works, I think, are his portraits, which he executed entirely with the graver; and fome of them do him great credit. That of Francis Maria Rhima, an ecclefiaftic, a fmall upright oval plate, is executed in a very clear, good ftyle. His hiftorical plates and fubjects, with figures, are by no means equally meritorious. They are dark and heavy, without effect, and in general very incorrectly drawn. He engraved fome few of the plates for a work entituled "Les Tableaux de la Pénitence," in fmall folio fize, from the defigns of Chauveau.

GRIMALDI (JOHN FRANCIS), a painter of Bologne, was born at Bologna in 1606, and ftudied under the Caracci, to whom he was related. He was a good defigner of figures, but became chiefly diftinguifhed for his landfcapes. When he arrived at Rome, Innocent X. did juftice to his merit, and employed him to paint in the vatican and other places. This pontiff ufed to fee him work, and talk familiarly with him.

His

His reputation reached cardinal Mazarine at Paris, who sent for him, settled a large pension on him, and employed him for three years in embellishing his palace and the Louvre, by the order of Lewis XIII. The troubles of the state, and the clamours raised against the cardinal, whose party he warmly espoused, put him so much in danger, that his friends advised him to retire among the jesuits. He was of use to them, for he painted them a decoration for the exposition of the sacrament during the holy days, according to the custom of Rome. This piece was mightily relished at Paris: the king honoured it with two visits, and commanded him to paint such another for his chapel at the Louvre. Grimaldi after that returned to Italy, and at his arrival at Rome found his great patron Innocent X. dead; but his two successors Alexander VII. and Clement IX. honoured him equally with their friendship, and found him variety of employment. His colouring is vigorous and fresh, his touch beautiful and light, his sites are pleasant, his fresco admirable, his leafing enchanting, and his landschapes, though sometimes too green, may serve as models to those who intend to apply themselves to that branch of painting. He understood architecture, and has engraved in aqua fortis forty-two landschapes in an excellent manner, five of which are after Titian. Grimaldi was amiable in his manners, as well as skilful in his profession: he was generous without profusion, respectful to the great without meanness, and charitable to the poor. The following instance of his benevolence may serve to characterise the man. A Sicilian gentleman, who had retired from Messina with his daughter, during the troubles of that country, was reduced to the misery of wanting bread. As he lived over-against him, Grimaldi was soon informed of it; and in the dusk of the evening, knocking at the Sicilian's door, without making himself known, tossed in money and retired. The thing happening more than once raised the Sicilian's curiosity to know his benefactor; who finding him out, by hiding himself behind the door, fell down on his knees to thank the hand that had relieved him: Grimaldi remained confused, offered him his house, and continued his friend till his death. He died of a dropsy at Rome in 1660, and left a considerable fortune among six children; of which the youngest, named Alexander, was a pretty good painter.

GRIMAREST (LEONARD), a french writer of no great merit. He published the "History of Charles the twelfth," and a "Life of Moliere," which Voltaire represents as full of misrepresentations both of Moliere and his friends. He wrote also "Eclaircissements sur la langue Françoise," in which are some sensible observations. He died in 1720.

GRIMOUX, a french painter, who flourished about the middle of this century. There was something so very curious, so original, in his portraits, that they are sought after as cabinet pieces. He was a whimsical and capricious character, and affected to make no distinction between the night and the day. He died in 1740.

GRINDAL (EDMUND), archbishop of Canterbury, was born, in 1519, at Hinsingham, a small village in Cumberland. After a suitable foundation of learning at school, he was sent to Magdalen-college in Cambridge, but removed thence to Christ's, and afterwards to Pembroke-hall; where, having taken his first degree in arts, he was chosen fellow in 1538, and commenced M. A. in 1541. In 1549, he became president [vice-master] of his college; and being now B. D. was unanimously chosen lady Margaret's public preacher at Cambridge; as he was also one of the four disputants in a theological extraordinary act, performed that year for the entertainment of king Edward's visitors.

Thus distinguished in the university, his merit was observed by Ridley, bishop of London, who made him his chaplain in 1550; perhaps by the recommendation of Bucer, the king's professor of divinity at Cambridge, who soon after his removal to London, in a letter to that prelate, styles our divine "a person eminent for his learning and piety." And thus, a door being opened to him into church-preferments, he rose by quick advances. His patron the bishop was so much pleased with him, that he designed for him the first preferments which should fall; and, in 1551, procured him to be made chaplain to the king. July 2, 1552, he obtained a stall in Westminster-abbey; which however he resigned to Dr. Bonner, whom he afterwards succeeded in the bishopric of London. In the mean time, there being a design, on the death of Dr. Tonstall, to divide the rich see of Durham into two, Grindal was nominated for one of these, and would have obtained it, had not one of the courtiers got the whole bishopric dissolved, and settled as a temporal estate upon himself.

In 1553, he fled from the persecution under queen Mary into Germany; and, settling at Strasbourg, made himself master of the german tongue, in order to preach in the churches there: in the disputes at Frankfort about a new model of government and form of worship, varying from the last liturgy of king Edward, he sided with Cox and others against Knox and his followers. Returning to England on the accession of Elizabeth, he was employed, among others, in drawing up the new liturgy to be presented to the queen's first parliament; and was also one of the eight protestant divines, chosen to hold a public dispute with the popish prelates

lates about that time. His talent for preaching was likewife very ferviceable, and he was generally appointed to that duty on all public occafions. At the fame time, he was appointed one of the commiffioners in the North, on the royal vifitation for reftoring the fupremacy of the crown, and the proteftant faith and worfhip. This vifitation extended alfo to Cambridge, where Dr. Young being removed, for refufing the oath of fupremacy, from the mafterfhip of Pembroke-hall, Grindal was chofen by the fellows to fucceed him in 1559.

July the fame year, he was nominated to the bifhopric of London, vacant by the depofition of Bonner. The juncture was very critical, and the fate of the church-revenues depended upon the event. An act of parliament had lately paffed, whereby her majefty was empowered to exchange the ancient epifcopal manors and lordfhips for tithes and impropriations; a meafure extremely regretted by thefe firft bifhops, who fcrupled whether they fhould comply in a point fo injurious to the revenue of their refpective fees. In this important point, our new-nominated bifhop confulted Peter Martyr: nor did he accept of the bifhopric, till he had received an opinion in favour of it from that divine, who faid, that the queen might provide for her bifhops and clergy in fuch manner as fhe thought proper, that being none of Grindal's concern. In 1560, he was made one of the ecclefiaftical commiffioners, in purfuance of an act of parliament to infpect into the manners of the clergy, and regulate all matters of the church; and the fame year he joined with Cox, bifhop of Ely, and Parker, archbifhop of Canterbury, in a private letter to the queen, perfuading her to marry. In 1561, he held his primary vifitation. In 1563, he affifted the archbifhop of Canterbury, together with fome civilians, in preparing a book of ftatutes for Chrift-church, Oxford, which as yet had none fixed. This year he was alfo very ferviceable, in procuring the Englifh merchants, who were ill ufed at Antwerp, and other parts of the fpanifh Netherlands, a new fettlement at Embden, in Eaft-Friefland; and the fame year, at the requeft of Sir William Cecil, fecretary of ftate, he wrote animadverfions upon a treatife intituled "Chriftiani Hominis Norma, &c." "The Rule of a Chriftian Man," the author whereof, one juftice Velfius, a dutch enthufiaft, had impudently, in fome letters to the queen, ufed fome menaces to her majefty; and, being at laft cited before the ecclefiaftical commiffion, was charged to depart the kingdom.

April 15, 1564, he took the degree of D. D. at Cambridge, and the fame year executed the queen's exprefs command for exacting uniformity in the clergy; but proceeded fo tenderly and flowly, that the archbifhop thought fit to excite and quicken him; whence the puritans fuppofed him inclined to their party. However, he brought feveral Nonconformifts to comply; to which end he publifhed a letter of Henry Bullinger, minifter of Zurich in Switzerland, to prove the lawfulnefs thereof, which had a very good effect. The fame year, October 3, on the celebration of the emperor Ferdinand's funeral, he preached a fermon at St. Paul's, afterwards printed. In 1567, he executed the queen's orders in proceeding againft the prohibited unlicenfed preachers; but was fo treated by fome with reproaches and rude language, that it abated much of his favourable inclinations towards them. May 1, 1570, he was tranflated to the fee of York. He owed this promotion to fecretary Cecil and archbifhop Parker, who liked his removal from London, as not being refolute enough for the government there. The fame year he wrote a letter to his patron Cecil, that Cartwright the famous nonconformift might be filenced; and in 1571, at his metropolitical vifitation, he fhewed a hearty zeal, by his injunctions, for the difcipline and good government of the church. In 1572, he petitioned the queen to renew the ecclefiaftical commiffion. In 1574, he held one for the purpofe of proceeding againft papifts, whofe number daily diminifhed in his diocefe, which he was particularly careful to provide with learned preachers, as being in his opinion the beft method of attaining that end. Upon the death of Parker, he was tranflated to Canterbury; in which fee he was confirmed, February 15, 1575. May 6, 1576, he began his metropolitical vifitation, and took meafures for the better regulation of his courts; but the fame year fell under her majefty's difpleafure, by reafon of the favour he fhewed to what was called the exercife of propheiying.

These prophefyings had been ufed for fome time, the rules whereof were, that the minifters of a particular divifion at a fet time met together at fome church, and there each in their order explained, according to their abilities, fome portion of fcripture allotted to them before: this done, a moderator made his obfervations on what had been faid, and determined the true fenfe of the place, a certain time being fixed for difpatching the whole. The advantage was the improvement of the clergy, who hereby confiderably profited themfelves in the knowledge of the fcripture; but this mifchief enfued, that at length there happened confufions and difturbances at thofe meetings, by an oftentation of fuperior parts in fome, by advancing heterodox opinions, and by the intrufion of fome of the filenced feparatifts, who took this opportunity of declaiming againft the liturgy and hierarchy, and hence even fpeaking againft ftates and particular perfons. The people alfo, of whom there was always a great conflux as hearers, fell to arguing and difputing much
about

about religion, and sometimes a layman would take upon himself to speak. In short, the exercises degenerated into factions.

Grindal laboured to redress these irregularities by setting down rules and orders for the management of these exercises; however, the queen still disapproved of them, as seeing probably how very apt they were to be abused. She did not like that the laity should neglect their secular affairs by repairing to those meetings, which she thought might fill their heads with notions, and so occasion dissentions and disputes, and perhaps seditions in the state. And the archbishop being at court, she particularly declared herself offended at the number of preachers as well as the exercises, and ordered him to redress both; urging, that it was good for the church to have few preachers, that three or four might suffice for a county, and that the reading of the Homilies to the people was sufficient. She therefore required him to abridge the number of preachers, and put down the religious exercises. This did not a little afflict him. He thought the queen infringed upon his office, to whom, next to herself, the highest trust of the church of England was committed; especially as this command was preremptory, and made without at all advising with him, and that in a matter so directly concerning religion: he wrote a letter to her majesty, declaring, that his conscience would not suffer him to comply with her commands.

This refusal was dated December 20, 1576. The queen therefore having given him sufficient time to consider well his resolution, and he continuing unalterable therein, she sent letters next year to the bishops, to forbid all exercises and prophesyings, and to silence all preachers and teachers not lawfully called, of which there were no small number; and in June the archbishop was sequestered from his office, and confined to his house, by an order of the court of Star-chamber. In November, the lord-treasurer wrote to him about making his submission, with which he not thinking fit to comply, his sequestration was continued; and in January there were thoughts of depriving him, which however were laid aside. June 1579, his confinement was either taken off, or else he had leave to retire to his house at Croydon; for we find him there consecrating the bishop of Exeter in that year, and the bishops of Winchester, and Lichfield and Coventry, the year following. This part of his function was exercised by a particular commission from the queen, who in council appointed two civilians to manage the other affairs of his see, the two of his nomination being set aside. Yet sometimes he had special commands from the queen and council to act in person, and issued out orders in his own name; and in general was as active as he could be, and vigilant in the care of his diocese as occasion offered. The precise time of his being restored does not clearly appear; but several of his proceedings shew, that he was in the full possession of the metropolitical power in 1582, in which year it is recorded, that he had totally lost his eye-sight. Towards the latter end of it, he resigned his see, and obtained a pension for his life from the queen, though in no degree of her majesty's favour. With this provision he retired to Croydon, where he died two months after, July 6, 1583, and was interred in that church.

GRINGONNEUR (JACQUEMIN), a french painter of the fourteenth century, and by some believed to have been the inventor of cards. This, however, is much disputed, perhaps he might invent the painting upon cards. He died about 1392.

GRINGORE (PETER), herald at arms to the duke of Lorrain, died in 1544. He was the author of "Moralities," in verse, which are remarkably scarce. They are very unentertaining to peruse; but are useful to mark and ascertain the progressive improvement of theatrical representations.

GRISAUNT (WILLIAM), a physician, astronomer, and mathematician, and like his countryman, frier Bacon, violently suspected of magic. He studied at Merton-college, Oxford; and, probably to escape the disagreeable effects concomitant with those suspicions, went into France, where he devoted himself entirely to the study of medicine, first at Montpelier, and then at Marseilles. In this city he fixed his residence, and lived by the practice of his profession, in which he acquired much skill and eminence. There is no greater proof of his genius, besides the imputations he laboured under in his youth, than his assiduously pursuing the method instituted by the greek physicians, of investigating the nature and cause of the disease and the constitution of the patient. The time of his death is not known; but we are told that he was an old man in 1350, and that he had a son, who was first an abbot of canons regular at Marseilles, and at length arrived at the pontificate under the name of Urban V. The list of his works may be found in Aikin's Biog. Memoirs of Medicine.

GRIVE (JOHN DE LA), a french geographer. He published the "Topography of Paris," which was remarkably accurate. He produced also "Plans of Versailles, Marly, the Environs of Paris, and a Tract on Spherical Trigonometry."

GROCYN (WILLIAM), a man eminently learned in his days, was born at Bristol in 1442, and educated at Winchester-school. He was elected thence to New-college, Oxford, in 1467; and in 1479, presented by the warden and fellows to the rectory of Newton-Longville in Berkshire. But his residence being mostly at Oxford, the society of Magdalen-college made him their divinity reader, about the beginning of

Richard

Richard the IIId's regn; and that king coming foon after to Oxford, he had the honour to hold a difputation before him, with which his majefty was fo pleafed, that he rewarded him gracioufly. In 1485, he was made a prebendary of Lincoln. In 1488, he quitted his readers's place, at Magdalen-college, in order to travel into foreign countries; for though he was reckoned a great mafter of the greek and latin languages here in England, where the former efpecially was then fcarcely underftood at all, yet he well knew that there was room enough for far greater perfection; and accordingly he went into Italy, and ftudied there fome time under Demetrius Chalcondylas and Politian. He returned to England, and fixed himfelf in Exeter-college, at Oxford, in 1491; where he publicly taught the greek language, and was the firft who introduced a better pronunciation of it than had been known in this ifland before. In this fituation he was, when Erafmus came to Oxford; and if he was not this great man's tutor, yet he certainly affifted him in attaining a more perfect knowledge of the greek. He was however very friendly to Erafmus, and did him many kind offices, as introducing him to archbifhop Warham, &c. and Erafmus fpeaks of him often in a ftrain, which fhews, that he entertained the moft fincere regard for him, as well as the higheft opinion of his abilities, learning, and integrity. About 1590, he refigned his living, being then made mafter of Allhallows-college at Maidftone, in Kent, though he continued ftill to live moftly at Oxford. Grocyn had no efteem for Plato, but applied himfelf intenfely to Ariftotle, whofe whole works he had formed a defign of tranflating, in conjunction with William Latimer, Linacre, and More, but did not purfue it. While his friend Colet was dean of St. Paul's, he read the divinity-lecture in that cathedral. He died at Maidftone in 1522, aged 80, of a ftroke of the palfy, which he had received a year before, and which made him, fays Erafmus, "fibi ipfi fuperftitem;" that is, outlive his fenfes. Linacre, the famous phyfician juft mentioned, was his executor, to whom he left a confiderable legacy, as he did a fmall one to Thomas Lilly the grammarian, who was his godfon. His will is printed in the appendix to Knight's " Life of Erafmus." A latin epiftle of Grocyn's to Aldus Manutius is prefixed to Linacre's tranflation of " Proclus de Sphæra," printed at Venice in 1449. Erafmus fays, that " there is nothing extant of his but this epiftle; indeed a very elaborate and acute one, and written in good latin." His publifhing nothing more feems to have been owing to too much delicacy; for, Erafmus adds, " he was of fo nice a tafte, that he had rather write nothing than write ill." Some other things, however, of his writing are mentioned by Bale and Leland, as

" Tractatus

"Tractatus contra hostiolum Joannis Wiclevi," "Epistolæ ad Erasmum & alios," "Grammatica," "Vulgaria puerorum," "Epigrammata," &c.

GRODITIUS (STANISLAUS), a polonese jesui. Died at Cracow in 1613. He left eight volumes of latin sermons, with many other polemic writings.

GRONOVIUS (JOHN FREDERIC), an eminent civilian, historian, and critic, was born at Hamburgh in 1611. Nature had given him a strong inclination to learning, so that he applied to books with indefatigable diligence from his infancy; and, having made a great progress in his own country, e travelled into Germany, Italy, and France, where h. searched all the treasures of literature that cou d be found in those countries, and was returning h me by the way of the United Provinces, when he was stopt at Daventer in the province of Over-Issel, and there made professor of polite learning. In this chair having acquired a great reputation, he was promoted to that of Leyden in 1658, vacant by the death of Lonne Heinsius. He published several works, and has given editions of a great number of the classics more correct than before; as Plautus, Sallust, Livy, Seneca, Pliny, Quintilian, Aulus Gellius, Statius, &c. He died at Leyden in 1672, much regretted.

He married a gentlewoman of Daventer, who brought him two sons that survived him, and were both eminent in the republic of letters: James, who is the subject of the ensuing article; and Theodore Laurent, who died young, having published "Emendationes Pandectarum, &c. Leyden, 1605," 8vo. and "A Vindication of the Marble Base of the Colossus erected in honour of Tiberius Cæsar. ibid. 1697," folio.

GRONOVIUS (JAMES), son of the preceding, was born October 20, 1645 at Daventer, and learned the elements of the latin tongue there; but, going with the family in 1658 to Leyden, he carried on his studies in that university with incredible industry under the eye of his father, who had the greatest desire to make him a complete scholar. In this view he not only read to him the best classic authors, but instructed him in the civil law. About 1670, he made the tour of England, and visited both the universities consulting their MSS; where he formed an acquaintance with several great men, particularly with Dr. Edward Pocock, Dr. Pearson, and Dr. Meric Casaubon, which last died in his arms. He was much pleased with the institution of the Royal Society, and addressed a letter to them, expressly testifying his approbation of it. After some months stay in England, he returned to Leyden, where he published an edition of Macrobius that year in 8vo, and another of Polybius the same year at Amsterdam, in 2 vols. 8vo. The same year he was also offered the professorship of Hogersius; but, not having finished the plan of his tracts, he declined,

though

though the professor, to engage his acceptance, proposed to hold the place till his return.

He had apparently other views in his head: he had felt the advantage of his visit to England, and he resolved to see France. In his tour thither, he passed through the cities of Brabant and Flanders; and arriving at Paris, was received with all the respect due to his father's reputation and his own merit, which presently brought him into the acquaintance of Chaplain, d'Herbelot, Thevenot, and several other persons of distinguished learning. This satisfaction was somewhat damped by the news of his father's death: soon after which he left Paris to attend Mr. Pointz, ambassador extraordinary from the States-general to the court of Spain. They set out in the spring of 1672; and our author went thence into Italy, where, visiting Tuscany, he was entertained with extraordinary politeness by the Great Duke, who, among other marks of esteem, gave him a very considerable stipend, and the professor's place of Pisa, vacant by the death of Chimantel. This nomination was the more honourable, both as he had the famous Henry Norris, afterwards a cardinal, for his colleague; and as he obtained it by the recommendation of Magliabecchi, whom he frequently visited at Florence, which gave him an opportunity of consulting the MSS. in the Medicean library.

Having finished his designs in Tuscany, he quitted his professorship; and visiting Venice and Padua, he passed through Germany to Leyden, whence he went to take possession of an estate left him by his mother's brother, at Daventer. Here he sat down closely to his books, and was employed in preparing an edition of Livy in 1679, when he was nominated to a professor's place at Leyden, which he accepted; and by his inaugural speech obtained an augmentation to the salary of 400 florins a year, which was continued to his death. He was particularly pleased with the honour shewn to his merit; and Leyden being the city most affected by him, as the place of his education and his father's residence, he fixed here as at home, and resolved never to leave it for the sake of any other preferment. In this view he refused the chair of the celebrated Octavio Ferrari at Padua, and declined an invitation made him by Frederic duke of Sleswick to accept a considerable stipend for a lecture at Kell, in Holstein. This post was offered him in 1696, and two years afterwards the venetian ambassador at the Hague made him larger offers to engage him to settle at Padua; but he withstood all attempts to draw him from Leyden, as his father had done before him; and, to engage him firmer to them, the curators of that university gave him the lecture of geography, with the same augmentation to the stipend as had been given to his predecessor Philip Cluver.

He

He was revising Tacitus in order to a new edition, when he lost his youngest daughter: this happened September 12, 1716, and he survived her not many weeks. The loss proved insupportable; he fell sick a few days after it, and died of grief, October 21, aged seventy-one. He left two sons, both bred to letters; the eldest being a doctor of physic, and the youngest, Abraham, professor of history at Utrecht. It is remarked of James Gronovius, that he fell short of his father, in respect of modesty and moderation, as he exceeded him in literature: in his disputes, he treated his antagonists with such a bitterness of style as procured him the name of the second Scioppius. The justness of this censure appears throughout his numerous works, which indeed are too many to give their titles a place here. It is sufficient to observe, that most of the variorum editions of the classics are owing to him and Grævius; in emulation also of whom, he published, which is his *chef d'œuvre*, " Thesaurus Antiquitatum Græcarum," 13 vols.

GROPPER (JOHN), born in Westphalia, an able polemic. He was remarkably well skilled in the history and discipline of the church of Rome, and had the honour of refusing a cardinal's cap, offered him by Paul IV. He published " Enchiridion Christianæ Religionis," which is thought an excellent abridgement of " Dogmatic Theology." Some ridiculous things are related of his abhorrence of women. He saw a maid-servant making his bed, at which he was so exasperated, that he severely reprehended the woman, and threw the bed into the street. He died at Rome, in 1559.

GROSE (PETER), an eminent sculptor, born at Paris, 1666. He contributed greatly to the ornament of Paris by ingenious performances, by many excellent models and original designs.

GROSE (NICHOLAS), a celebrated theologian of Rheims, where he was born, in 1675. From his opposition to the Bull Unigenitus he was obliged to become an exile, and among other places visited England. He wrote a great number of books, chiefly on temporary subjects. He was the principal support of the Jansenist church in Holland.

GROSE (FRANCIS), an eminent antiquary and ingenious and entertaining writer. He illustrated the "Antiquity of England and Wales," in four volumes, and those of Scotland in two; and was pursuing the same design, with respect to Ireland, when he was cut off by death in the year 1791, at Dublin, aged fifty-two. He wrote also a great number of works, among the principal of which are, " Military Antiquities respecting a History of the English Array," a " Treatise on Ancient Armour," a " Classical Dictionary of the Vulgar Tongue," a " Provincial Dictionary with various Dissertations" in the " Archæologia." Mr. Grose had an excellent talent

lent for drawing; and was of a very agreeable and communicative difpofition. After his death the following epitaph was inferted in the St. James's Chronicle:

" Here lies Francis Grofe.
On Thurfday, May 12, 1791,
Death put an end to
His *Views* and *Profpects.*"

GROSLEY (PETER JOHN), member of the Academy of Infcriptions, and Belles Lettres, at Paris, was born at Troyes in 1718. He appears not only to have been refpectable as a fcholar but very amiable as a man. His principal work is entituled, " Recherches pour l'Hiftoire du Droit François," a book full of erudition and found argument. He alfo wrote "The Lives of the Two Pithous," and " Obfervations of two Swedifh Gentlemen on Italy," and " An Account of his Travels in England." The French " Encyclopedia" was much indebted to this author, as were alfo the compilers of the " Dictionnaire Hiftorique." He died at his native city, in 1785.

GROSSETESTE (ROBERT), bifhop of Lincoln, and a man of great learning and endowments. He was probably, as his hiftorian Dr. Pegge informs us, born about 1175.

He was of obfcure birth, and where and how he received his education is uncertain; but we know that he completed his ftudies at Oxford. At a time when Greek was hardly known in this country, he became, by application, a proficient in that language. From Oxford he went to Paris, which feems at this period to have been as cuftomary, among fuch of our countrymen as defired improvement, as it was formerly for the gentlemen of Rome to go to Athens. From Paris he returned to Oxford, where he read lectures on philofophy and divinity with great applaufe. His firft preferment was given him by Hugh de Welles, bifhop of Lincoln. He was foon afterwards made archdeacon of Chefter, which was the more honourable to him, as this ftation was always filled by great and learned men; this, however, he exchanged for the archdeaconry of Wiltfhire. To be minute in his progrefs towards other and higher preferments would be ufelefs; he was, in 1234-5, elected by the chapter of Lincoln to be bifhop of that fee, which choice was readily confirmed by the king. Matthew Paris, who was not in many inftances at all favourable to our bifhop, does not refufe him the higheft encomiums with regard to his learning, his integrity, and piety. He died at Buckden, in 1253. An accurate account of his works may be found in bifhop Tanner's " Bibliotheque," from which it appears that he was a moft voluminous writer, both on fubjects of divinity, philofophy, &c. He left alfo fome " Commentaries on Ariftotle," and was author of fome some tranſlations from the Greek. He was a man of ſtrong and clear intellect, but his ſtyle correſponds with the language of the time, and is turgid, verboſe, and inharmonious:

GROSTESTE (CLAUDE), a French refugee clergyman. He came to London, in 1685, after the revoking of the Edict of Nantz. He was miniſter of the Savoy, and was remarkable for his learning, his piety, and his benevolence. He wrote a treatiſe on the inſpiration of the ſacred books, and numbers of Sermons. He died in 1713.

GROTIUS (HUGO), or HUGO DE GROOT, was deſcended from a family of the greateſt diſtinction in the Low Countries: his father, John de Groot, was burgomaſter of Delft, and curator of the univerſity of Leyden. He was born at Delft on Eaſter-day, April 10, 1583, and came into the world with the moſt happy diſpoſitions; a profound genius, a ſolid judgement, and a wonderful memory. Theſe extraordinary natural endowments had all the advantages that education could give them: he was ſo happy as to find in his own father a pious and an able governor, who formed his mind and his morals. He was ſcarce paſt his childhood, when he was ſent to the Hague, and boarded with Mr. Utengobard, a celebrated clergyman among the Arminians, who took great care of his truſt; and, before he had completed his 12th year, was removed to Leyden, under the learned Francis Junius. He continued three years at this univerſity, where Joſeph Scaliger was ſo ſtruck with his prodigious capacity, that he condeſcended to direct his ſtudies; in 1597, he maintained public theſes in the mathematics, philoſophy, and law, with the higheſt applauſe.

At this early age he ventured to form plans which required very great learning; and he executed them with ſuch perfection, that the republic of letters were ſtruck with aſtoniſhment. But theſe were not publiſhed till after his return from France. He had a ſtrong inclination to ſee that country, and an opportunity offered at this time of gratifying it. The States-General came to a reſolution of ſending, on an embaſſage to Henry IV. in 1598, count Juſtin of Naſſau, and the grand penſioner Barnevelt: and Grotius put himſelf into the train of thoſe embaſſadors, for the latter of whom he had a particular eſteem. The learned youth was advantageouſly known in France before. M. de Buzanval, who had been ambaſſador in Holland, introduced him to the king, who preſented him with his picture and a gold chain. After almoſt a year's ſtay in France he returned home, much pleaſed with his journey; only one thing was wanting to complete his ſatisfaction, a ſight of the celebrated M. de Thou, or Thuanus, the perſon among all the French whom he moſt eſteemed.

esteemed. He had sought an acquaintance with that great man, but did not succeed: he resolved to repair this ill luck by opening a literary correspondence, and presenting him with the first-fruits of his studies in print, which he had just dedicated to the prince of Condé. This was his edition of "Martianus Capella." He had formed the plan of this work, and almost finished it, before he left Holland; and he published it presently after his return in 1569. M. de Thou was extremely well pleased with this address, and from this time to his death there subsisted an intimate correspondence between them.

Grotius, having chosen the law for his profession, had taken an opportunity before he left France of procuring a doctor's degree in that faculty; and upon his return he attended the law-courts, and pleaded his first cause at Delft with universal applause, though he was scarcely seventeen; and he maintained the same reputation as long as he continued at the bar. This employment, however, did not fill up his whole time; on the contrary, he found leisure to publish the same year, 1599, another work, which discovered as much knowledge of the abstract sciences in particular as the former did of his learning in general. Stevin, a mathematician to prince Maurice of Nassau, composed a small treatise for the instruction of pilots in finding a ship's place at sea; in which he drew up a table of the variations of the needle, according to the observations of Plancius, a famous geographer, and added directions how to use it. Grotius translated into Latin this work, which prince Maurice had recommended to the college of admiralty, to be studied by all officers of the navy; and, because it might be equally useful to Venice, he dedicated his translation to that republic. In 1600, he published his " Phænomena of Aratus." This book discovers a great knowledge in physics, and especially, astronomy. The corrections he made in the Greek are very judicious: the notes shew that he had reviewed several of the rabbies, and had some insight into the Arabic tongue; and the verses made to supply those of Cicero that were lost have been thought equal to them. In the midst of these profound studies, this prodigy of a young man found time to cultivate the Muses, and with such success, that he was esteemed one of the best poets in Europe. The prosopopœia, in which he makes the city of Ostend speak, after having been three years besieged by the Spaniards, is reckoned a masterpiece. It was translated into French by Du Vaër, Rapin, Pasquier, and Malherbe; and Casaubon turned it into Greek. Neither did our youth content himself with writing small pieces of verse; he rose to tragedy. We have three written by him;

the firſt, called "Adamus Exul," was printed in Leyden, in 1601. He was indeed diſſatisfied with this performance, and would not let it appear in the collection of his poems publiſhed by his brother. "Chriſtus patiens" was his ſecond tragedy; it was printed at Leyden, 1608, and much approved: Caſaubon greatly admires its poetical fire. Sandys tranſlated it into Engliſh verſe, and dedicated it to Charles I. It was favourably received in England, and in Germany propoſed as the model of perfect tragedy. His third was the ſtory of Joſeph, and its title "Sophomphanœus," which, in the language of Egypt, ſignifies the Saviour of the World; he finiſhed this in 1633, and the following year, at Hamborough.

In 1603, the glory which the United Provinces had obtained by their illuſtrious defence againſt the whole power of Spain, after the peace of Vervins, determined them to tranſmit to poſterity the ſignal exploits of that memorable war; and for this purpoſe they ſought out a proper hiſtorian. Several made great intereſt for the place; among others Baudius, the profeſſor of eloquence at Leyden. But the States thought young Grotius, who had taken no ſteps to obtain it, deſerved the preference; and, what is ſtill more ſingular, Baudius himſelf did not blame their choice, becauſe he looked upon Grotius to be already a very great man. All this while his principal employment was that of an advocate, and he acquired great honour therein. However, upon the whole, the profeſſion did not pleaſe him, though the brilliant figure he made at the bar procured him the place of advocate-general of the fiſc for Holland and Zealand, which, becoming vacant, was immediately conferred on him by thoſe provinces. He took poſſeſſion of this important office in 1607, and filled it with ſo much reputation, that the States augmented his ſalary, and promiſed him a ſeat in the court of Holland. Upon this promotion, his father began to think of a wife for him, and fixed upon Mary Reigeſberg, a lady of great family in Zealand, whoſe father had been burgomaſter of Veer. The marriage was ſolemnized in July, 1608. At the time of his marriage he was employed in writing his "Mare liberum, i. e. the Freedom of the Ocean, or the Right of the Dutch to trade to the Indies." The work was printed, in 1609, without his knowledge, and publiſhed without his conſent. Indeed he appears not to have been quite ſatisfied with it: and though there came out ſeveral anſwers, particularly that of Selden, intituled, "Mare clauſum, ſeu de dominio maris," yet, being ſoon after diſguſted with his country, he took no farther concern in the controverſy. The enſuing year, he publiſhed his piece, "De antiquitate Reipublicæ Batavæ." His deſign is, to ſhew the original independence of Holland and Frieſland againſt the Spaniſh claim;

he

he dedicated it to thofe States, March 16, 1610. They were extremely pleafed with it, returned thanks to the author, and made him a prefent.

Elias Olden Barnevelt, penfionary of Rotterdam, and brother to the grand penfionary of Holland, dying in 1613, the city of Rotterdam offered that important place to Grotius; but it was fome time before he yielded to the offer. By the ferment of men's minds he forefaw that great commotions would fpeedily fhake the republic; this made him infift, that he fhould never be turned out; and, upon a promife of this, he accepted of the poft, which gave him a feat in the affembly of the States of Holland, and afterwards in that of the States-General. Hitherto he had but very little connexion with the grand penfionary Barnevelt; but from this time he contracted an intimate friendfhip with him, infomuch that it was reported that Barnevelt defigned to have his friend fucceed him as grand penfionary of Holland.

At this time a difpute arofe between the Englifh and the Dutch, concerning the right of fifhing in the Northern feas. Two Amfterdam veffels, having caught 22 whales in the Greenland ocean, were met by fome Englifh fhips bound to Ruffia; who, finding that the Dutch had no paffports from the king of England, demanded the whales, which the Dutchmen, unable to refift, were obliged to deliver. On their arrival in Holland, they made their complaint; and the affair being laid before the States, it was refolved that Grotius, who had written on the fubject, and was more mafter of it than any one, fhould be fent to England to demand juftice: but he could obtain no fatisfaction. Hereupon the Dutch determined not to fend to Greenland for the future without a force fufficient to revenge themfelves on the Englifh, or at leaft to have nothing to fear from them. The difpute growing ferious, to prevent any acts of hoftility, a conference was held, in 1615, between the commiffioners of England and Holland, in which the debate turned chiefly on the whale-fifhery: but, the Englifh infifting on the right to Greenland, which the Dutch refufed, the conference broke up without any fuccefs. Grotius, who was one of the commiffioners from Holland, gives the hiftory of this conference, in a letter to Du Maurier, dated at Rotterdam, June 5, 1615. However, he had reafon to be well fatisfied with the politenefs of king James, who gave him a gracious reception, and was charmed with his converfation. But the greateft pleafure he received in this voyage was the intimate friendfhip he contracted with Cafaubon.

Hitherto this great man went on fmoothly in the paths of glory without any material interruption; but fortune had now refolved resolved to put his virtue to the trial. The United Provinces had been kindled into a warm dispute about grace and predestination, from the year 1608, when Arminius first broached his opinions. His doctrines, being directly opposite to those of Calvin, gave great offence to that party, at the head of which appeared Gomar, who accused his antagonist before the synod of Rotterdam. Gomar's party prevailing there, Arminius applied to the States of Holland, who promised the disputants to have the affair speedily discussed in a synod. The dispute still continuing with much bitterness, in 1611, the States ordered a conference to be held between twelve ministers on each side: but the consequence of this, like that of most other disputes, especially in matters of religion, was, that men's minds were the more inflamed. Arminius died October 19, 1609, some time before this conference; and Grotius made his eulogium in verse. He had hitherto applied little to these matters, and ingenuously owns he did not understand a great part of them, being foreign to his profession; but, upon a farther enquiry, he embraced the Arminian doctrine. In 1610, the partisans of Arminius drew up a remonstrance, setting forth their belief; first negatively against their adversaries, and then positively their own sentiments, each comprehended in six articles. This remonstrance was drawn up by Utengobard, minister at the Hague, and was probably made in concert with Grotius, the intimate friend of that minister. To this the Gomarists opposed a contra-remonstrance: the former proposed a toleration, the latter a national synod; and, the disputes increasing, the States, at the motion of the grand pensionary, in the view of putting an end to them, revived an obsolete law made in 1591, placing the appointment of ministers in the civil magistrates. But this was so far from answering the purpose, that the Contra-remonstrants resolved not to obey it. Hence grew a schism, which occasioned a sedition, and many riots.

It was at this time that Grotius was nominated pensionary at Rotterdam, as mentioned above; and ordered to go to England, with secret instructions, as is thought, to get the king and principal divines of that kingdom to favour the Arminians, and approve the conduct of the States. He had several conferences with king James on that subject. On his return to Holland, he found the divisions increased: Barnevelt and he had the direction of the States' proceedings in this matter; and he was appointed to draw up an edict which might restore tranquillity. He did so, and the draught was approved by the States; but it was so favourable to the Arminians that it gave great offence to the Contra-remonstrants, who determined to pay no regard to it. Hence this
edict

edict serving to increase the troubles, by driving the Gomarists to despair, the grand pensionary Barnevelt, in hourly expectation of fresh riots, proposed to the States of Holland, that their magistrates should be empowered to raise troops for the suppression of the rioters, and the security of their towns. Dort, Amsterdam, and three others of the most favourable to the Gomarists, protested against this step, which they regarded as a declaration of war against the Contra-remonstrants. Barnevelt's motion however was agreed to, and August 4, 1617, the States issued a placart accordingly. This fatal decree occasioned the death of the grand pensionary, and the ruin of Grotius, by incensing prince Maurice of Nassau against them, who looked upon the resolution of the States, taken without his consent, to be derogatory to his dignity, as governor and captain-general.

Amsterdam, almost as powerful singly as all Holland, favoured the Gomarists, and disapproved the toleration which the States wanted to introduce. These resolved therefore to send a deputation to that city, in order to reconcile them to their sentiments. Grotius was one of these deputies: they received their instructions April 21, 1616; and, arriving at Amsterdam next day, met the town-council on the 23d, when Grotius was their spokesman. But neither his speech nor all his other endeavours could avail any thing. The burgomasters declared their opinion for a synod, and that they could not receive the cachet of 1614, without endangering the church, and risquing the ruin of their trade. The deputies wanted to answer, but were not allowed. Grotius presented to the States on his return an account in writing of all that had passed at this deputation, and he flattered himself for some time with the hopes of some good effects from it; the disappointment whereof chagrined him so much, that he was seized with a violent fever, which had well nigh carried him off. He was removed to Delft, where he found himself better; but, being forbid to do any thing which required application, he wrote to Vossius, desiring his company, as the best restorative of his health. The time of his recovery he employed in examining the part he had acted in the present disputes; and, the more he reflected on it, the less reason he had for blushing or repentance: he foresaw the danger he incurred, but his resolution was, not to change his conduct, and to refer the event to Providence. The States of Holland, wholly employed in seeking ways to compound matters, came to a resolution, February 21, 1617, to make a rule or formula, to which both parties should be obliged to conform; and such an instrument was accordingly drawn up at their request by Grotius, who presented it to prince Maurice. But the project did not pleafe him; he wanted a national fynod, which was at length determined by the States-General, and to be convoked in Holland at Dort. In the mean time the prince, who faw with the utmoft difpleafure feveral cities, agreeably to the permiffion given them by the particular States, levy a new militia, under the title of Attendant Soldiers, without his confent, engaged the States-General to write to the provinces and magiftrates of thofe cities, enjoining them to difband the new levies. This injunction not being complied with, he confidered the refufal as a rebellion; concerted with the States-General, that he fhould march in perfon with the troops under his command, to get the attendant foldiers difbanded, depofe the Arminian magiftrates, and turn out the minifters of their party. He accordingly fet out, accompanied by the deputies of the States-General, in 1518; and, having reduced the province of Gueldres, he was proceeding to Utrecht, when the States of Holland fent thither Grotius, with Hoogarbetz, penfionary of Leyden, to put that city into a pofture of defence againft him. But, their endeavours proving ineffectual, the prince reduced the place; and foon afterwards fent Grotius and Hoogarbetz to prifon in the caftle at the Hague, where Barnevelt alfo was confined, Auguft 29th this year. After this, the States of Holland confented to the national fynod, which was opened at Dort, Nov. 15, 1618, which, as is well known, ended in a fentence, condemning the five articles of the Arminians, and in imprifoning and banifhing their minifters. This fentence was approved by the States-General, July 2, 1619.

After the rifing of that fynod, our three prifoners were brought in order to their trial, the iffue whereof was the execution of Barnevelt, May 13, 1619. Five days after came on the trial of Grotius. He had been treated, as well as his fellow-prifoner, with inconceivable rigour during their imprifonment, and alfo while their caufe was depending. He tells us himfelf, that, when they were known to be ill, it was concerted to examine them; that they had not liberty to defend themfelves; that they were threatened and teazed to give immediate anfwers; and not fuffered to have their examinations read over to them. Grotius, having afked leave to write his defence, was allowed only five hours, and one fheet of paper; he was alfo perfuaded that, if he would own he had tranfgreffed and afk pardon, he might obtain his liberty; but, as he had nothing to reproach himfelf with, he would never take any ftep that might infer confcioufnefs of guilt. His wife, his father, brother, and friends, all approved this refolution. His fentence, after reciting the feveral reafons thereof, concludes thus: " For thefe caufes, the judges, appointed

appointed to try this affair, adminiftering juftice in the name of the States-General, condemn the faid Hugo Grotius to perpetual imprifonment, and to be carried to the place appointed by the States-General, there to be guarded with all precaution, and confined the reft of his days; and declare his eftate confifcated. Hague, May 18, 1619." In purfuance of this fentence, he was carried from the Hague to the fortrefs of Louveftein near Gorcum in South Holland, June 6, 1619, and 24 fols per day affigned for his maintenance, and as much for Hoogarbetz; but their wives declared they had enough to fupport their hufbands, and that they chofe to be without an allowance, which was looked upon as an affront. Grotius's father afked leave to fee his fon, but was denied; they confented to admit his wife into Louveftein, but, if fhe came out, not to be fuffered to return. However, in the fequel, it was granted that fhe might go abroad twice a week.

Grotius now became more fenfible than ever of the advantage of ftudy; which became his bufinefs and confolation. December 5, 1619, he writes to Voffius, that the Mufes, which were always his delight, even when immerfed in bufinefs, were now his confolation, and appeared more amiable than ever. He wrote fome fhort notes on the New Teftament, which he intended to fend Erpenius, who was projecting a new edition of it; but a fit of illnefs did not fuffer him to finifh them. When he was able to refume his ftudies, he compofed, in Dutch verfe, his "Treatife of the Truth of the Chriftian Religion," and fent it to Voffius, who thought fome places obfcure. In 1620, he promifes his brother to fend him his obfervations on Seneca's tragedies; thefe he had written at Voffius's defire. In 1621, Du Maurier lofing his lady, Grotius writes him, February 27, a very handfome confolatory letter, in which he deduces with great eloquence every topic of fupport that philofophy and religion can fuggeft on that melancholy occafion. The only method he took to unbend himfelf, was to go from one work to another. He tranflated the "Pheniffæ of Euripides," wrote his "Inftitutions of the Laws of Holland in Dutch," and compofed fome fhort "Inftructions for his Daughter" Cornelia, in the form of a catechifm, &c.

He had been above 18 months fhut up at Louveftein, when, January 11, 1620, Muys-van-Halli, his declared enemy, who had been one of his judges, informed the States-General, that he had advice from good hands their prifoner was feeking to make his efcape. Some perfons were fent to examine into this matter; but, notwithftanding all the enquiry that could be made, they found no reafon to believe that he had laid any plot

plot to get out. His wife, however, was wholly employed in contriving it: he had been permitted to borrow books of his friends, and, when he had done with them, they were carried back in a chest with his foul linen, which was sent to Gorcum to be washed. The first year his guards were very exact in examining the chest; but, being used to find nothing in it besides books and linen, they grew tired of searching, and even did not take the trouble to open it. His wife, observing their negligence, proposed to take advantage of it. She represented to her husband, that it was in his power to get out of prison when he pleased, if he would put himself into this chest. However, not to endanger his health, she caused holes to be bored opposite where his face was to be, to breathe at; and persuaded him to try if he could continue shut up in that confined posture, as long as it would require to go from Louvestein to Gorcum. Finding it might be done, she resolved to seize the first favourable opportunity; which very soon offered. The commandant of Louvestein going to Heusden to raise recruits, she made a visit to his lady, and told her in conversation, that she was desirous of sending away a chest of books; for, her husband was so weak, that it gave her great uneasiness to see him study with such application. Having thus prepared the commandant's wife, she returned to her husband's apartment, and in concert with a valet and a maid who were in the secret, shut him up in the chest; and at the same time, that the people might not be surprised at not seeing him, she spread a report of his being ill. Two soldiers carried the chest; which was brought down, and put into the boat; and Grotius's maid, who was in the secret, had orders to go to Gorcum with it. There it was put on a horse, and carried by two chairman to David Dazelaer's, a friend of Grotius, and brother-in-law to Erpenius; and, when every body was gone, the maid opened the chest. Grotius had felt no inconvenience in it, though its length was not above three feet and a half. He got out, dressed himself like a mason with a rule and a trowel; and was secretly conveyed in this disguise to Valvic in Brabant. Here he made himself known to some Arminians, and hired a carriage to Antwerp; and, at Antwerp, he alighted at the house of Nicolas Grevincovius, who had been formerly a minister at Amsterdam, and made himself known to nobody else. It was March 22, 1621, that he thus received his liberty.

Mean while, his wife's account, that he was ill, gained credit at Louvestein; and, to give him time to get off, she gave out that his illness was dangerous: but as soon as she learnt by the maid's return that he was at Brabant, and consequently in safety, she told the guards the bird was flown. They informed the commandant, by this time returned from Heusden, who, finding it true, confined Grotius's wife more closely; but upon her petition to the States General, April 5. 1621, she was discharged two days after, and suffered to carry away every thing that belonged to her in Louvestein. From Antwerp, Grotius wrote to the States General, March 30, that, in procuring his liberty, he had employed neither violence nor corruption with his keepers; that he had nothing to reproach himself with in what he had done; that he gave those counsels which he thought best for appeasing the troubles that had arisen in public business; that he only obeyed the magistrates of Rotterdam his masters, and the States of Holland his sovereigns; and that the persecution he had suffered would never diminish his love for his country, for whose prosperity he heartily prayed. He continued some time at Antwerp, deliberating what course to take; and at length determined to go to France, where he had many friends. He arrived at Paris, April 13, 1621; his wife in October following: and, after some difficulties, obtained a pension of 3000 livres. But, notwithstanding the king's grant, he could not touch the money; they had forgot to put it on the civil list, and the commissioners of the treasury found daily some new excuse for delaying the payment. At length, however, by the solicitation of some powerful friends, he received it; but it continued to be paid as grants were paid at that time, that is to say, very slowly.

Having collected some materials in prison for his Apology, he printed it in the beginning of 1622; it was translated into latin, and published the same year at Paris. It was sent to Holland immediately, where it caused so much disgust, that the States-General proscribed it as slanderous, tending to a perse by falshoods the sovereign authority of the government of the United Provinces; the person of the prince of Orange, the States of the particular provinces, and the towns themselves; and forbad all persons to have it in their custody on pain of death. Grotius presented a petition to the king of France, to be protected against this edict, which imported, that he should be apprehended wherever found; whereupon his majesty took him into his special protection, the letters for that purpose being issued at Paris, February 25, 1623. The malevolence of those who were thence in place made no change in Grotius. In the height of this new persecution, he wrote to his brother, that he would still labour to promote the interest of Holland; and that, if the United Provinces were desirous of entering into a closer union with France, he would assist them with all his credit. In reality, he still preserved many friends, who ardently wished for his return; though

P 3   they

they were not able in any wife to facilicate it. In 1623, he publifhed at Paris his edition of Stobæus.

He had now lived a year in the noife of Paris, and began to think of retiring into the country, when the prefident de Meme offered him one of his feats at Bologne, near Senlis. Grotius accepted the offer, and paffed there the fpring and fummer of the year 1623. In this caftle he began his great work, which fingly is fufficient to render his name immortal; I mean his "Treatife of the Rights of Peace and War." He had vifited the moft diftinguifhed men of learning; among others Salmafius and Rigault, and had the free ufe of de Thou's library: he fometimes alfo made excurfions to St. Germains, where the court was; but, having learned that de Meme wanted to refide himfelf at Bologne, he returned to Paris in October. April 23, 1625, prince Frederic Henry fucceeding to the poft of Stadtholder on the death of his brother Maurice, Grotius's friends conceived great hopes of obtaining leave for his return to Holland: and, at their requeft, he wrote to the new Stadtholder for this purpofe, but without effect; as he had before conjectured. However, he was now in the height of his glory by the prodigious fuccefs of his book, "De Jure Belli & Pacis," which was publifhed this year. In the mean time he began to grow tired of that city. His penfion was ill-paid, and his revenue infufficient to keep him decently with a wife and a family. He had an offer of being profeffor of law in a college at Denmark; but, though he was fatisfied with the falary, he thought the place beneath his acceptance.

His heart was ftrongly bent upon returning to his native country; and in thefe wifhes he fent his wife into Holland in the fpring of 1627, that fhe might enquire how matters ftood: but, as he continued in the refolution to make no folicitations for leave, all the endeavours of his friends were fruitlefs. However, they obtained a caufe of fome confequence to him. He reclaimed his effects which were confifcated, and his demand was granted. In fine, notwithftanding the inefficacy of his friend's folicitations, he refolved, by his wife's advice, to go thither; and accordingly fet out, October 1631. The fentence paffed againft him being ftill in force, his friends advifed the concealing of himfelf. This ftep appeared to him fhameful and ill-timed. He went to Rotterdam, as thinking it the fafeft, becaufe, having filled the place of penfionary with much honour, he was greatly beloved in the town; but the magiftrates giving him to underftand, that they did not approve his appearing in public, he left Rotterdam, and, paffing to Amfterdam, he was extremely well received there; and Delft alfo, where he was born, fhewed him a fincere refpect.

But

But no city ventured publicly to protect him; and the States-General, thinking themselves affronted by this boldness in continuing in the country without their leave, and by the repugnance he shewed to ask them pardon, issued an ordnance, December 10, 1631, enjoining all bailiffs of the country to seize his person, and give them notice: but nobody would execute it; and, to employ himself till his fate should be determined, he resolved to follow the business of a chamber-council. With this view he desired his brother, in a letter dated February 16, 1632, to send him what law books he had, such as he might want for that office. He could make no use of these books; for, the States General, on March 10, renewed their ordinance, upon pain to those, who would not obey, of losing their places, and with a promise of 2000 florins to any one who should deliver him into the hands of justice. Upon this he thought proper to seek his fortune elsewhere; and, March 17, he set out from Amsterdam on his way to Hamburgh, and passed the fine season at an agreeable seat called Okenhuse, near the Elbe, belonging to William Morth, a dutchman. On the approach of winter, he went to Hamburgh, and lodged with one Van Sorgen, a merchant: but the town did not prove agreeable to him, and he passed his time but heavily till the return of his wife from Zealand in autumn 1633. She had always been his consolation in adversity, and rendered his life more agreeable. Her business at Zealand was to pick up the remains of their fortune, which she probably brought with her to Hamburgh. While he continued here, some advantageous proposals were made him from Spain, Poland, Denmark, the duke of Holstein, and several other princes; but he still entertained the thought of a reconciliation with his native country. At length, however, he was determined.

He had always entertained a very high opinion of Gustavus king of Sweden; and that prince having sent to Paris Benedict Oxenstiern, a relation of the chancellor, to bring to a final conclusion the treaty between France and Sweden, this minister made acquaintance with Grotius, and resolved, if possible, to draw him to his master's court: and Grotius writes, that if that monarch would nominate him ambassador, with the proper salary for the decent support of the dignity, the proposal should merit his regard. In this situation Salvius, vice-chancellor of Sweden, a great statesman, and a man of learning, being then at this city, Grotius made acquaintance with him, and saw him frequently. Polite literature was the subject of their conversation. Salvius conceived a great esteem for Grotius, and the favourable report he made of him to the high-chancellor Oxenstiern determined the latter to write to

Grotius to come to him, that he might employ him in affairs of the greatest importance. Grotius accepted of this invitation; and setting out for Frankfort on the Maine, where that minister was, arrived there in May 1634. He was received with the greatest politeness by Oxenstiern, but without explaining his intentions. However, in confidence of the high-chancellor's character, he sent for his wife; and she arrived at Francfort with his daughters and son, in the beginning of August. The chancellor continued to heap civilities upon him, without mentioning a word of business; but ordered that he should follow him to Mentz, and at length declared him counsellor to the queen of Sweden, and her ambassador to the court of France.

As soon as he could depend upon an establishment, he resolved to renounce his country, and to make it known by some public act, that he considered himself as no longer a dutchman. In this spirit he sent his brother letters for the prince of Orange and the dutch to that purport, July 13, this year: he likewise wrote to Rotterdam, which had deferred nominating a pensionary after the sentence passed against him, that they might proceed to an election, since they must no longer look upon him as a dutchman. He set out from Mentz on his embassy to France in the beginning of 1636, and always supported with great firmness the rights and honours belonging to the rank of an ambassador. He continued in that character in France till 1644, when he was recalled at his own request. In order to his return, having obtained a passport through Holland, he embarked at Dieppe, and arrived at Amsterdam in 1645, where he was extremely well received and entertained at the public expence. That city fitted out a vessel to carry him to Hamburgh, where he was, May 16, this year. He went next day to Lubeck, and thence to Wismar, where count Wrangle, admiral of the Swedish fleet, gave him a splendid entertainment, and afterwards sent a man of war with him to Calmer, whither the chancellor sent a gentleman with his coach to bring him to Suderacher. He continued there about a fortnight with the chancellor and other embassadors, who treated him with great honours. Returning to Calmer, he went by land to Stockholm, whither queen Christina came from Upsal to see him.

Her majesty had, before his departure from France, assured him that she was exremely satisfied with his services; and she now gave him several audiences, and made him dine with her, and he appeared to be abundantly pleased with the honours he received: but as he saw they were in no haste to do any thing for him, and only rewarded him with compliments, he grew uneasy, and asked leave to retire. He was confirmed

confirmed in this resolution, by finding the court filled up
with persons that had conceived a jealousy against him; besides,
the air of Sweden did not agree with him. The queen several times refused to grant him his dismission, and signified
that if he would continue in her service in quality of counsellor
of state, and bring his family into Sweden, he should have no
reason to repent it: but he excused himself on account of his
own health, who could not bear the cold air of that kingdom.
He asked a passport, which they delaying to grant, he grew so
uneasy, that he resolved to be gone without it. Leaving
Stockholm, therefore, he went to a sea-port two leagues
distant, in order to embark for Lubeck. The queen, being
informed of his departure, sent a gentleman to tell him she
wanted to see him once more, otherwise she should think he
was displeased with her. He returned therefore to Stockholm,
and explained himself to the queen, who seemed satisfied with
his reasons, and made him a large present in money; adding
to it some silver plate which was not finished sooner, and
which he was assured had delayed the granting of his passport.
That was afterwards issued; and the queen gave him a vessel,
on-board which he embarked, August 12, for Lubeck.

But the vessel was scarce sailed when a violent storm arose,
which obliged her after three days tossing to put in, August 17,
on the coast of Pomerania, fourteen miles from Dantzick.
Grotius set out in an open waggon for Lubeck, and arrived at
Rostock, August 26, very ill, having travelled about sixty miles
through wind and rain. He lodged with Balleman, and sent
for Stochman the physician, who, from the symptoms, judged
he could not live long. On the 28th he sent for Quistorpius,
minister of that town, who gives the following account of his
last moments: "You are desirous of hearing how that phœnix
of literature, Hugo Grotius, behaved in his last moments: I
am going to tell you." He then proceeds to give an account
of his voyage, and his sending for Stochman, a scotch physician, after which he goes on as follows: "he sent for me
about nine at night, I went, and found him almost at the point
of death. I said, ' There was nothing I desired more than to
have seen him in health, that I might have had the pleasure of
his conversation;' he said, ' God hath ordered it otherwise.'
I desired him 'to prepare himself for a happier life, to acknowledge he was a sinner, and repent of his faults;' and happening
to mention the publican, who acknowledged he was a sinner,
and asked God's mercy, he answered, ' I am that publican.' I
went on, and told him that ' he must have recourse to Jesus
Christ, without whom there is no salvation.' He replied, ' I
place my hope in Jesus Christ.'. I began to repeat aloud in
german

german the prayer that begins Herr Jefu[p]; he followed me in a very low voice with his hands clafped. When I had done, I afked him if he underftood me; he anfwered, 'I underftand you very well.' I continued to repeat to him thofe paffages of the word of God, which are commonly offered to the remembrance of dying perfons; and, afking if he underftood me, he anfwered me, 'I heard your voice, but did not underftanc what you faid.' Thefe were his laft words; foon after he expired, juft at midnight. His body was delivered to the phyficians, who took out his bowels, and eafily obtained leave to bury them in our principal church, dedicated to the Virgin Mary."

Thus died this extraordinary perfon, Auguft 28, at night, 1645. His corpfe was carried to Delft, and depofited in the tomb of his anceftors. He wrote this modeft epitaph for himfelf.

" Grotius hic Hugo eft Batavûm captivus & exul,
  Legatus regni, Suecia magna, tui."

Among his works thefe are the principal, firft, his "Anthologia." 2. " Via ad Pacem Ecclefiafticam." 3. " Hiftoria Gothorum, &c." 4. "Remarks on Juftinian's Laws." 5. "Commentary on the Old and New Teftament, with feveral Pieces annexed." 6 "Differtatio Hift. & Politic. de Dogmatis, Ritibus, & Gubernatione Ecclefiæ, &c." 7. "De Origine Gentium Americanarum, &c." with two anfwers to Dr. Laets in its defence. 8. " An Introduction to the Laws of Holland." 9. "Notes to Tacitus," publifhed in Lipfius's edition, 1640. 10. " Notes upon Lucian," publifhed in 1614. In 1652, there came out a fmall collection in 12mo, with this title, "Hugonis Grotii quædam inedita, aliaque ex Belgicè editis Latinè verfa argumenti theolog. jurid. politic." and in 1687, an edition of his " Epiftles."

GROTIUS (WILLIAM), a native of Delft, and a younger brother of Hugo Grotius, was an eminent lawyer, and wrote feveral books; in particular, " Enchiridion de Principis Juris Naturæ," printed at the Hague. He wrote alfo, and which were publifhed after his death "Vitas Juris confultorum quorum in Pandectis extant nomina." He died in 1662.

• GROTIUS (PETER), the fecond fon of Hugo Grotius, was eminent both for his knowledge as a lawyer and his acutenefs as a philologift. He died in 1678.

[p] It is a prayer addreffed to Jefus Chrift, and fuited to the condition of a dying perfon, who builds his hopes on the Mediator. Le Clerc has recited it at length, in Sentimens de quelq. Theolog. lett. xvii. p. 397.

GROVE (HENRY), a learned divine among the englifh prefbyterians, was defcended from the Groves of Wiltfhire, and the Rowes of Devonfhire. His grandfather Grove was ejected from a living in Devonfhire for nonconformity in 1662: his father fuffered much in the fame caufe for lay-nonconformity under Charles and James II. The eminent piety of Mr. Rowe, his grandfather by the mother's fide, may be known by the account of his life by Mr. Theophilus Gale. His father, in particular, filled a life of eighty years honourably and ufefully, and died univerfally efteemed and lamented. From fuch parents our author was born at Taunton in Somerfetfhire, January 4, 1683; and, at fourteeen years of age, being poffeffed with a fufficient ftock of claffical literature, he went through a courfe of academical learning under the Rev. Mr. Warren of Taunton, who was for many years at the head of a flourifhing academy. Having finifhed his courfe of philofophy and divinity under Mr. Warren, he removed to London; and ftudied fome time under the Rev. Mr Rowe, to whom he was nearly related. At this time he contracted a friendfhip with feveral perfons of merit, and particularly with Dr. Watts, which continued till his death, though they differed in their judgement upon feveral points warmly controverted among divines.

After two years fpent in London, he returned into the country; and, being now twenty-two years of age, began to preach with great reputation. The fpirit of devotion which prevailed in his fermons early procured the friendfhip of Mrs. Singer, afterwards Mrs. Rowe, which fhe expreffed in an "Ode on Death," addreffed to Mr. Grove. Soon after his beginning to preach, he married; and at the age of twenty-three, upon the death of his tutor, Mr. Warren, was chofen to fucceed him in the academy at Taunton. The province firft affigned him, was ethics and pneumatology; and he compofed fyftems in each. His concern in the academy obliging him to a refidence in Taunton, he preached for eighteen years to two fmall congregations in the neighbourhood. In 1708, he commenced author, by a piece intituled, "The Regulation of Diverfions," drawn up for the ufe of his pupils; and about the fame time, Dr. Samuel Clarke publifhed his "Difcourfe on the Being and Attributes of God;" and the proof therein from the neceffary ideas of fpace and duration not convincing our author, he wrote to the doctor for information and fatisfaction upon that head. This occafioned their exchanging feveral letters; when, not being able to convince each other, the debate was dropped with expreffions of great mutual efteem. The next offering he made to the public was feveral papers in the eighth volume of the "Spectator," viz.

No.

No. 588, 601, 626, 635. In 1718, he published "An Essay towards a Demonstration of the Soul's Immortality." About 1719, when those angry disputes upon the Trinity unhappily divided the presbyterians, and when the animosities were carried so high as to produce excommunications, &c. Mr. Grove's moderate conduct was such, as drew on him the censures and displeasure of some of his own persuasion: the reasons for this moderate conduct are mentioned in his "Essay on the Terms of Christian Communion."

In 1725, he lost his partner in the academy, the Rev. Mr. James; and was now obliged to take the students in divinity under his direction. He confined himself to no system in divinity, but directed his pupils to the best writers on natural and revealed religion, and an impartial consideration of the chief controversies therein. He likewise succeeded Mr. James in his pastoral charge at Fullwood near Taunton, in which he continued till his death. In 1730, he published, "The Evidence of our Saviour's Resurrection considered;" and, the same year, "Some Thoughts concerning the Proof of a future State from Reason," in answer to the Rev. Mr. Hallet, junior, which drew him into a dispute on the point with that divine. In this controversy, he was thought to disparage the necessity of revelation, in regard to that proof. In 1732, he printed "A Discourse concerning the Nature and Design of the Lord's Supper," where he set that institution in the same light with bishop Hoadly. In 1734, he published, without his name, "Wisdom the first Spring of Action in the Deity," which was animadverted on, as to some particulars, by Mr. Balguy, who, however, allowed the discourse in general to abound with solid remarks and sound reasonings. In 1736, he published "A Discourse on saving Faith." The same year he met with an affliction, which gave him an opportunity of shewing the strength of his christian patience and resignation; this was the death of his wife: and, a little more than a year after this, he died himself: for, having preached on February 19, 1737-8, and with such an uncommon flow of spirits as he said he could hardly govern he was violently seized at night with a fever, which carried him off upon the 27th. His friends erected a handsome monument over his grave, on which is a latin inscription composed by the late Dr. Ward, rhetoric-professor at Gresham-college, who hath also obliged the world with an english version of it. Besides the works already mentioned, he published many sermons upon several occasions, and also a volume of "Miscellanies in Prose and Verse." After his death came out by subscription his "Posthumous Works, 1740," in 4 vols. 8vo.

GRUCHIUS

GRUCHIUS (NICOLAS) of a noble family of Rouen, was, as the compilers of the Dictionnaire Historique affirm, "le Premier qui expliqua Aristote en Grec." He was author of various works. He translated Castanedo's "History of the Indies," and he published a treatise "De Comitiis Romanorum."

GRUDIUS (NICOLAS EVERARD), treasurer of Brabant, wrote poetry, sacred and prophane, in latin. He died in 1571.

GRUE (THOMAS), a frenchman, celebrated for his various translations of english works into french. Among others he published Ross's "History of all Religions," and Abraham Roger's "Gate opened to the Knowledge of Paganism."

GRUGET (CLAUDE), lived in the sixteenth century. He was famous for his translations from italian and spanish into french, in particular an edition of the "Heptameron of the Queen of Navarre."

GRUNER (JOHN FREDERIC), an eminent theologian and excellent scholar; was author of many useful and important works, a catalogue of which is given by Harles in his book "De vitis Philologorum." His talents are represented to have been very various, and his diligence indefatigable. He published a new edition of "Cælius Sedulius," with various commentaries, "An Introduction to Roman Antiquities," "Miscellanea Sacra," "Various critical Remarks on the Classics," new editions of "Eutropius and Velleius Paterculus, &c." He was born at Coburg in 1723, and died in 1778.

GRUTERUS (JANUS), a celebrated philologer, was born December 3, 1560, at Antwerp in Brabant. He was the son of John Walter Gruter, burgo-master of Antwerp; who, being one of those who signed the famous petition to the duchess of Parma, the governess of the Netherlands, which gave rise to the word *Gueux* [Beggars], was proscribed his country. He crossed the sea to Norwich in England taking his wife (who was an english woman) and family along with him. Young Gruter was then but an infant; he had the peculiar felicity of imbibing the elements of learning from his mother, Catharine Tishem; who, besides french, italian, and english, was complete mistress of latin, and so well skilled in greek that she could read Galen in the original; which, Bayle says, is more than one physician in a thousand can do. The family, being persecuted on account of the protestant religion, found an asylum in England, where they resided several years, and at a proper age sent their son to complete his education at Cambridge. His parents, after some time, repassing the sea to Middleburg, the son followed them to Holland; and, going to Leyden, studied the civil law, and
took

took his doctor's degree there in that faculty; but, applying himself at the same time to polite literature, he became an early author in that way, as appears by some latin verses which he published, under the title of "Ocelli," at twenty years of age.

After taking his degree, he went to Antwerp, to his father, who had returned thither as soon as the States had possessed themselves of it; but, when the city was threatened with a siege by the duke of Parma in 1584, was sent to France, where he resided some years, and then visited other countries. The particular route and circumstances of his travels afterwards are not known; only it is certain that he read public lectures upon the classics at Rostock, particularly on Suetonius. He was in Prussia, when Christian, duke of Saxony, offered him the chair of history-professor in the university of Wittemburg; which place he enjoyed but a few months: for, upon the death of that prince, his successors desiring the professors to subscribe the act of concord on pain of forfeiting their places, Gruterus chose rather to resign than subscribe any confession of faith against his conscience. He was treated with particular severity on this occasion; for, though two others were deprived on the same account, yet half a year's salary was allowed them by way of gratification, according to the custom of those countries, with regard to persons honourably discharged: whereas this present was so far from being made to Gruterus, that they did not defray even the expences of his journey. The truth is, he was the worst courtier in the world; and he judged that, all things considered, it would be more advantageous to him to give up all thoughts of that present than to trifle away his time in tedious solicitations. We do not know whither he directed his steps next; only we are told, that, being at Padua at the time of Riccoboni's death, that professor's place was offered to him, together with liberty of conscience: the salary too was very considerable, but he refused all these advantages. He was apprehensive that so profitable and honourable employ would expose him to the attacks of envy, and he would not submit to the bare exercise of his religion in private. He was much better pleased with his invitation to Heidelberg, where he filled the professor's chair with great reputation for many years; and, in 1502, had the direction of that famous library, which was afterwards carried to Rome.

This employ suited his genius, and soon after he published the most useful of his works, his large collection of inscriptions, which is dedicated to the emperor Rodolphus II. who bestowed great encomiums upon it, and gave Gruterus the choice,

choice of his own reward. He anfwered that he would leave it to the emperor's wifdom, only begged it might not be pecuniary. In the fame temper, upon hearing there was a defign to give a coat of arms, in order to raife the dignity of his extraction, he declared, that, fo far from deferving a new coat of arms, he was too much burthened with thofe which had devolved to him from his anceftors. The emperor was then defired to grant him a general licence for all the books of his own publifhing. The emperor not only confented to it, but alfo granted him a privilege of licenfing others. The emperor intended to create him a count of the facred palace; and the affair was carried fo far, that the patent was drawn, and brought back to be ratified by his fign manual; but, the emperor happening to die in the interim, it was left without the fignature, and fo the affair came to nothing. Neverthelefs Gruterus beftowed the fame encomiums on the good emperor as if it had been completed. His privilege, however, of licenfing books was of great advantage to him, fince he publifhed a vaft number, being one of the moft laborious writers of his age. This tafk he was the better enabled to execute by the help of his library, which was large and curious, having coft him no lefs than 1200 crowns in gold. Imagine, then, how deep his affliction muft be, when it was deftroyed and plundered, together with the city of Heidelberg, in 1622. Ofwald Smendius, his fon-in-law, endeavoured to fave it, but in vain. For this purpofe, he wrote to one of the great officers of the duke of Bavaria's troops; but the wild licentioufnefs of the foldiers could not be reftrained. Afterwards he went to Heidelberg, and faw the havock that had been made at his father's houfe; he then tried to fave at leaft what Gruterus's amanuenfis had lodged in the elector's library, and brought the Pope's commiffion to give him leave to remove them. He received for anfwer, that as to the MSS. the Pope had ordered them all to be fought for carefully, and carried to Rome; but as to the printed books, leave would be given to reftore them to Gruterus, provided it was approved by Tilly under his hand. However, this pretended favour proved of no effect, becaufe Tilly could not be fpoken with.

Gruterus had left Heidelberg before it was taken, and retired to his fon-in-law's at Bretten, whence he went to Tubingen, where he ftayed fome time. He made feveral removes afterwards, and received invitations to read lectures at feveral places, particularly one from Denmark. The curators alfo of the univerfity of Franecker offered him the profefforfhip of hiftory in 1624; but, when the affairs of the Palatinate were a little fettled, he returned to Bretten; where, however,

ever, he found himself very much teazed by some young jesuits, who were fond of disputing. In reality, Gruterus never loved controversy, especially upon religious subjects. Nor indeed was it the business of a critic of his fame to dispute about controverted points with young jesuits just fresh plumed with the subtleties of the schools; and he found no other way of getting rid of their importunities than to go and live at a distance from them. He retired therefore to a country-house, which he purchased near Heidelberg, where he used to make visits occasionally. He came from one of these, September 10, 1527; and going to Bernhelden, a country-seat belonging to his son-in-law 'mendius, about a league's distance from Heidelberg, he fell sick the same day, and expired there ten days afterwards. His corpse was carried to Heidelberg, and interred in St. Peter's church.

GRUTERUS (PETER), was a practitioner of physic in several parts of Flanders. In 1609 he published at Leyden a "Century of Latin Letters," in which he affected old words and obsolete phrases. In 1629, he published a "New Century of Letters," at Amsterdam, at which place he died in 1634.

GRYNÆUS (SIMON), a very learned german, was the son of a peasant of Suabia, and born at Veringen in the county of Hohenzollern in 1493. He pursued his studies in Pfortsheim at the same time with Melancthon, which gave rise to a friendship between them which lasted long. He continued them at Vienna, and there taking the degree of master in philosophy, was appointed greek professor. Having embraced the protestant religion, he was exposed to many dangers; and particularly in Baden, where he was some years rector of the school. He was thrown into prison at the instigation of the friers; but at the solicitation of the nobles of Hungary, was set at liberty, and retired to Wittemberg, where he had a conference with Luther and Melancthon. Being returned to his native country, he was invited to Heidelberg, to be greek professor in that city, in 1523. He exercised this employment till 1529, when he was invited to Basil to teach publicly in that city. In 1531, he took a journey into England, and carried with him a recommendatory letter from Erasmus to William Montjoy, dated Friburg, March 18, 1531. After desiring Montjoy to assist Grynæus as much as he could, in shewing him libraries, and introducing him to learned men, Erasmus adds: "Est homo Latinè Græcèquè ad unguem doctus, in philosophia & mathematicis disciplinis diligenter versatus, nullo supercilio, pudore pene immodico. Pertraxit hominem istuc Britanniæ visendæ cupiditas, sed præcipue Bibliothecarum vestrarum amor. Rediturus est ad nos, &c." Erasmus recommended him also to Sir Thomas More, from whom he received

ceived the highest civilities. In 1534, he was employed, in conjunction with other persons, to reform the church and school of Tubingen. He returned to Basil in 1536, and in 1540 was appointed to go to the conferences of Worms, with Melancthon, Capito, Bucer, Calvin, &c   He died of the plague at Basil in 1541.

He did great service to the commonwealth of learning, and we are obliged to him for editions of several ancient authors. He was the first who published the "Almagest" of Ptolemy in greek, which he did at Basil in 1538, and added a preface concerning the use of that author's doctrine. He also published a greek "Euclid," with a preface, in 1533, and Plato's works with some commentaries of Proclus, in 1534. His edition of Plato was addressed to John More, the chancellor's son; as a testimony of gratitude for favours received from the father; and as the following passage in the dedication shews Sir Thomas, as well as Grynæus, in a very amiable light, we think it not amiss to insert it here. "It is, you know, three years, since arriving in England, and being recommended most auspiciously by my friend Erasmus to your house, the sacred seat of the Muses, I was there received with great kindness, was entertained with greater, was dismissed with the greatest of all.   For that great and excellent man your father, so eminent for his high rank and noble talents, not only allowed me, a private and obscure person, (such was his love of literature) the honour of conversing with him in the midst of many public and private affairs, gave me a place at his table, though he was the greatest man in England, took me with him when he went to court or returned from it, and had me ever by his side, but also with the utmost gentleness and candour enquired, in what particulars my religious principles were different from his; and though he found them to vary greatly, yet he was so kind as to assist me in every respect, and even to defray all my expences.   He likewise sent me to Oxford with one Mr. Harris, a learned young gentleman, and recommended me so powerfully to the university, that at the sight of his letters all the libraries were open to me, and I was admitted to the most intimate familiarity with the students."

GRYNÆUS (Thomas), nephew of Simon, was born at Syringen, in Suabia, in 1512. He pursued his studies under the auspices of his uncle, and taught the latin and greek languages at Berne. He also read public lectures at Basil, and was a respectable and amiable character. He left four sons, all of whom were eminent for their learning.

GRYPHIARDER (John), was professor of poetry and history in the university of Jena. He died in 1612, and was author of several books.

GRYPHIUS (SEBASTIAN), a celebrated printer of Lyons in France, was a german, and born at Suabia near Augſburg in 1493. He performed the duties of his profeſſion with ſo much honour, that he was publicly applauded for it by very learned men. Conradus Geſner has even dedicated one of his books, namely, the twelfth of his pandects, to him; and takes occaſion to beſtow the following praiſes on him. "You, moſt humane Gryphius, who are far from meriting the laſt place among the excellent printers of this age, came firſt into my mind: and eſpecially on this account, becauſe you have not only gained greater fame than any foreigner in France, by a vaſt number of moſt excellent works, printed with the greateſt beauty and accuracy, but becauſe, though a german, you ſeem to be a countryman, by your coming to reſide among us." Baillet ſays, that Julius Scaliger dedicated alſo to him his work, "De Cauſis Linguæ Latinæ:" but he is miſtaken. Scaliger wrote a kind letter to Gryphius, in the ſame manner as Quintilian wrote to Trypho, a bookſeller, which is indeed printed at the head of the work: but the dedication is to Silvius Scaliger, his eldeſt ſon, to whom he alſo addreſſed his "Ars Poetica." Scaliger was too proud to dedicate a book to a printer.

Gryphius is allowed to have reſtored the art of printing at Lyons, which was before exceedingly corrupted; and the great number of books printed by him are valued by the connoiſſeurs. He printed many books in Hebrew, Greek, and Latin, with new and very beautiful types; and his editions are no leſs accurate than beautiful. The reaſon is that he was a very learned man, and perfectly verſed in the languages of ſuch books as he undertook to print. Thus a certain epigrammatiſt has obſerved, that Robert Stephens was a very good corrector, Colinæus a very good printer, but that Gryphius was both an able printer and corrector. This is the epigram:

"Inter tot norunt libros qui cudere, tres ſunt
Inſignes: languet cætera turba fame.
Caſtigat Stephanus, ſculpit Colinæus, utrumque
Gryphius edocta mente manuque facit."

He died, 1556, in his 63d year: and his trade was carried on honourably in the ſame city by his ſon, Anthony Gryphius. One of the moſt beautiful books of Sebaſtian Gryphius is a "Latin Bible;" it was printed, 1550, with the largeſt types that had then been ſeen, in 2 vols. folio.

GRYPHIUS (ANDREW), born at Glogaw in 1616, died in 1664. He was called the Corneille of Germany, and acquired conſiderable reputation by his compoſitions for the theatre, and is among the very firſt writers of tragedy in the catalogue

catalogue of german writers. He alſo wrote in a fine vein of irony a "Critique on the Ancient Comedies of the Germans."

GRYPHIUS (CHRISTIAN), ſon of the preceding, born in 1649. He was profeſſor of eloquence at Breſlaw, and a man of various and excellent talents. He was a great improver of his native language, and wrote many eſteemed works, the principal of which are a "Treatiſe on the Origin and Progreſs of the German Language," "A Diſſertation on the Writers who principally illuſtrated the Hiſtory of the ſeventeenth Century," and a "Collection of Poems." He was alſo a contributor to the "Journal de Leipſic."

GUADAGNOLO (PHILIP), a great orientaliſt of Italy, was born about 1596 at Magliano. After going through his ſtudies, he entered among the "Clerici regulares minores," and made his profeſſion at Rome in 1612. His genius prompted him to the ſtudy of languages, to which he devoted himſelf entirely; ſo that he acquired the greek, hebrew, chaldean, ſyriac, perſian, and arabic languages, but excelled chiefly in the arabic. He ſpent the greateſt part of his life in tranſlating books from that language, and in writing books in it, to facilitate the learning of it to others. He taught it many years in the college della Sapienza at Rome; and was indeed ſo perfect a maſter of it, that he ſpoke an oration in it, before Chriſtina queen of Sweden, in 1656. The eaſtern prelates preſented a petition to Urban VIII. to have the bible tranſlated into arabic; and, the congregation "de propaganda fide" complying with their deſires, Guadagnolo was immediately pitched upon as the propereſt and beſt qualified perſon to undertake this great work. He began it in 1622, and finiſhed it in 1649; having, however, aſſiſtants under him, and ſometimes only acting the part of a corrector. During the time that he was employed in it, he gave an account twice a week of what progreſs he had made to a congregation aſſembled for that purpoſe. It was publiſhed at Rome, 1671, in 3 vols. folio, with this title, "Biblia Sacra Arabica Sacræ Congregationis de propaganda fide juſſu edita ad uſum eccleſiarum orientalium. Additis è regione Bibliis Vulgatis Latinis." In 1631, he publiſhed a latin work, intituled, "Apologia pro Chriſtiana Religione, qua reſpondetur ad objectiones Ahmed filii Zin Alabedin Perſæ Aſphaenſis contentas in libro inſcripto Politor Speculi," 4to. The occaſion of this work was as follows. A ſpaniard had publiſhed a religious book, intituled, "The true Looking-glaſs;" which falling into the hands of a learned perſian, he wrote an anſwer to it in his native tongue, intituled, "The Poliſher of the Lookingglaſs;" and added theſe words at the end of it, "Let the Pope anſwer it." This book being brought to Rome in 1625, Urban VIII. ordered Guadagnolo to refute it; which he did so effectually, that the persian, to whom it was sent, renounced the mahometan faith, and became as zealous a defender of christianity as he had been before an opposer of it. Guadagnolo published his apology in arabic in 1637, 4to. He wrote another work in arabic, intituled, "Considerations against the Mahometan religion;" in which he shews, that the Koran is a mere rhapsody of falsehood and imposture. He published also at Rome, in 1642, "Breves Institutiones Linguæ Arabicæ," folio: a very methodical grammar. He had also compiled a dictionary in that language, but the publication of it was prevented by his death, which happened in 1656. The MS. is preserved in the convent of San Lorenzo in Lucina.

GUAGNIN (ALEXANDER), born at Verona 1538, and died at Cracow, at the age of seventy-six. He was naturalized in Poland, and published some typographical works which are highly esteemed, in particular "Sarmatiæ Europeæ Descriptio," printed at Spires in 1581. He also published "Rerum Polonicarum Scriptores," in 3 vols 8vo.

GUALBERT (S. JOHN), a florentine gentleman, who founded a monastery in the celebrated retirement of Vallombrosa, among the Apeninnes, thus mentioned by our Milton:

" Thick as autumnal leaves that strow the brooks
In Vallombrosa, where the Etrurian shades
High overarched imbower, &c."

GUALDUS (PRIORATUS, alias GALEAZZO). a native of Vicenza, where he died in 1678. He was historiographer to the emperor, and has left many historic works, written in italian; of these the principal are the "History of Ferdinand the second, and Ferdinand the third;" "An Account of the Ministry of Cardinal Mazarin;" "History of the Emperor Leopold," which last is the most esteemed, and was published at Venice in three volumes, folio, with plates.

GUALTERUS (RODOLPHUS), born at Zurich in 1529, wrote many works, and particularly "Commentaries on the Bible." He also published a translation of "Julius Pollux," at Basil, concerning which, see Fabricius. Saxius says he was born in 1519.

GUARIN, (PETER), a Benedictine, born at Rome in 1678, was eminently skilled in the greek and hebrew languages. He published a "Heb-ew Grammar," in two volumes, quarto; a "Hebrew Lexicon." He was also tutor to the abbé Bleterie, celebrated for his lives of Julian and Jovian.

GUARINI, was of an illustrious family of Verona, and merits a place in our volumes, as being the first who, after the reftoration of letters, taught greek in Italy, which he went to Conftantinople to learn. He was alfo author of various tranflations and notes on ancient authors, at the command of pope Nicolas the fifth. He tranflated "Strabo." He died in 1460.

GUARINI (JOHN BAPTIST), a celebrated italian poet, was great-grandfon of the former. In the courfe of his education he fpent fome time at Pifa, and at Padua; where he was much efteemed by the rector of the univerfity, but at an early age he went to Rome. He was apparently bred for the court and public affairs, and foon taken notice of by Alphonfus II. who firft fent him on an embaffy to Venice, and then to Piedmont, where he refided five years. The nuptials of the duke of Savoy with the princefs Catharine, fifter to Philip III. king of Spain, being celebrated about the time of his refidence at the court of Turin, he had an opportunity of prefenting that prince with his "Paftor Fido," which was then, Guarini himfelf being prefent, exhibited for the firft time with the greateft magnificence, as it was afterwards in other parts of Italy. In 1571, he went to Rome to congratulate, on the part of the duke of Ferrara, Gregory XIII. on his elevation to the pontificate. Returning to Ferrara, he fpoke the funeral oration, when the fervice was folemnized there for the emperor Maximilian and Lewis cardinal of Effe. He afterwards carried his prince's compliments to Henry of Valois upon his election to the crown of Poland; and, paffing through Germany, he had on this occafion an interview with the emperor; and on his return home was made fecretary and counfellor to the duke of Ferrara. He executed all thefe negociations with great integrity and prudence; and when the throne of Poland became vacant by the refignation of Henry Valois, who quitted it in the view of fucceeding to the crown of France, after the death of Charles IX. May 1374, Guarini was fent a fecond time to Poland, together with Galengui, by Alphonfo duke of Ferrara, to manage his intereft for that crown. Thefe deputies negociated the affair with great prudence, though without fuccefs, on account of a variety of obftacles which ftood in the way.

At length, however, not meeting with the return he thought his fervices deferved, he grew difgufted; and, in 1582, applied to the duke for leave to retire, upon pretence of attending to his private concerns. During his retreat, he fpent the winters in Padua, and the fummers at a delightful country-feat of his called La Guarina, fituate in Polefine de Rovigo, which duke Borfo had prefented to Battifta Guarini his grandfather, as a reward for his fervices performed in France, where he had been his envoy. He had fpent three years in his retirement, when he was recalled by duke Alphonso, restored to the office of secretary of state, and employed in various negociations; but, meeting with some vexations, he again quitted the court. Alessandro Guarini, his eldest son, who, in 1587, had married a rich heirefs, niece to cardinal Canani, being weary of living under the subjection of his father, and disgusted with the imperious treatment he met with from him, resolved to leave his house, and live apart with his wife. Battista was so highly offended at their departure, that he immediately seized their income, on pretence of debts due to him for money expended at their marriage. His son, deprived of his income for nine months, at last applied to the duke of Ferrara to interpose his authority, which he did; when commanding the chief judge to take cognizance of the affair, that magistrate immediately decided it in favour of Alessandro. This sentence exasperated the father still more; so that, looking on it as a proof that the duke had no regard for him, he addressed a letter to him in the most respectful but strongest terms, to be dismissed the service; which the duke granted, though not without intimating some displeasure at Guarini, for shewing so little regard to the favours he had conferred on him.

In this ill humour, 1588, he offered his services to the duke of Savoy, and was immediately employed; but, not continuing long there, he went to Padua, where he had the affliction to lose his wife in 1589. This loss inspired him with different thoughts from those he had hitherto entertained; it is even presumed by his letters, that he intended to go to Rome, and turn ecclesiastic. However, he was diverted from this step by an invitation, received in 1592 from the duke of Mantua, who sent him to Infpruck to negociate some affairs at the archduke's court. But he afterwards was dismissed this service, as he had been that of Ferrara, by the solicitations of duke Alphonso; who, it is said, could not bear that a subject of his, of Guarini's merit, should serve other princes. Thus persecuted, he went to Rome apparently with the design just mentioned, but was again prevented from executing it by a reconciliation with Alphonso, which brought him back to Ferrara in 1595. This reconciliation was obtained by his son Alessandro, who was very much beloved at court. However, fresh quarrels between father and son soon broke out again, which were afterwards carried to a great height; and, great changes happening upon the death of Alphonso in 1597, Guarini thought himself ill used, and left Ferrara to go to Ferdinand de Medicis, Grand duke of Tuscany, who expressed a great esteem for him.

But here again an unlucky accident cut short his hopes; he carried with him to Florence Guarino Guarini, his third son,

son, but fifteen years of age, and sent him to Pisa to complete his studies in that city. There the youth fell in love with a noble but poor widow, named Cassandra Pontaderi, and married her. Guarini no sooner heard the news, but suspecting the Grand Duke was privy to the marriage, and even promoted it, he left his service abruptly; and, returning to Ferrara, went thence to the prince of Urbino, but in a year's time came back to Ferrara. This was in 1604; he was sent the same year by the magistrates of the city of Rome, to congratulate Paul V. on his elevation to the papal chair. This was probably his last public employ. He resided at Ferrara till 1609, going occasionally to Venice to attend his law-suits, which carried him in 1610 to Rome, where they were determined in his favour. Passing through Venice on his return home, he was seized, in his inn there, with the distemper which put a period to his life, October 1612, when he was seventy years of age.

He was a knight of St Stephen, and member of several academies, besides other societies; as that of the Ricouvrati of Padua, the Intrepidi of Ferrara, and the Umoristi of Rome. Notwithstanding the reputation he had gained by his "Pastor Fido," he could not endure the title of poet, which he thought was so far from bringing any honour to the bearers, that it rather exposed them to contempt. He wrote other things, a complete catalogue of which may be seen in Niceron; but this was his favourite work, as appears from the warmth of his resentment against a critic who censured it.

GUARINI, a celebrated architect born at Modena in 1624. His talents were principally exercised in the sacred buildings which adorn Turin, and not only Turin, but various parts of Italy and even of Paris He seems to have had more knowlege than true taste. His posthumous works in architecture shew the extent of his skill, while his performances are marked with irregularities, and what the french call Bizarreries.

GUASCO (OCTAVIAN), born at Turin, and died at Verona in 1783. He was member of the Royal Society of London, and of the Academy of Inscriptions, &c. of Paris. He possessed considerable talents and much learning, which he made appear by various publications. Many of them are well esteemed, in particular "A Treatise on Asylums," "Literary Dissertations," "An Essay on the Statues of the Ancients." He was also the intimate friend of the president Montesquieu, and translated his great work into italian.

GUAZZI (STEPHEN), secretary to the duchess of Mentz, died at Pavia in 1565. He published "Poems," "A Tract on Polite Conversation," and "Dialogues;" all of which were much esteemed in their time.

GUAZZI (MARK), native of Padua, was eminent both in arms and learning. He died in 1556, and published a "History of Charles the Eighth," a history of his own time, and "An Abridgment of the Wars of the Turks againſt the Venetians." He was alſo the author of ſome poetical pieces.

GUDIUS (MARQUARD), a learned critic, was of Holſtein in Germany; but we know nothing of his parents, nor in what year he was born. He laid the foundation of his ſtudies at Renſburg under Jonſius, and went afterwards to Jena, where he was in 1654. He continued ſome years in this city, manifeſting a ſtrong inclination for letters, and making diligent ſearch after ancient inſcriptions. He was at Francfort in July 1658, when the emperor Leopold was crowned; and went thence to Holland, where John Frederic Gronovius recommended him to Nicolas Heinſius, as a young man of uncommon parts and learning, who had already diſtinguiſhed himſelf by ſome publications, and from whom greater things were to be expected. His parents in the mean time wanted to have him at home, and offered at any price to procure him a place at court, if he would but abandon letters, which they conſidered as a frivolous and unprofitable employment. But he remained inexorable; preferring a competency with books to any fortune without them; and, above all, was particularly averſe from a court, where "he ſhould," he ſaid, " be conſtantly obliged to keep the very worſt of Company."

His learned friends all this while were labouring to ſerve him. Grævius tried to get him a place at Duiſburg, but could not. The magiſtrates of Amſterdam ſoon after offered him a conſiderable ſum to digeſt and reviſe Blondel's "Remarks upon Baronius's Annals," and gave him hopes of a profeſſorſhip: but receiving a letter from Gronovius, which propoſed to him a better offer, he declined the undertaking. Gronovius propoſed to him the making the tour of France, Italy, and other countries of Europe, in quality of tutor to a rich young gentleman, whoſe name was Samuel Schas: and this propoſal he readily embraced, though he had another letter from Alexander Morus, with the offer of a penſion of Saumur, and a lodging in the houſe of the celebrated profeſſor Amyrault, if he would read lectures upon ancient hiſtory to ſome french noblemen.

He ſet out with Schas, November 1659; and, April 1660, got to Paris, where he found Menage at work with Diogenes Laërtius, and communicated to him ſome obſervations of his own. He eaſily found admittance to all the learned wherever he came, being furniſhed from Holland with inſtructions and recommendations for that purpoſe. The two travellers arrived at Touloufe, October 1661, where they both

fell

fell fo ill, that they were expected to die: but recovering, the went to Italy, where they ſtayed all 1662, and part of 1663. At Rome, at Florence, and at Capua, they found ſeveral of the learned, ſuch as Leo Allatius, Carolus Dati, &c. In 1663, they returned to France, and continued there the remaining part of the year. Gudius, who ſeems to have been a provident man, had defired his friends at parting, to keep a look-out for ſome place of ſettlement for him at his return: and accordingly Heinſius, Gronovius, and Grævius, were very attentive to his intereſt. But his pupil Schas wiſhed to make another tour, and Gudius thought it better to attend him than to accept of any thing that the others could get him. The truth is, Gudius found himſelf at preſent in a condition to make his fortune: for, Schas was a lover of letters; and, though immenſely rich, reſolved to ſpend his life in ſtudious purſuits. He was withal very fond of Gudius, whom he diſſuaded from accepting any place; and preſſed to accompany him through the libraries of Germany, as he had aready done through thoſe of France and Italy.

Before they ſet out for Germany, Iſaac Voſſius, moved with envy upon ſeeing in the hands of Gudius ſo many valuable monuments of literature, which they had collected in their firſt tour, is ſaid to have acted a double part, neither becoming a ſcholar nor an honeſt man. On the one hand, he affected to hold them light, when he talked with Gudius; whom alſo he did not ſcruple to treat with an air of contempt, even in the preſence of his friend Gronovius, ſaying, that Gudius had never collated any MS. but always uſed a copyiſt for that purpoſe, and that he did not know the value of them, but was ready to ſell them for a trifle to the firſt purchaſer. On the other hand, when he talked to Schas, he repreſented to him what an eſtimable treaſure he was in poſſeſſion of, exhorted him not to be the dupe of Gudius, but invited him to join his MSS. with his own; alleging, that they would enjoy them in common during their lives, and after their deaths bequeath them to the public; which unuſual act of generoſity would gain them great honour. But Voſſius miſtook his man, who loved books, and underſtood MSS. perhaps as well as Gudius: and Grævius, in the preface to his edition of "Florus," makes his acknowledgements to Schas, whom he calls *vir eximius*, for having collated three MSS. of that author in the king of France's library. Voſſius uſed other ungenerous and diſhoneſt means to ſet Gudius and Schas at variance; he cauſed a quarrel between Schas and his brother, by inſinuating, that Gudius had too great a ſhare in the poſſeſſions as well as the affections of Schas; and he did what he could to ruin Gudius's character with the States of Holland. It was all in vain; but it

ſhews

shews to what terrible paffions even learned men are fometimes fubject.

Gudius and Schas fet out for Germany, July 1664; but their excurfion was fhort, for they returned to the Hague in December. They went over to England, fome time before they went to Germany: but no particulars of this journey are recorded. He continued at the Hague till 1671, refufing to accept any thing, though a profefforfhip or two were offered him; and then went to fettle in his own country, yet without difuniting himfelf from his pupil, with whom he had lived long as an intimate friend. Heinfius te'ls Ezekiel Spanheim in a letter, Auguft 1671, that Gudius was made librarian and counfellor to the duke of Holftein; and in another to Falconieri, June 1672, that he was married. In 1674, he was fent by that prince to the court of Denmark: and, December 1675, was informed at the Hague, that Schas was dead at Holftein. He was fo, and had left his eftate to Gudius, with legacies to Grævius, Gronovius, Heinfius, and other learned men: which legacies, however, were revoked in a codicil. There was a conteft about the will, fet on foot by the relations of Schas; but Gudius carried the eftate, and, as Heinfius relates in a letter, 1676, from that time thought proper to break off his correfpondence with his learned friends in Holland. What a picture of ingratitude! thofe very friends, to whom he owed his firft rife, and who laid the foundation of all his grandeur.

In 1678, he was irretrievably difgraced with his prince, which created him much affliction. One would think, that a man, who loved books fo well as he did, far from being afflicted with an accident of this nature, might have been pleafed to be thus fet at liberty, and in full power to purfue his humour: but his learning had not freed his mind from avarice and ambition. However, he was a little comforted afterwards, by being made counfellor to the king of Denmark. He died, fomewhat immaturely, in 1689; Burman calls his death immature; and he could not be old. Though it was conftantly expected from him, yet he never publifhed any thing of confequence. At Jena, in 1657, came out a thefis of his "De Clinicis, five Grabatariis veteris Ecclefiæ:" and in 1661, when he was at Paris, he publifhed "Hippolyti Martyris de Antichrifto librum, Græcè," a piece never printed before. His MSS. however, with his own collations, he communicated to Gronovius, Grævius, Heinfius, and others, who all confidered him as excellent in philology and criticifm. "Ingenio & doctrina r. condita in primis hujus fæculi confpicuus Marquardus Gudius," are the words of Grævius, in his preface to "Florus:" and Burman, who was far from giving people more than their due, speaks of him in the higheft terms, in the preface to
"Phædrus,"

"Phædrus," which he published at Amsterdam, 1698, merely for the sake of Gudius's notes. To this edition are added four new fables, which Gudius extracted from a MS. at Dijon. Burman had published in quarto, the year before at Utrecht, "A Collection of Epistles of Gudius and his Friends," whence these memoirs of him are taken: and, in 1731, came out "Antiquæ Inscriptiones, cum Græcæ tum Latinæ, olim à Marquardo Gudio collectæ, nuper à Joanne Koolio digestæ, hortatu consilioque Joannis Georgii Grævii; nunc à Francisco Hesselio editæ, cum annotationibus eorum. Leuwardiæ." folio.

GUDIUS (Gottlob Frederic), a Lutheran minister, who wrote many works worthy of being remembered. Among others, we have from his pen a "Treatise of the Difficulties of Learning the Hebrew Tongue," various "Theological Compositions," "Remarks on the Conduct of the Emperor Julian," and a "Life of the learned Hoffman."

GUERCHEVILLE (Antoinette de Pons Marchioness of), remarkable for her spirited answer to Henry the Fourth, who made some attempts upon her chastity. If, said she, I am not noble enough to be your wife, I am too much so to be your mistress. When Henry the Fourth married Mary of Medicis, he made this lady dame d'honneur to that princess. Since, said he, you are really dame d'honneur, be so to the queen my wife.

GUERCINO, so called from a cast he had in one of his eyes, for his true name was Francesco Barbieri da Cento, was a celebrated Italian painter, and born near Bologna in 1590. He learned the principles of his art under a Bolognian painter, whose capacity was not extraordinary: but conversing afterwards with the works of Michael Angelo and the Caracci, into whose academy he entered, he made a vast progress. He designed gracefully, and with correctness: he was an admirable colourist: he was, besides, very famous for a happy invention and freedom of pencil, and for the strength, relievo, and becoming boldness, of his figures. While he was forming a manner of designing, he consulted that of his contemporary artists. Guido's and Albani's seemed to him too weak; and therefore he resolved to give his pictures more force. He painted for a long time in this strong way, but began, in the decline of life, to alter his style; and took up another more gay, neat, and pleasant, yet by no means so grand and natural as his former gusto. This however he did, not to please himself, for it was against his judgement, but the undiscerning multitude, who were drawn by Guido's and Albani's great reputation to approve no manner but theirs. He was invited to Rome by Gregory
XV,

XV. and, after two years spent there with universal applause, returned home: whence he could not be drawn by the most powerful allurements from either the kings of England or France. Nor could Christina, queen of Sweden, prevail with him to leave Bologna, though in her passage through it she made him a visit, and would not be satisfied till she had taken him by the hand; "that hand," said she, "which had painted 106 altar-pieces, 144 pictures for people of the first quality in Europe, and had, besides, composed ten books of designs." He received the honour of knighthood from the duke of Mantua. He died a bachelor in 1666, very rich, notwithstanding vast sums of money, which he had expended in building chapels, founding hospitals, and other acts of charity: for, it is remarkable, and much to this painter's honour, that he was every where as illustrious and as much venerated for his exemplary piety, prudence, and morality, as he was for his knowledge and skill in his profession.

GUERET (GABRIEL), born in 1641, was eminent both at the bar as an advocate, and in the "Republic of Letters" as an author. He left a number of works which do honour to his memory. Among others are, "Parnassus reformed," and "The War of Authors," a satirical but very witty performance. He published also many facetious works in conjunction with Blondeau.

GUERIN (FRANCIS), professor of the college de Beauvais, at Paris, translated Tacitus and Livy into French. The latter performance is by learned men preferred to the former; and has been printed at the elegant press of Barbou, in ten volumes, 12mo.

GUERINIERE (FRANCIS ROBICHON), author of two works, "l'Ecole de Cavalerie," and "Elémens de Cavalerie," which have passed through numerous editions, and are in considerable esteem.

GUESCLIN (BERTRAND DU), constable of France, and one of the greatest warriors of his time. His life has been written by many of his countrymen, all of whom agree in declaring that his person and appearance were as mean as his mind was noble. He rendered very important services to France, although by birth a Breton. His education was so much neglected that he could neither read nor write; though it must be confessed that at the period when he lived this was not uncommon, even in families of the highest rank. He died in 1380.

GUETTARD (JOHN STEPHEN), a physician, and skilful botanist, in which character he was honourably employed by the duke of Orleans. He published "Memoirs on different Parts of the Sciences and Arts," in three volumes, quarto.

quarto. We have also from his pen "Observations on Plants," in two volumes, in twelves. He was a man of exemplary probity, and was brought, by extreme attention to literary pursuits, to a too early grave in 1786.

GUEVARA (ANTONY DE), a spanish writer, was born in the province of Alaba, towards the end of the 15th century, and was brought up at court. After the death of Isabella, queen of Castile, he turned Franciscan monk; but afterwards, having made himself known at court, became preacher and historiographer to Charles V. He was much admired for his politeness, eloquence, and great parts; but, pretending to write books, he made himself ridiculous to good judges. His high-flown figurative style, full of antitheses, is not the greatest of his faults: an ill taste, and a wrong notion of eloquence, might lead to this error. This however was trifling, compared with his extravagant way of handling history. The liberty he took to falsify whatever he pleased, and to advance, as matter of fact, the inventions of his own brain, approaches near that of romance-writers. He broke the most sacred and essential laws of history with a boldness that cannot be sufficiently detested: and, when he was censured for it, alleged by way of excuse, that no history, excepting the Holy Scripture, is certain enough to be credited. Being in the emperor's retinue, he saw a great part of Europe, and was made bishop of Guadix, in the kingdom of Granada, and then bishop of Mandonedo in Galicia. He died in 1544 He was the author of several works in Spanish; the most famous of which is his "Dial of Princes, or Life of Marcus Aurelius Antoninus;" for, it has been translated into all the languages of Europe. Vossius has passed the following judgement upon this performance, "which," says he, "has nothing in it of Antoninus, but is all a fiction, and the genuine offspring of Guevara himself; who scandalously imposes upon the reader, plainly against the duty of an honest man, but especially of a bishop In the mean time he has many things not unuseful nor unpleasant, especially to a prince; whence it is entitled, 'The Dial of Princes'." Those, who may be supposed to have spoken of Guevara in the most indulgent manner, have yet been forced to set him in a most scandalous light. "It deserves our pity rather than our censure," says Nicolas Antonio, "that a writer of such fame should think himself at liberty to forget ancient facts, and to play with the history of the world, as with Æsop's Fables or Lucian's Monstrous Stories." Among Guevara's works must be ranked his Epistles, with which some have been so charmed, that they have not scrupled to call them Golden Epistles; but, says Montaigne in his dry manner,
" Whoever

"Whoever gave them this title, had a very different opinion of them from what I have, and perhaps faw more in them than I do." Bayle had fuch a contempt for Guevara as an author, that he thinks " the eagernefs of foreigners, in tranflating fome of his works into feveral languages, cannot be fufficiently admired."

GUEVARA (Louis Velez de), a fpanifh comic poet, who recommended himfelf at the court of Philip IV. by his humour and pleafantries. He is faid to have poffeffed in the higheft degree the talent of turning the moft ferious things into ridicule, and even of diffipating, in an agreeable manner, the deepeft and the jufteft grief. He was the author of feveral comedies, which were printed at different places in Spain; and of an humorous piece, intituled, "El diabolo cojuelo, novella de la otra vida," printed at Madrid in 1641. He was born at Icija in Andalufia, we know not in what year; but he died in 1646. His being a contemporary with Lopez de Vega did not hinder him from acquiring a great reputation.

GUEULETTE (Thomas Simon), was the author of many works of the gay and lighter kind, which difcovered a warm fancy and confiderable ingenuity. Among many others are the " Sultans of Guzerat, and the marvellous Adventures of the Mandarin Fum-Ho-Hum." He was alfo the author of many pieces in Italian, and edited feveral popular works in his own language. He died in 1766.

GUGLIELMINI, a native of Bologna, and moft eminent mathematician. He wrote many valuable works on fubjects of Philofophy, and Natural Hiftory, particularly that which is his greateft performance, a "Treatife on the Nature of Rivers." He was elected into the academy of Paris in 1662, and partook of the liberality of Louis XIV. He wrote alfo a "Tract on the Nature of Comets," which has not been fo favourably received by the learned. The whole of his works were printed at Geneva, in two volumes, quarto, in 1719, and he himfelf died in 1710.

GUICHARD (Claude de). He was hiftoriographer to the duke of Savoy, and author of a tranflation of Livy; and a curious work on "The Funerals of the Ancients," printed in quarto, at Lyons, in 1581.

GUICCIARDINI (Francesco), the celebrated hiftorian of Italy, was defcended of an ancient and noble family at Florence, where he was born March 6, 1482. His father, Piero Guicciardini, being an eminent lawyer, bred up his fon in his own profeffion; in which defign he fent him, in 1498, to attend the lectures of M. Jacobo Modefti, of Carmignano, who read upon Juftinian's Inftitutes at Florence. Francefco
submitted

submitted to this resolution of his father with some reluctance. He had an uncle, who was archdeacon of the metropolitan church of Florence, and bishop of Cortona; and the prospect of succeeding to these benefices, which yielded near 1500 ducats a year, had fired the ambition of the nephew. He had hopes of rising from such a foundation through richer preferments by degrees to the highest, that of a cardinal; and the reversion of the uncle's places might have been easily obtained. But, though his father had five sons, he could not think of placing any of them in the church, by reason of the neglect which he observed in the discipline. Francesco proceded therefore with great vigour in the study of the law: he took his degrees at Pisa, in 1505; but, looking upon the canon law as of little importance, he chose to be doctor of the civil law only. The same year he was appointed a professor of the institutes at Florence, with a competent salary for those times. He was now no more than twenty-three years of age, yet soon established a reputation superior to all the lawyers his contemporaries, and had more business than any of them. In 1506, he married Maria, daughter of Everardo Salviati, by far the greatest man in Florence; and, in 1507, was chosen standing counsellor to several cities of the republic. Two years after he was appointed advocate of the Florentine chapter, a post of great honour and dignity, which had been always filled with the most learned counsellors in the city; and, in 1509, he was elected advocate of the order of Calmaldoli.

He continued thus employed in the proper business of his profession till 1511; but that year the crisis of the public affairs gave occasion to call forth his abilities for more important matters. The Florentines were thrown into great difficulties by the league, which the French and Spaniards had entered into against the Pope. Perplexed about their choice to remain neuter or engage in the league, they had recourse to our advocate, whom they sent ambassador to Ferdinand king of Spain, to treat of this matter; and at the same time charged him with other affairs of the highest importance to the state. With this character he left Florence, 1512, and arriving safely at Bruges, where his Spanish majesty then resided, remained two years at that court. Here he had an opportunity of exerting and improving his talents as a statesman. Many events happened in that time, the consequences whereof came within his province to negociate; such as the taking and plundering Ravenna and Prato by the Spaniards, the deposing of Piero Soderini, and the restoration of the family of Medici. The issues of these and several other occurrences, which happened at that time, were conducted by him with such a happy address, that the republic found no occasion

occasion to employ any other minister; and the king testified the satisfaction he found in him by a great quantity of fine wrought plate, which he presented to him at his departure. On his arrival at Florence, in 1514, he was received with uncommon marks of honour; and, in 1515, constituted advocate of the consistory by Leo X. at Cortona. The Pope's favours did not stop here. Guicciardini's extraordinary abilities, with a hearty devotion to the interest of the church, were qualifications of necessary use in the ecclesiastical state. Leo therefore, that he might reap the full advantage of them, sent for him not long after to Rome, resolving to employ him where his talents might be of most service. In 1518, when Modena and Reggio were in great danger of being lost, he was appointed to the government of those cities, and approved himself equal to the charge.

His merit in this government recommended him, in 1521, to that of Parma, whence he drove away the French, and confirmed the Parmesans in their obedience; and this at a time, when the holy see was vacant by the death of Leo, and the people he commanded full of fears, disheartened, and unarmed. He retained the same post under Adrian VI. to whom he discovered the dangerous designs of Alberto Pio da Carpi, and got him removed from the government of Reggio and Rubiera. Clement VII. on his exaltation to the pontificate, confirmed him in that government. This Pope was of the house of Medici, to which Guicciardini was particularly attached; and, in return, we find him presently raised to the highest dignities in the ecclesiastical state. For instance, having on his part, in 1523, prevented the duke of Ferrara from seizing Modena, the Pope, in acknowledgement thereof, not only made him governor of that city, but constituted him president of Romagna, with unlimited authority. This was a post of great dignity and power, yet as factions then ran very high, the situation was both laborious and dangerous. However, he not only by his prudence overcame all these difficulties, but found means, in the midst of them, to improve the conveniences and delight of the inhabitants. Their towns which lay almost in rubbish, he embellishe[1] with good houses and stately buildings; a happiness, of which they were so sensible, that it rendered the name of Guicciardini dear to them, insomuch that they were overjoyed, when, after a farther promotion of Francesco, they understood he was to be succeeded in his government by his brother. This happened in 1526, when the Pope, by a brief, declared him lieutenant-general of all his troops in the ecclesiastical state, with an authority over his forces in other parts also, that were under the command of any captain-general. It

has

has been observed, that he was the chief favourite of Pope Clement, and his present situation is a most illustrious proof of that remark. This post of lieutenant-general of the forces, added to what he held in the civil government, were the highest dignities which his holiness could bestow: but this honour was yet more increased by the command of the confederate army, which was given him soon after: for, in 1527, he led these joint forces to Ravenna, and relieved that country, then threatened with entire destruction. The same year he also quelled a dangerous insurrection in Florence, when the army of the league was there under the command of the constable of Bourbon.

In 1531, the Pope made him governor of Bologna, contrary to all former precedents, that city having never before been committed to the hands of a layman. He was in this post when his holiness met Charles V. there, in December, 1532; and he assisted at the pompous coronation of the said emperor, on St. Matthias's day following. This solemnity was graced with the presence of several princes, who all shewed our governor particular marks of respect, every one courting his company, for the sake of his instructive conversation. He had at this time laid the plan of his history, and made some progress in it; which coming to the ears of the emperor before he left Bologna, his imperial majesty gave orders, when Guicciardini should attend his levee, to have him admitted into his dressing-room, where he conversed with him on the subject of his history. So particular a distinction gave umbrage to some persons of quality and officers of the army, who had waited many days for an audience. The emperor, being informed of the pique, took Francesco by the hand, and, entering thus into the drawing-room, addressed the company in these terms: " Gentlemen, I am told you think it strange that Guicciardini should have admission to me before yourselves; but I desire you would consider, that in one hour I can create a hundred nobles, and a like number of officers in the army; but I shall not be able to produce such an historian in twenty years. To what purpose serve the pains you take to discharge your respective functions honourably, either in the camp or cabinet, if an account of your conduct is not to be transmitted to posterity for the instruction of your descendants? Who are they that have informed mankind of the heroic actions of your great ancestors, but historians? It is necessary then to honour them, that they may be encouraged to convey the knowledge of your illustrious deeds to futurity. Thus, gentlemen, you ought neither to be offended nor surprised at my regard for Guicciardini, since you have as much interest in his province as myself."

Our governor did not remain continually at Bologna, but divided his time between that city and Florence. February this year, he sent a letter of instructions to Florence; and in April received orders from the Pope to reform the state there, and to put Alessandro in the possession of the government. Wise and prudent, however, as he was, discontents and faction at length arose. As long as Clement sat in the papal chair, the murmurers grumbled only in private; but upon that Pope's death, in 1534, the disgust shewed itself openly: two noblemen in particular, who till then had been fugitives, entered the city at noon-day, with a retinue of several of their friends, and some outlawed persons, well armed. The governor, looking upon this as done in contempt of his person, meditated how to revenge the affront. One evening two proscribed felons, under Pepoli's protection, were taken up by the officers as they were walking the streets, and carried to prison: and Guicciardini, without any farther process, ordered them to be immediately executed. Pepoli, who was one of these noblemen, highly incensed, assembled a number of his friends, and was going in quest of the governor to seek his revenge, when the senate sent some of their members to desire him to return home, and not to occasion a tumult, which, for fear of disobliging that body, he complied with.

It was this good disposition of the senate towards him, which prevailed with Guicciardini to keep the reins in his hands after the death of Clement. He foresaw that the people would no longer submit to his commands, and therefore had resolved to quit the government; but the senate, considering that many disorders might happen, if they were left without a governor in the time of the vacant see, begged him to continue, promising that he should have all the assistance requisite. To this he at last consented; and, with true magnanimity and firmness of mind, despising the danger that threatened him, remained in the city, till he understood that a new governor was appointed, when he resolved to quit the place. Some time after his arrival in Florence, upon the death of the duke, he had influence enough in the senate to procure the election of Cosmo, son of Giovanni de Medici, to succeed in the sovereignty. But though he had interested himself so much in the election, yet he soon quitted the court, and meddled in public affairs no farther than by giving his advice occasionally, when required. He was now past fifty, an age when business becomes disgusting to persons of a reflecting turn. His chief wish was, that he might live long enough,

enough, in a quiet recefs, to finifh his hiftory. In this refolution he retired to his delightful country-feat at Emmæ, where he gave himfelf up entirely to the work; nor could he be drawn from it by all the intreaties and advantageous offers that were made him by Pope Paul III. who, in the midft of his retirement, paffing from Nice to Florence, was at the pains to folicit our hiftorian, firft in perfon, then by letters, and at laft by the mediation of cardinal Ducci, to come to Rome. But he was proof againft all folicitations, and, excufing himfelf in a handfome manner to his holinefs, ftuck clofe to his great defign; fo that, though he enjoyed this happy tranquillity a few years only, yet in that time he brought his hiftory to a conclufion; and had revifed the whole, except the four laft books [Q], when he was feized with a fever, which carried him out of this world, May 27, 1540. He died in his fifty-ninth year.

As to the productions of his pen, his hiftory claims the firft place. It would be tedious to produce all the encomiums beftow'd upon it by perfons of the firft character: it is fufficient to obferve, that lord Bolingbroke calls him " The admirable hiftorian;" and fays, he " fhould not fcruple to prefer him to Thucydides in every refpect." In him are found all the tranfactions of that æra, wherein the ftudy of hiftory, as that lord fays, ought to begin; as he wrote in that point of time when thofe events and revolutions began, that have produced fo vaft a change in the manners, cuftoms, and interefts, of particular nations; and in the policy, ecclefiaftical and civil, of thofe parts of the world. And, as Guicciardini lived in thofe days, and was employed both in the field and cabinet, he had all the opportunities of furnifhing himfelf with materials for his hiftory: in particular, he relates at length the various caufes, which brought about the great change in religion by the reformation; fhews by what accidents the French kings were enabled to become mafters at home, and to extend themfelves abroad; difcovers the origin of the fplendor of Spain in the 15th century, by the marriage of Ferdinand and Ifabella; the total expulfion of the Moors, and the difcovery of the Weft-Indies. Laftly, in refpect to the empire, he gives an account of that change which produced the rivalfhip between the two great powers of France and Auftria; whence arofe the notion of a balance of power, the prefervation whereof has been the principal care of all the wife councils of Europe, and is fo to this day. As foon as his hiftory appeared in public, it was tranf-

---

[Q] This is the reafon why we fee no more than 16 books in all the firft editions of his hiftory, publifhed by his nephew.

lated into Latin, and has had several editions in most of the European languages. Our author wrote several other pieces, as " The Sacking of Rome ;" " Considerations on State-Affairs ;" " Councils and Admonitions."

Besides, there are extant several of his " Law-Cases," with his opinion, preserved in the famous library of Signior Carlo Tomaso Strozzi; and an epistle in verse, which has given him a place among the Tuscan poets, in the account of them by Crescimbeni. It were to be wished, that we could look into his correspondence; but all his letters, by fatal negligence, have perished; our curiosity in that point can only be satisfied by some written to him: part of these are from cardinal Pietro Bembo, secretary to Pope Leo X. and are to be seen in his printed letters; and others from Barnardo Tasso, among which is that famous sonnet in his works,

" Arno ben puoi il tuo natio soggierno,
" Lasciar nel Appeninno, &c."

Bembo's letters shew, that his correspondent possessed the agreeable art of winning the affections both of private persons and princes.

Guicciardini was survived by his wife (who lived till 1559) and three daughters. Two married into the family of Capponi and the third into that of Ducci.

GUICCIARDINI (LOUIS), was nephew of the preceding, and an historical writer of approved fidelity. He wrote different works, the principal of which is a " Description of the Pays Bas," in folio. The original is in Italian, but was translated into French by Belleforêt. It is a very interesting and curious performance. He left also other performances; and, though in some respects eclipsed by the fame of his uncle, he was equal to him in knowledge if not in talents.

GUICHERON (SAMUEL), advocate at Bourge in Bresse, deserves an illustrious place among the writers of history in the seventeenth century. He published among other things the " Genealogical History of the House of Savoy." He is much commended by Bayle.

GUIDI (ALEXANDER), an Italian poet, was born at Pavia, in Milan, 1650; and sent to Parma at sixteen years of age. His uncommon talents for poetry recommended him so powerfully at court, that he received encouragement from the duke himself. He composed some pieces at that time, which, though they favoured of the bad taste then prevailing, yet shewed genius, and a capacity for better things. He had afterwards a desire to see Rome, and, in 1683, went thither by the permission of the duke of Parma. He was already known by his poems, which were much sought after; so that he found no difficulty in being introduced to perfons of the firſt diſtinction there. The queen of Sweden, Chriſtina, wiſhed to ſee him; and was ſo pleaſed with a poem, which he compoſed at her requeſt, that ſhe had a great deſire to retain him at her court. The term allowed him by the duke being expired, he returned to Parma; but the queen having ſignified her defire to that prince's reſident at Rome, and the duke being acquainted with it, Guidi was ſent back to Rome in May, 1685.

His abode in this city was highly advantageous to him; for, being received into the academy, which was held at the queen of Sweden's, he became acquainted with ſeveral of the learned, who were members of it. He began then to read the poems of Dante, Petrarch, and Chiabrara; which reformed the bad taſte he had contracted. The reading of theſe and other good authors entirely changed his manner of writing; and the pieces he wrote afterwards were of quite a different ſtyle and taſte. Though the queen of Sweden was very kind to him, and obtained a good benefice for him from Innocent XI. yet he did not ceaſe to feel the eſteem of his maſter the duke of Parma, but received from him a penſion, which was paid very punctually. The death of his royal patroneſs happened in 1689, but he did not leave Rome; for the duke of Parma gave him an apartment in his palace there, and his loſs was abundantly recompenſed by the liberality of many perſons of quality. July 1691, he was made a member of the academy of Arcadians at Rome, under the name of Erilo Cleoneo, nine months after its foundation, and was one of its chief ornaments. Clement XI. who knew him well, and did him kind offices while he was a cardinal, continued his favours to him after he was raiſed to the pontificate.

In 1709, he took a journey to his own country, to ſettle ſome private affairs. He was there when the emperor made a new regulation for the ſtate of Milan, which was very grievous to it; and being capable of any thing as well as poetry, was pitched upon to repreſent to prince Eugene of Savoy the inconveniences and burden of this regulation: for, prince Eugene, being then governor of the country, was deputed by the emperor to manage the affair. For this purpoſe Guidi drew up a memorial, which was thought ſo juſt and ſo well reaſoned, that the new regulation was immediately revoked. The ſervice he did his country, in this reſpect, procured him a mark of diſtinction from the council of Pavia; who, in 1710, enrolled him in the liſt of nobles and decurions of the town. He was now ſolely intent upon returning to Rome; but made his will firſt, as if he had foreſeen

foreseen what was shortly to happen to him. Upon his arrival there, he applied himself to a versification of six homilies of the Pope, which he caused to be magnificently printed, and would have presented it to the pontiff, who was then at Castel-Gandolfe. With this view he set out from Rome in June 1712, and arrived at Frescati, where he was seized with an apoplectic fit, of which he died in a few hours, aged almost sixty-two. His body was carried back to Rome, and interred in the church of St. Onuphrius, near Tasso.

Though nature had been very kind to his inner man, yet she had not been so to his outer; for he was deformed both before and behind; his head, which was unreasonably large, did not bear a just proportion to his body, which was small; and he was blind of his right eye. In recompence, however, for these bodily defects, he possessed very largely the faculties of the mind. He was not learned, but he had a great deal of wit and judgement. His taste lay for heroic poetry, and he had an aversion to any thing free or satirical. His goût is original, though we may sometimes perceive that of Dante, Petrarque, and Chiabrara, who were his models.

Though the writers of his life tell us of some prose piece before it, yet the first production we know of is, " Poësie Liriche, in Parma, 1681;" which, with " L'Amalasunta," an opera, printed there the same year, he afterwards made no account of, they being written during the depravity of his taste. In 1687, he published at Rome, " Accademia per musica;" written by order of Christina of Sweden, for an entertainment, which that princess made for the earl of Castlemain, whom James II. of England sent embassador to Innocent XI. to notify his accession to the throne, and to implore his holiness's assistance in reconciling his three kingdoms to Popery. " L'Endimione di Erilo Cleoneo, pastor Arcade, con un discorso di Bione Crateo al cardinale Albano. In Roma, 1692." The queen of Sweden formed the plan of this species of pastoral, and furnished the author with some sentiments, as well as with some lines, which are marked with commas to distinguish them from the rest. The discourse annexed, by way of pointing out the beauties of the piece, was written by John Vincent Gravina. " Le Rime. In Roma, 1704." He takes an opportunity of declaring here, that he rejects all his works, which had appeared before these poems, except his " L'Endimione." " Sei Omelie di M. S. Clemente XI. Spiegate in versi. In Roma, 1712," folio. This edition is very magnificent, and adorned with cuts. It is not properly either a version or a paraphrase, the author having only taken occasion, from some passages in these thefe homilies, to compofe fome verfes according to his own genius and tafte.

In 1726, was publifhed at Verona, in 12mo, "Poëfie d'Aleffandro Guidi non piu raccolte. Con la fua vita novamente fcritta dal fignor Canonico Crefcimbeni. E con due Ragionamenti di Vincenzo Gravina, non piu divulgati." This is a collection of his printed poems and MSS, and it confifts of pieces which he had recited before the academy of Arcadians upon various fubjects.

GUIDO (RENI), an Italian painter, was born at Bologna in 1575, and learned the rudiments of painting under Denis Calvert, a flemifh mafter, who taught in that city, and had a good reputation. But, the academy of the Carracci beginning to be talked of, Guido left his mafter, and entered himfelf of that fchool, in order to be polifhed and refined. He chiefly imitated Ludovico Carracci, yet always retained fomething of Calvert's manner. He made the fame ufe of Albert Durer as Virgil did of old Ennius, borrowed what he pleafed from him, and made it afterwards his own; that is, he accommodated what was good in Albert to his own manner. This he executed with fo much gracefulnefs and beauty, that he alone got more money and more reputation in his time than his own mafters, and all the fcholars of the Carracci, though they were of greater capacity than himfelf. He was charmed with Raphael's pictures; yet his own heads are not at all inferior to Raphael's. Michael Angelo, moved probably with envy, is faid to have fpoken very contemptuoufly of his pictures; and his infolent expreffions might have had ill confquences, had not Guido prudently avoided difputing with a man of his impetuous temper. Guido acquired fome fkill alfo in mufic, by the inftruction of his father, who was an eminent profeffor of that art.

Great were the honours this painter received from Paul V. from all the cardinals and princes of Italy, from Lewis XIII. of France, Philip IV. of Spain, and from Udiflaus, king of Poland and Sweden, who, befides a noble reward, made him a compliment, in a letter under his own hand, for an Europa he had fent him. He was extremely handfome and graceful in his perfon; and fo very beautiful in his younger days, that his mafter Ludovico, in painting his angels, took him always for his model. Nor was he an angel only in his looks, if we may believe what Giofeppino told the Pope, when he afked his opinion of Guido's performances in the Capella Quirinale, "Our pictures," faid he, "are the works of men's hands, but thefe are made by hands divine." In his behaviour he was modeft, gentle, and very obliging; lived in great fplendor both at Bologna and Rome; and was only unhappy in his immoderate love of gaming. To this in his latter days he abandoned himself to entirely, that all the money he could get by his pencil, or borrow upon interest, was too little to supply his losses: and he was at last reduced to so poor and mean a condition, that the consideration of his present circumstances, together with reflexions on his former reputation and high manner of living, brought a languishing distemper on him, of which he died in 1642. His chief pictures are in the cabinets of the great. The most celebrated of his pieces is that which he painted in concurrence with Domenichino, in the church of St. Gregory. It is obfervable, that there are several designs of this great master, in print, etched by himself.

GUIDOTTI (PAUL), a painter, engraver, and architect, in each of which arts he attained some degree of eminence. He was also a good anatomist; but he made himself too ridiculous by pretending to construct wings by which he was to fly through the air. He made the attempt at Lucia, and the event need hardly be told—he fell and broke his limbs.

GUIGNARD (JOHN), a Jesuit, born at Chartres, and professor of divinity in the college of Clermont, was executed at Paris, January 7, 1595, for high treason, that is, for having written a book filled with rebellion and fury against Henry III. and Henry IV. of France. As the parliament were carrying on the prosecution against Chastel, some of them, deputed for that purpose, went to the college of Clermont, and seized several papers: and among these were found a book in the hand-writing of Guignard, containing propositions to prove, that it was lawful to kill the king; with inferences, to advise the murther of his successor also. As the juncture of things at that time required the government to prosecute with the strictest severity a doctrine, which not long before had exposed the king's life to the wicked attempt of John Chastel, it was not thought proper to shew the least favour to the Jesuit. He refused to make the *Amende Honorable*, and obstinately persisted till his death in not acknowledging Henry IV. for king of France: for which he has been placed in the Jesuits martyrology. The whole kingdom of France abounded then with seditious preachers and persons, who both in their conversation and writings hinted at the assassination of princes like Henry IV. whom they suspected to favour the enemies of Popery; and this, perhaps, was one of the reasons, which induced the parliament of Paris to involve all the Jesuits of France in the cause of Chastel and Guignard.

GUILD (WILLIAM), D. D. He was born near Dundee, 1602, and educated in the Marischal college, Aberdeen, where he took his degrees, and was successively professor of philosophy,

philosophy, divinity, church-history, and one of the ministers of that city. When the troubles broke out in 1638, he opposed the covenant, but afterwards complied with the Presbyterian form of church-government, and was continued professor of divinity. In times when the passions of men were generally heated by controversy he conducted himself with great moderation, so as to be esteemed by both parties. In 1657 he wrote a learned answer to a Roman Catholic book concerning innovations; and when the restoration took place he was sent over to Breda as one of the commissioners from the church of Scotland to congratulate king Charles II. He died 1662, aged sixty, much esteemed both by the Presbyterians and Episcopalians.

GUILLANDIUS (MELCHIOR), a famous physician and eminent botanist, native of Konigsberg in Prussia. He was taken prisoner by the Algerines, in an expedition to the coast of Africa, made solely to accomplish himself in botanical knowledge. After passing some time in slavery he returned to his country of Prussia, and published different works. His principal performance is one named Papyrus, which is a commentary on three chapters of Pliny on the same subject, and is full of erudition and acuteness. He died at Padua, in 1589.

GUILLEMEAU (JAMES), an eminent french surgeon, who published many important chirurgical works, anatomical tables, and accounts of chirurgical operations. He died at Paris in 1612.

GUILLET (DE ST. GEORGE), first historiographer to the academy of painting and sculpture at Paris. He was author of various works, among the principal of which are, the "History of Mahomet the Second," "Ancient and Modern Sparta," and "Ancient and Modern Athens," concerning which latter place he was engaged in a serious dispute with Spon.

GUILLIM (JOHN), was son of John Guillim of Westburg in Gloucestershire, yet born in Herefordshire about 1565. He was sent to a grammar-school at Oxford, and apparently entered a student of Brazen-nose-college in 1581. Having completed his pursuit of literature in the university, he returned to Minsterworth in Gloucestershire; and had been there only a short space, when he was called to London, and made a member of the Society of the college of Arms, by the name of Portsmouth; and hence promoted to the honours of Rouge-Croix Pursuivant of arms in ordinary in 1617; in which post he continued till his death, which happened in 1621. His claim to a place in this work arises from his celebrated book, intituled, "The Display of Heraldry," published by him in 1610, folio, which has gone through many

many editions. To the fifth, which came out in 1679, was added a treatise of honour, civil and military, by captain John Loggan. The last was published, with very large additions, in 1724, and is generally esteemed the best book extant upon the subject.

GUISE (HENRY), of Lorraine, (eldest son of François of Lorraine, duke of Guise), memorable in the history of France as a gallant officer; but an imperious, turbulent, seditious, subject, who placed himself at the head of an armed force, and called his rebel band The League; the plan was formed by the cardinal, his younger brother, and, under the pretext of defending the Roman catholic religion, the king Henry III. and the freedom of the state, against the designs of the Huguenots, or french protestants; they carried on a cival war, massacred the Huguenots, and governed the king, who forbade his appearance at Paris; but Guise now became an open rebel, entered that city against the king's express order, and put to the sword all who opposed him; the streets being barricaded to prevent his progress, this fatal day is called, in the french history, The Day of the Barricades. Masters of Paris, the policy of the Guises failed them; for they suffered the king to escape to Blois, though he was deserted in his palace at Paris by his very guards. At Blois, Henry convened an assembly of the states of France; the duke of Guise had the boldness to appear to a summons sent him for that purpose; a forced reconciliation took place between him and the king, by the advice of this assembly; but it being accidentally discovered, that Guise had formed a plan to dethrone the king, that weak monarch, instead of resolutely bringing him to justice, had him privately assassinated, December 23, 1558, in the thirty-eighth year of his age. His brother the cardinal shared the same fate the next day. Vide Henault's history of France.

GUISCARD (ROBERT), a famous norman knight one of the sons of Tancred de Hauteville, the father of a race of heroes, originally of Coutance in Normandy, was one of the conquerors of Naples and Sicily, from the Saracens in the eleventh century. The right of conquest gave him the sovereignty, or rather dukedom, of Apulia and Calabria. He made himself master of the person of Pope Gregory VII. when besieged by the emperor Henry IV. in the castle of St. Angelo, and carried him with him to Salerno, where this pope, who had deposed so many kings, died the captive of a norman gentleman, who was at the same time his protector. Princess Anna Comena, daughter of the emperor Alexius, in her history of these times, looks upon Guiscard in no better light than a free booter, and expresses much indignation at
his

his presuming to marry his daughter as he did to Constantine, the son of the emperor Michael Ducas; she ought to have recollected that power confers a right to titles and honours, and that every thing in this world must yield to force. Died in the isle of Corfu about 1085.

GUISCARD (CHARLES), a prussian officer, who managed with equal skill his pen and his sword. He published "Military Memoirs of the Greeks and Romans," of which it is observed that although it too much depresses the celebrated Folard, it is distinguished by much sagacity and learning.

GUISE (WILLIAM), an english divine, was born at Ablond's Court, near Gloucester, in 1653; and was entered, in 1669, a commoner of Oriel college, Oxford, which he changed for All-Souls, where he was chosen fellow, a little before he took his first degree in arts, April 4, 1674. He commenced M. A. in 1677, and entered into orders; but, marrying, he resigned his fellowship. However, he still continued at Oxford; he took a house in St. Michael's parish, resolving not to leave the university, on account of his studies, which he prosecuted with indefatigable industry, and soon became a great master of the oriental learning and languages. In that way, he translated into english, and illustrated with a commentary, "Misnæ pars ordinis primi Zeraim Tituli septem;" and was preparing an edition of Abulfeda's Geography, when he was seized with the small-pox, which carried him off in 1638. Thomas Smith gives him the title of "Vir longe eruditissimus;" and observes, that his death was a prodigious loss to the republic of letters. Foreigners style him a "person of great learning, and the immortal ornament of the university of Oxford." He was buried at St. Michael's church in that city, where a monument was erected to his memory by his widow, with a latin inscription. He left issue a son John, who, being bred to the army, raised himself to the highest posts there, and was well known in the military world, by the title of General Guise.

GUITTON (d'AREZZO), an eminent italian poet, who flourished about the year 1250. Many of his performances are to be found in a "Collection of the Ancient Italian Poets," published at Florence, in 8vo. in 1527.

GULDENSTAEDT (JOHN ANTONY), a celebrated traveller, of whose various performances a list is given in "Cox's Travels," Vol. I. p. 162. On account of his great skill in natural history and knowledge of foreign languages, he was invited to Petersburg, where he was made professor. He was absent three years on his travels. He first went to Astracan and Kislar, and afterwards to the eastern extremity of Caucasus. Here he collected vocabularies of the language spoken in those parts,

parts, and difcovered fome traces of chriftianity among the people. He next procceded to Georgia, was introduced to prince Heraclius, and carefully examined the adjacent country. He then explored the fouthern diftricts, inhabited by the Turcoman Tartars, and, penetrating into the middle chain of Mount Caucafus, vifited Mingrelia, Middle Georgia and Eaftern and the Lower Jmeretia. It was his intention next to have journeyed to Crim Tartary, but was recalled to Peterfburg, where he died of a fever. He was a native of Riga, and was a man poffeffed of every requifite for the accomplifhment of the purpofes which he had in view.

GUNDLING (NICOLAS JEROME), a native of Nuremburg, and profeffor of eloquence, philofophy, and civil law, at Hall. He was in great eftimation at Berlin, where he was often fent for and confulted on affairs of ftate. He left a great number of literary works on fubjects of jurifprudence, hiftory, and politics. His writings difcover much fpirit and various knowledge, and are withal very numerous. Thofe in moft efteem are his "Hiftory of Moral Philofophy;" "A Courfe of Literary Hiftory;" "A Courfe of Philofophy," in 3 vols. 8vo. He had alfo a principal fhare in the "Obfervationes Hallenfes," an excellent work in eleven volumes, octavo.

GUNNING (PETER), bifhop of Ely, was the fon of Peter Gunning of Hoo in Kent, and born there in 1613. He had his firft education at the king's fchool in Canterbury, where he commenced an acquaintance with Somner, the antiquary, his fchool-fellow. At fifteen, he was removed to Clare-hall, in Cambridge, and promoted to a fellowfhip in 1633: he became an eminent tutor in the college. Soon after he commenced M. A. and had taken orders, he had the cure of Little St. Mary's from the mafter and fellows of Peterhoufe. He became an eminent preacher, and was licenfed as fuch by the univerfity in 1641; when he diftinguifhed himfelf by his zeal for the church and king. About the fame time, making a vifit to his mother at Tunbridge, he exhorted the people, in two fermons, to make a charitable contribution for the relief of the king's forces there: which conduct rendered him obnoxious to the powers then in being, who firft imprifoned him; and, on his refufing to take the covenant, deprived him of his fellowfhip. This obliged him to leave the univerfity, but not before he had drawn up a treatife againft the covenant, with the affiftance of fome of his friends, who took care to publifh it.

Being thus ejected, he removed to Oxford, where he was incorporated M. A. July 10, 1644; and kindly received by Dr. Pink, warden of New-college, who appointed him one of the chaplains of that houfe. During his refidence there, he he officiated two years at the curacy of Caffingdon, under Dr. Jasper Mayne, near Oxford; and sometimes preached before the court, for which service he was complimented, among many other Cambridge-men, with the degree of B. D. June 23, 1646. Soon after this, he became tutor to the lord Hatton and Sir Francis Compton, and then chaplain to Sir Robert Shirley, who was so much pleased with his behaviour, in some disputations with a romish priest, as well as with his great worth and learning in general, that he settled upon him an annuity of £100. Upon the decease of Sir Robert, he held a congregation at the chapel of Exeter-house, in the Strand, where he duly performed all the parts of his office according to the liturgy of the church of England; yet he met with no other molestation, from the usurper Cromwell, than that of being now and then sent for and reproved by him. On the return of Charles II. he was restored to his fellowship, and created D. D. by the king's mandate September 5, 1660; having been first presented to a prebend in the church of Canterbury; soon after which he was instituted to the rectories of Cotesmore in Rutland, and of Stoke-Bruen in Northamptonshire. But this was not all; for, before the expiration of the year, he was made master of Corpus-Christi college in Cambridge, and also lady Margaret's professor of divinity: nor did he stop even here, for in a few months he succeeded to the regius professorship of divinity, and the headship of St. John's-college, upon the resignation of Dr. Tuckney, who had been obliged, June 12, 1661, to give way for Gunning; he being looked upon as the properest person to settle the university on right principles again, after the many corruptions that had crept into that body.

All the royal mandates indeed, for his several preferments, were grounded upon his sufferings and other deserts; for he was reckoned one of the most learned and best-beloved sons of the church of England: and as such was chosen proctor both for the chapter of the church of Canterbury, and for the clergy of the diocese of Peterborough, in the convocation held in 1661; one of the committee upon the review of the liturgy, when it was brought into that state of sufficiency where it has rested ever since; and was principally concerned in the conference with the dissenters at the Savoy the same year. In 1669, he was promoted to the bishopric of Chichester, which he held with his regius professorship of divinity till 1674, when he was translated to Ely; where, after ten years enjoying it, he died a bachelor, in his 71st year, July 6, 1684. His corpse was interred in the cathedral of Ely, under an elegant monument of white marble, the inscription upon which has been often printed. As to his character, he has been so variously

riously drawn by writers of different principles and parties, that we shall not take upon us to determine what is so warmly disputed among them, viz. Whether his head was as good as his heart.

However, all agree in allowing him to be a profound divine, as well as a person of great erudition, of a most unblamable life and conversation, and of most extensive and exemplary charity. To the former, his writings bear testimony; and to the latter, his many extraordinary benefactions to the public.

GUNTER (EDMUND), an english mathematician, was of welsh extraction from a family at Gunter's-town in Brecknockshire; but his father, being settled in the county of Hereford, had this son born to him there in 1581. As he was a gentleman possessed of a handsome fortune, he thought proper to breed him up in a liberal way: to which end he was placed by Dr. Busby at Westminster school, where he was admitted a scholar on the foundation; and, in consequence thereof, elected student of Christ-church, Oxford, in 1599. Having taken both his degrees in arts at the regular times, he entered into orders, and became a preacher in 1614, and proceeded B. D. November 23, 1615. But, genius and inclination leading him chiefly to mathematics, he applied early to that study; and, about 1606, merited the title of an inventor by the new projection of his sector, which he then described, together with its use, in a latin treatise; and several of the instruments were actually made according to his directions. These being greatly approved, as being more extensively useful than any that had appeared before, on account of the greater number of lines upon them, and those better contrived, spread our author's fame universally: their uses also were more largely and clearly shewn than had been done by others; and, though he did not print them, yet many copies, being transcribed and dispersed abroad, carried his reputation along with them, recommended him to the patronage of the earl of Bridgewater, brought him into the acquaintance of the celebrated Mr. Oughtred, and Mr. Henry Briggs, professor of geometry at Gresham; and thus, his fame daily increasing the more he became known, he was preferred to the astronomy-chair at Gresham-college on March 6, 1619.

He had invented a small portable quadrant, for the more easy finding of the hour and azimuth, and other solar conclusions of more frequent use, in 1618; and, in 1620, he published his latin " Canon Triangulorum, or Table of artificial Sines and Tangents to the Radius of 10 000,000 Parts to each Minute of the Quadrant." This was a great improvement to astronomy, by facilitating the practical part of that science in the resolution of spherical triangles without the use of secants

or

or verfed fines, the fame thing being done here (by addition and fubtraction only) for performing which the former tables of right fines and tangents required multiplication and divifion. This admirable help to the ftudious in aftronomy was gratefully commemorated, and highly commended, by feveral of the moft eminent mathematicians who were his contemporaries, and who at the fame time did juftice to his claim to the improvement, beyond all contradiction.

The ufe of aftronomy in navigation unavoidably draws the aftronomer's thoughts upon that important fubject; and, as great genii can hardly look into any art without improving it, we find Gunter difcovering a new variation in the magnetic needle, or the mariner's compafs, in 1622. Gilbert, in the beginning of that century, had inconteftibly eftablifhed the firft difcovery of the fimple variation; after which the whole attention of the ftudious in thefe matters was employed in fettling the rule obferved by nature therein, without the leaft apprehenfion or fufpicion of any other; when our author, making an experiment this way at Deptford, in 1622, found that the direction of the magnetifm there had moved no lefs than five degrees within two minutes, in the fpace of forty-two years. Indeed the fact was fo furprifing, and fo contrary to the opinion then univerfally received of a fimple variation only, which had fatisfied and bounded all their curiofity, that our author dropt the matter apparently, expecting, through modefty, an error in his obfervation to have efcaped his notice in his experiment. But afterwards, what he had done induced his fucceffor at Grefham to purfue it; and, the truth of Gunter's experiment being confirmed by a fecond, farther enquiry was made, which ended in eftablifhing the fact. We have fince feen Halley immortalize his name, by fettling the rule of it in the beginning of this century.

The truth is, Gunter's inclination was turned wholly the fame way with his genius; and it cannot be denied that he reached the temple of fame by treading in that road. To excite a fpirit of induftry in profecuting mathematical knowledge, by leffening the difficulties to the learner; to throw new light into fome things therein, which before appeared fo dark and abftrufe as to difcourage people of ordinary capacities from attempting them; and by that means to render things of wonderful utility in the ordinary employment of life fo eafy and practicable as to be managed by the common fort; is the peculiar praife of our author, who effected this by that admirable contrivance of his famous rule of proportion, now called the line of numbers, and the other lines laid down by it, and fitted in his fcale, which, after the inventor, is called "Gunter's scale;" the description and use of which he published in 1624, 4to. together with that of his sector and quadrant already mentioned. It is no wonder that his fame by this time had reached the ears of his sovereign, or that prince Charles should give directions, that he should draw the lines upon the dials in Whitehall garden, and give a description and use of them; or that king James should order him to print the book the same year, 1624. There was, it seems, a square stone there before of the same size and form, having five dials upon the upper part, one upon each of the four corners, and one in the middle, which was the principal dial, being a large horizontal concave; besides these, there were others on the sides, east, west, north, and south; but the lines on our author's dial, except those which shewed the hour of the day, were greatly different. And Dr. Wallis tells us, that one of these was a meridian, in fixing whereof great care was taken, a large magnetic needle being placed upon it, shewing its variation from that meridian from time to time. If the needle was placed there with that intention by our author, it is a proof that his experiment at Deptford had made so much impression upon him, that he thought it worth while to pursue the discovery of the change in the variation, of which the world would doubtless have reaped the fruits, had his life been continued long enough for it.

But he was taken off December 10, 1626, about his 45th year, the prime of life for such studies. He died at Gresham-college, and was buried in St. Peter the Poor, Broad-street, without any monument or inscription; but his memory will always be preserved in the mathematical world as an inventor, which entitles him to the honour of being the parent of instrumental arithmetic. The 5th edition of his works was published by Mr. Leybourn in 1674, 4to.

GUNTHER, a german poet of great genius, but whose talents proved his destruction. A rival mixed some drugs in his drink just as he was about to be presented to the king of Poland. The consequence of which was, that, at the moment he was preparing to address the monarch in a complimentary speech, he staggered and fell down. His vexation was so extreme that it caused his death at the age of twenty-eight. Among other elegant pieces he wrote "An Ode on the Victory of Prince Eugene over the Turks," a subject which has also been handled by Rousseau.

GURTLER (NICOLAS), born at Basle in 1654, and died in 1711. He was author of a "Greek, German, and French, Lexicon." He wrote also "Historia Templariorum, Origines Mundi," a work of prodigious learning; but in which the

writer

writer has been too fanciful in his etymologies, and sometimes abfurd in his ideas of mythology.

GUSMAN (LEWIS), a fpanifh jefuit, known in "Ribandeneira's Catal. Script. Sac. Jef." as the author of the "Hiftory of the Jefuits in the Indies, and the fuccefs of their Miffions in Japan," in fourteen books, in Spanifh. He was rector of feveral colleges in his fociety, and afterwards provincial of Seville and Toledo. He died at Madrid in 1605.

GUSSANVILLAN (PETER), a native of Chartres. He publifhed a good edition of the works of Gregory the great, the beft before that which was edited by the Benedictines.

GUSTAVUS (VASA), or more properly Guftavus Ericfon, was the fon of Eric Vafa, and defcended from the ancient kings of Sweden, His great paffion was the love of glory, and this difficulties and dangers increafed rather than diminifhed. He lived at a time when the greateft part of the wealth of Sweden was in the hands of the clergy, when every nobleman was, in his own territories, a fovereign; and, laftly, when Steeno was adminiftrator of the realm. In the war, which was originally profecuted betwixt the Swedes and Chriftian king of Denmark, this laft, having got Guftavus into his power, kept him a prifoner many years in Denmark. He at length made his efcape, and through innumerable dangers got back to his native country, where, for a long period, he ufed every effort to roufe his countrymen to refift and repel their invaders and victorious enemies. In this, however, he was not fuccefsful; and Chriftian of Denmark having got poffeffion of Stockholm, and Colmar, exercifed the crueleft tyranny on all ranks, and in one day put to death ninety-four nobles, among whom was the father of Guftavus. Guftavus at length prevailed on the Dalecarlians to throw off the yoke; and, at the head of a refpectable body of forces, entered the provinces of Halfingia, Geftricia, and fome others. After a feries of defperate adventures, temporary fucceffes, and frequent defeats, he recovered Upfal; for which difappointment Chriftian put to death the mother and fifter of Guftavus in cruel torments. Having overcome Eaft Gothland and blockaded Stockholm, he convened the States-General, and was by them offered the title of king. This he refufed, and was fatisfied with the regal power and title of adminiftrator. A fhort time afterwards, a revolution took place in Denmark; Chriftian was driven from his dominions; and Frederic duke of Holftein, uncle of Chriftian, was made his fucceffor. There was now nothing to oppofe Guftavus; he therefore fummoned the ftates to meet at Stregnez, filled up the vacancies of the fenate, and was proclaimed king with the ufual forms of election. He alfo prevailed on the ftates to render the crown hereditary

hereditary to the male heirs; and, to make the life of Gustavus yet more remarkable in history, it was in his reign that Lutheranism was established as the natural religion of Sweden. The latter part of his life was spent in cultivating the arts of peace, and in decorating his metropolis with noble edifices. He died at Stockholm, of a gradual decay, on the 9th of September 1560, in the 70th year of his age, and was quietly succeeded in his throne by his eldest son Eric.

GUSTAVUS (ADOLPHUS), king of Sweden, commonly called the Great, a title, which, if great valour, united with great wisdom; great magnanimity with regard to himself, and great consideration of the wants and infirmities of others, have any claim, he seems well to have deserved. He was born at Stockholm in 1594. His name Gustavus he inherited from his grandfather Gustavus Vasa, and he was called Adolphus from his grandmother Adolpha. His education was calculated to form a hero, and seems, in all respects, to have resembled that bestowed on Henry the fourth of France. He had a great genius, a prodigious memory, and a docility and desire of learning almost beyond example. He ascended the throne of Sweden in 1611, being then no more than fifteen; but the choice he made of ministers and counsellors proved him fully adequate to govern. His valour in the field was tried first against Denmark, Muscovy, and Poland. He made an honourable peace with the two first, and compelled the last to evacuate Livonia. He then formed an alliance with the protestants of Germany against the emperor, and what is commonly called the league. In two years and a half he overran all the countries from the Vistula as far as the Danube and the Rhine. Every thing submitted to his power, and all the towns opened to him their gates. In 1631, he conquered Tilly, the imperial general, before Leipsic, and a second time at the passage of the Lech. In the following year, he fought the famous battle in the plains of Lutzen, where he unfortunately fell at the immature age of thirty-eight. Besides his other noble qualities he loved and cultivated the sciences. He enriched the university of Upsal; he founded a royal academy at Abo, and an university at Dorp in Livonia. Before his time there were no regular troops in Sweden; but he formed and executed the project of having 80,000 men constantly well armed, disciplined, and cloathed. This he accomplished without difficulty, on account of the love and confidence which his subjects without reserve reposed in their king. Some historians have delighted to draw a parallel between Gustavus and the great Scipio, and it is certain that they had many traits of character in common. Scipio attacked the Carthaginians in their own dominions; and

Gustavus undertook to curb the pride of Austria by carrying the war into the heart of her country. Here indeed the advantage is with Gustavus; for, the Carthaginian power was already debilitated; but the emperors had before never received any check. He died literally, as it is said of him, with the sword in his hand, the word of command on his tongue, and victory in his imagination. His life has been well written by our countryman Harte; and he appears in all respects to have deserved the high and numerous encomiums which writers of all countries have heaped upon his memory. Some have suspected this exalted character to have lost his life from the intrigues of cardinal Richelieu; others from Lawemburgh, one of his generals, whom Ferdinand the emperor is said to have corrupted. He left an only daughter, whom he had by the princess Mary of Brandenburg, and who succeeded her father at the age of five. This princess was the celebrated Christina queen of Sweden.

GUTHRIE (WILLIAM), was born at Breichen, in Angusshire, 1701, and educated in King's college, Aberdeen; where he took his degrees, and removed to London in consequence of a love-affair, which created some disturbance in his family. As his fortune was small, he was obliged to write for the booksellers, and compiled the "History of England," in three volumes, folio, a work of considerable merit, but not generally known. He afterwards suffered his name to be prefixed to a "History of Scotland," in ten volumes; to an "Universal History," in twelve volumes, and to a "Peerage," in quarto. His last and most esteemed work is his "Geographical Grammar," in 8vo. and 4to. He died in 1769, aged 68. He was in the commission of the peace for the county of Middlesex, but never acted.

GUTTEMBERG (JOHN), one of those who disputes with Faust, Schoeffer, and others, the invention of the art of printing. He was of a noble family; and there is very good evidence for the assertion, that, if he did not absolutely invent the art, he was the first who conceived the idea of regularly printing a book. To enter into the arguments for or against his claims, would be to compose a dissertation on printing. Bowyer affirms, that the real inventor of printing was Laurentius of Haerlem. The types of Laurentius, he affirms, were stolen from him by an elder brother of Guttemberg. This man entered into a partnership with Faust, and they were afterwards joined by Guttemberg. Guttemberg it undoubtedly was who first invented cut metal types, which were used in the earliest edition of the bible. Guttemberg had endeavoured, but without success, to introduce printing into Strasburg, before he joined his brother and Faust at Mentz, which explains the circumstance that these three cities Haerlem,

lem, Mentz, and Strasburgh, severally claim the invention of the art. Guttemberg died at Mentz in 1468, aged about sixty years; and the circumstance of his claims, which are more or less valid, certainly justifies his having a place here.

GUY (THOMAS), founder of Guy's hospital, was the son of Thomas Guy, lighterman and coal-dealer in Horseley-down, Southwark. He was put apprentice, in 1660, to a bookseller, in the porch of Mercers chapel, and set up trade with a stock of about 200l. in the house that forms the angle between Cornhill and Lombard-street. The English Bibles being at that time very badly printed, Mr. Guy engaged with others in a scheme for printing them in Holland, and importing them; but, this being put a stop to, he contracted with the university of Oxford for their privilege of printing them, and carried on a great bible-trade for many years to considerable advantage. Thus he began to accumulate money, and his gains rested in his hands; for, being a single man and very penurious, his expences were next to nothing. His custom was to dine on his shop-counter, with no other table-cloth than an old newspaper; he was also as little nice in regard to his apparel. The bulk of his fortune, however, was acquired by purchasing seamen's tickets during queen Anne's wars, and by South-sea stock in the memorable year 1720.

To shew what great events spring from trivial causes, it may be observed, that the public are indebted to a most trifling incident for the greatest part of his immense fortune's being applied to charitable uses. Guy had a maid-servant, whom he agreed to marry; and, preparatory to his nuptials, he had ordered the pavement before his door to be mended so far as to a particular stone which he marked. The maid, while her master was out, innocently looking on the paviours at work, saw a broken place they had not repaired, and mentioned it to them; but they told her that Mr. Guy had directed them not to go so far. "Well," says she, "do you mend it: tell him I bade you, and I know he will not be angry." It happened, however, that the poor girl presumed too much on her influence over her wary lover, with whom the charge of a few shillings extraordinary turned the scale entirely against her: for, Guy, enraged to find his orders exceeded, renounced the matrimonial scheme, and built hospitals in his old age.

In 1707, he built and furnished three wards on the north side of the outer court of St. Thomas's hospital in Southwark; and gave 100l. to it annually for eleven years preceding the erection of his own hospital. Some time before his death, he erected the stately iron gate, with the large houses on each side, at the expence of about 3000l. He was seventy-six years of age when he formed the design of building the hospital near St. Thomas's which bears his name. The charge of of erecting this vast pile amounted to 18,793l. besides 219,499l. which he left to endow it: and he just lived to see it roofed in. He erected an alms-house with a library at Tamworth, in Staffordshire, (the place of his mother's nativity, and which he represented in parliament,) for fourteen poor men and women; and for their pensions, as well as for the putting out of poor children apprentices, bequeathed 125l. a year. To Christ's hospital he gave 400l. a year for ever; and the residue of his estate, amounting to about 80,000l. among those who could prove themselves in any degree related to him.

He died December 17, 1724, in the 81st year of his age, after having dedicated to charitable purposes more money than any one private man upon record in this kingdom.

GUY, a monk of Arezzo, famous for inventing music in several parts. Guy, being a born a musician, found out, by the powers of reflection, that, by observing certain proportions, several different voices might be made to sing together, and form a delightful harmony. He invented the times of the gamut and the six famous syllables, ut re mi fa sol la. He lived about the year 1026; and his invention was received with unbounded applause; for by means of it a child might learn in a few months what would have employed a man for many years.

GUYARD (DE BERVILLE), a poor french author of great merit, who encountered a fate similar to that of Otway and Chatterton, and died in prison at the age of seventy-three. He wrote the "Histories of Bertrand, Duguesclin," and in particular of the "Chevalier Bavard." He died in 1760.

GUYET (FRANCIS), an eminent critic, was born of a good family at Angers, in 1575. The circumstance of his life, however, came to be known only by his heirs; for, he never would tell in what year he was born, but concealed his age with as much solicitude as an ancient virgin who proposes to be married: though, indeed, it is said he had hardly a confidant in any other thing. He lost his father and mother when a child; and the small estate they left him came almost to nothing by the ill management of his guardians. Nevertheless, he applied himself intensely to books; and, being of opinion, that Paris would enable him to perfect his judgement and knowledge by the conversation of learned men, he took a journey thither in 1599. The acquaintance he soon got with the sons of Claudius du Puy proved very advantageous to him; for, the most learned persons in Paris frequently visited these brothers, and many of them met every day in the house of Thuanus, where Mess. du Puy received company. After the death of that president, they held those conferences in the same place; and Guyet constantly made one. He went to Rome in 1608, and applied himself to the italian tongue with such success that he could make good italian verses. He was much esteemed by cardinal du Perron and several great personages. He returned to Paris by the way of Germany, and was taken into the house of the duke d'Epernon, to teach the abbot de Granselve, who was made cardinal de la Valette in 1621. Being thoroughly skilled in greek and latin authors, he picked out of them what was most proper for his pupil; and explained it to him, not like a pedant, but with a view to the use which a man, designed for great employments would make of it. His noble pupil conceived so great an esteem for him, that he always entrusted him with his most important affairs. He took him with him to Rome, and procured him a good benefice; but Guyet, being returned to Paris, chose to live a private life rather than in the house of the cardinal, and pitched upon Burgundy-college to make his abode in. Here he spent the remainder of his life, minding nothing but his studies: and applied himself chiefly to a work, wherein he pretended to shew, that the latin tongue was derived from the greek, and that all the primitive words of the latter consisted only of one syllable. His work came to nothing; for, they found, after his death, only a vast compilation of greek and latin words, without any order or coherence, and without any preface to explain his project. But the reading of the ancient authors was his main business: for, as to the moderns, he meddled with nothing but histories and voyages. The margins of his classics were full of notes, many of which have been published. Those upon Hesiod were imparted to Grævius, who inserted them in his edition of that author, 1667. The most complete thing that was found among his papers was his notes upon Terence; and therefore they were sent to Bocclerus, and afterwards printed. He took great liberties as a critic: for he rejected as suppositious all such verses as seemed to him not to favour of the author's genius. Thus he struck out many verses of Virgil; discarded the first ode in Horace; and would not admit the secret history of Procopius. Notwithstanding the boldness of his criticisms, and his free manner of speaking in conversation, he was afraid of the public; and dreaded Salmasius in particular, who threatened to write a book against him, if he published his thoughts about some passages in ancient authors. He was so happy as to be accounted a man of great learning, though he had printed nothing; and was contented with the praises others bestowed upon him. He is said to have been a sincere and honest man. He was cut for the stone in 1636; abating which, his long life was hardly attended with any illness. He died of a catarrh, after three days illness, in the arms of James du Puy and Menage his countryman, April 12, 1655, aged 80. His life is written in latin, with great judgement and politeness,

by

by Mr. Portner, a senator of Ratisbon, who took the suppositious name of Antonius Periander Rhætus; and is prefixed to his notes upon Terence, printed with those of Boeclerus, at Strasburg, in 1657.

GUYON (JOHANNA-MARY BOUVIERS de la MOTHE), a french lady, memorable for her writings and for her sufferings in the cause of quietism, was descended from a noble family, and born at Montargis, in 1648. She discovered an anxiety to take the veil at a very premature age; but with this her friends refused to comply, and obliged her to marry a gentleman to whom they had betrothed her. At the early age of twenty-eight she was a widow; when giving up the care of her children to their other relations, she distinguished herself in and made many converts to what is called quietism. The author of this was Michael de Molinos, a spanish priest, who resided at Rome. Madam Guyon was doubtless eminent for goodness of heart; but she was as certainly of an inconstant and unsettled temper, and subject to be drawn away by the seduction of a warm and unbridled fancy. She was confined for some months by order of the king. She was, however, defended by Fenelon, who adopted many of her tenets, and who obtained her release. Bossuet, who was jealous of Fenelon, obtained the condemnation of what he had written on this subject; and Madame Guyon, who was involved with Fenelon, was again imprisoned. Her latter days were consumed in mystical reveries, covering not only her books and papers; but her furniture, walls, and cielings, with the wanderings of her spiritual fancies. Her verses were collected and published after her death, in five volumes, and were called "Canticles Spirituels." Her other publications were "Le moyen court et très facile de faire oraisons, et le Cantique des Cantiques de Solomon, interpreté selon le sens mystique;" but these last were condemned by the archbishop of Paris.

GUYON (CLAUDE), a french historian, who died at Paris in 1771. Although he did not satisfy the fastidious taste of Voltaire, he is an interesting and useful writer. He published a "Continuation of Echard's Roman History," "The History of Empires and Republics," of which it is said, that, if compared with Rollin's, it is less agreeable and elegant: but from which it is certain that Guyon drew his materials from the original sources of the ancients; whilst, on the contrary, Rollin has often copied the moderns. Guyon also wrote the "History of the Amazons;" a "History of the Indies;" and an "Ecclesiastical History," a very successful performance.

GWYNN (ELEANOR), better perhaps known by the familiar name of Nell, was, at her first setting out in the world, a plebeian of the lowest rank, and sold oranges at the playhouse.

Some affirm that she was born in a night-cellar; certain it is, that she rambled from tavern to tavern, entertaining the company with her songs. As early as the year 1667, she was admitted in the theatre-royal, and was mistress to Hart, to Lacy, and to Buckhurst. She became eminent in her profession as an actress, and performed the most spirited parts with admirable address. The pert prattle of the orange-wench by degrees refined into a wit, which pleased our Charles the second. She ingratiated herself into her sovereign's affection, in which she retained a place to the time of her death. Dryden was very partial to her, and greatly assisted her in her rise at the theatre; in return, when possessed of the power, she distinguished the poet by particular marks of gratitude. Many benevolent actions are recorded of her; and perhaps she was the only one of the king's mistresses who was never guilty of any infidelity towards him. It is ludicrous, perhaps, but it is nevertheless true, that Madam Gwynne (for so she was latterly called) piqued herself on her attachment to the church of England. She was low in stature, and careless of her dress; but her pictures represent her as handsome. She died in 1687.

GWYNNE (MATTHEW), a famous physician in his time, was born in London, and descended from an ancient welch family. He was educated at Oxford, of which he afterwards became perpetual fellow. He first practised physic in and about Oxford, and in 1593 was created doctor. In 1595, by leave of the college, he attended Sir Henry Unton, ambassador from queen Elizabeth to the french court, in quality of his physician. The date of his death, supposes him to have died after 1639, because his name was still in the edition of the "Pharmacopœia," printed in that year; but Dr. Aikin has made it appear that the "Pharmacopœia" of 1618 was many times reprinted by the booksellers without changing the names of the college members. Of his miscellaneous works, the latin ones do not stand very high in estimation, the style being formed upon a wrong taste.

H.

HABAKKUK, the eighth of the lefs prophets, whom fome affirm to have been a native of Belthraker, and of the tribe of Simeon. Some fuppofe him to have lived in the reign of Manaffeth, others in that of Jofiah, and fome have even placed him fo late as Zedekiah; it is, however, moft probable that he prophefied under Jehoiakim, who reigned A. M. 3395. Habakkuk is faid, as well as Jeremiah, to have chofen to remain amidft the fad fcenes of a deferted and defolate land rather than follow his countrymen into captivity. The ftyle of this prophet's book is poetical, and the conclufion is eminently beautiful; he is imitated by fucceeding prophets, and is cited as an infpired perfon by the evangelifts.

HABERT (HENRY LOUIS), a member of the French Academy, deferves a place in this work as having been the friend of Gaffendi, and, by his kindnefs to that philofopher, proving, which is often the cafe, that a friend may be better than a patron. He publifhed the works of Gaffendi, with an elegant latin preface; He alfo wrote fome epigrams and other pieces of poetry, and is reprefented to have been a man omnis doctrinæ & fublimioris & humanioris amantiffimus. He died in 1679.

HABICOT (NICOLAS), an eminent furgeon, who not only obtained confiderable reputation by his profeffional fkill, but alfo by his " Treatife on the Plague." He was born at Bonny, in Gatinois, and died in 1624.

HABINGTON (WILLIAM), an englifh poet and hiftorian, was defcended from an ancient family, and born at Hendlip, in Worcefterfhire, 1605. He received his education at St. Omer's and Paris, where he was earneftly preffed to take the habit of a jefuit; but, this fort of life not fuiting with his genius, he excufed himfelf, and left them. After his return from Paris he was inftructed in hiftory and other branches of polite literature, and became, fays Wood, a very accomplifhed gentleman. He died Nov. 30, 1654, leaving behind him, 1. "Poems," 1635, in 8vo. 2d edit. under the title of " Caftara." 2. " The Queen of Arragon," a tragi-comedy. 3. " Obfervations on Hiftory, 1641," 8vo. 4. " Hiftory of Edward IV, King of England, 1640," folio, Nicolfon, fpeaking of Edward the IVth's reign, fays, that Habington " has given us as fair a draught of it as the thing would bear; at leaft, he has copied this king's picture as agreeably as could be expected from one ftanding at fo great a diftance from the original." Our author, during the civil war, is faid by Wood to have run with the times, and not to have been unknown to Oliver Cromwell;

Cromwell; but there is no account of his being raised to any preferment during the protector's government.

HACKET (WILLIAM), an english fanatic in the reign of Elizabeth, was at first a gentleman's servant, and afterwards married a rich widow, whom he soon ruined by his extravagance. He was enormously vicious; being not only addicted to wine and women, but even to robbing upon the highway. He had never studied, but had a great memory, which he abused in repeating the sermons of ministers over his cups. At length he set up for a prophet, and declared, that England should feel the scourges of famine, pestilence, and war, unless it established the consistorial discipline; and that for the future there should be no more popes. He began to prophesy at York and Lincoln, where, for his boldness, he was publicly whipped and condemned to be banished. The people believed, nevertheless, that he had the extraordinary gift of the Holy Spirit; and he was so confident of his own favour with heaven as to affirm, that, if all England should pray for rain, and he should pray for the contrary, it would not rain. Coppinger and Arthington, two persons of learning, joined with him: the first by the title of "The Prophet of Mercy," the second by the title of "The Prophet of Judgement." These two visionaries pretended an extraordinary mission, and gave out, that Hacket was the sole monarch of Europe: and that, next to Jesus Christ, none upon earth had greater power than he. They afterwards went farther, and equalled him in all things to Jesus Christ, without being opposed by Hacket, who used to say in his prayers, "Father, I know thou lovest me equally with thyself." As they protested a most unreserved obedience to him, he ordered them to go and proclaim, through all the streets of London, that Jesus Christ was come to judge the world, and lodged in such an inn; and that nobody could put him to death. They did so; and, drawing together a vast concourse of people, discoursed of the important mission of William Hacket. They returned to him; and, when they saw him, Arthington cried out, "Behold the king of the earth!" They were prosecuted and tried: Hacket was sentenced, and executed accordingly July 28, 1592.

The blasphemies he uttered in his prayer upon the scaffold are so horrid, that we cannot transcribe them. He had an inconceivable hatred against queen Elizabeth, whom, as he confessed to the judges, he had stabbed to the heart in effigy: and he cursed her with all manner of imprecations a little before he was hanged. As for Coppinger and Arthington, the former famished in prison, and the latter, upon his repentance, was pardoned.—These instances serve to shew, that there is nothing too extravagant for the human heart to be capable

capable of; and might, one would hope, be of use to those, who would attentively contemplate them.

HACKET (JOHN), bishop of Lichfield and Coventry, was descended from an ancient family, and born in London, September 1, 1592. He was admitted very young into Westminster-school; and, in 1608, elected thence to Trinity-college, in Cambridge. His uncommon parts and learning recommended him to particular notice; so that, after taking the proper degrees, he was chosen fellow of his college, and became a tutor of great repute. One month in the long vacation, retiring with his pupil, afterwards lord Byron, into Nottinghamshire, he there composed a latin comedy, intituled, "Loyala," which was twice acted before James I. and printed in 1648. He took orders in 1618, and had singular kindness shewn him by bishop Andrews and several great men. But, above all others, he was regarded by Dr. Williams, dean of Westminster and bishop of Lincoln, who, being appointed lord-keeper of the great seal in 1621, chose Hacket for his chaplain, and ever loved and esteemed him above the rest of his chaplains. In 1623, he was made chaplain to James I. and also a prebendary of Lincoln; and the year following, upon the lord-keeper's recommendation, rector of St. Andrew's, Holborn, in London. His patron also procured him the same year the rectory of Cheam, in Surrey; telling him, that he intended Holborn for wealth, and Cheam for health.

In 1625, he was named by the king himself, to attend an ambassador into Germany; yet was dissuaded from the journey by being told, that, on account of his severe treatment of the jesuits in his "Loyola," he would never be able to go safe, though in an ambassador's train. In 1628, he commenced D. D. and, in 1631, was made archdeacon of Bedford. His church of St. Andrew being old and decayed, he undertook to rebuild it, and for that purpose got together a great sum of money in stock and subscriptions; but, upon the breaking out of the civil war, this was seized by the parliament, as well as what had been gathered for the repair of St. Paul's cathedral. March 1646, he was one of the sub-committee, appointed by the house of lords to consult of what was amiss and wanted correction in the liturgy, in hopes by that means to dispel the cloud hanging over the church. He made a speech against the bill for taking away deans and chapters, which is published at length in his life by Dr. Plume. March 1642, he was presented to a residentiary's place in St. Paul's London; but, the troubles coming on, he had no enjoyment of it, nor of his rectory of St. Andrew's. Besides, some of his parishioners there having articled against him, at the committee of plunderers, his friend Selden told him it

was

was in vain to make any defence; and advised him to retire to Cheam, where he would endeavour to prevent his being molested. He was disturbed here by the earl of Essex's army, who marching that way took him prisoner along with them; but he was soon after dismissed, and from that time lay hid in his retirement at Cheam, where we hear no more of him, except that, in 1648-9, he attended in his last moments Henry Rich, earl of Holland, who was beheaded for attempting the relief of Colchester.

After the restoration of Charles II. he recovered all his preferments, and was offered the bishopric of Gloucester, which he refused; but he accepted shortly after that of Litchfield and Coventry, and was consecrated December 22, 1661. The spring following he repaired to Litchfield, where, finding the cathedral almost battered to the ground, he set up in eight years a complete church again, better than ever it was before, at the expence of 20,000l. of which he had 1000l. from the dean and chapter: and the rest was of his own charge, and procuring from benefactors. He laid out 1000l. upon a prebendal house, which he was forced to live in, his palaces at Litchfield and Ecclefhall having been demolished during the civil war. He added to Trinity-college, in Cambridge, a building called Bishop's hostel, which cost him 1200l. ordering that the rents of the chambers should be laid out in books for the college-library. Besides these acts of munificence, he left several benefactions by will; as 50l. to Clare-hall, 50l. to St. John's college, and all his books, which cost him about 1500l. to the university library. He died at Lichfield, October 21, 1670, and was buried in the cathedral, under a handsome tomb, erected by his eldest son Sir. Andrew Hacket, a master in chancery; for he was twice married, and had several children by both his wives.

He published only the comedy of "Loyola" above mentioned, and "A Sermon preached before the King, March 22, 1660;" but, after his decease, "A Century of his Sermons upon several remarkable Subjects" was published by Thomas Plume, D. D. in 1675, folio; and, in 1693, "The Life of Archbishop Williams," folio, of which an improved abridgement was published in 1700, 8vo. by Ambrose Philips. He intended to have written the life of James I. and for that purpose the lord-keeper Williams had given him Camden's MS. notes or annals of that king's reign; but, these being lost in the confusion of the times, he was disabled from doing it. He was a man of great acuteness, and applied himself to all parts of learning, but could never make himself master of the oriental languages. He was deeply versed in ecclesiastical history, especially as to what concerned our own church. In the

the univerſity, when young, he was much addicted to ſchool-learning; but grew afterwards weary of it, as being full of ſhadows without ſubſtance, and containing horrid and barbarous terms, more fit, he would ſay, for incantation than divinity. He was a man of exemplary conduct, and as remarkable for virtue and piety as for parts and learning.

HACKSPAN (THEODORE), a Lutheran miniſter, who was a great proficient in the oriental languages, and the firſt profeſſor at A torf. He wrote a number of books, on theological ſubjects, which are much eſteemed in Germany. He died in 1659.

HADDOCK (SIR RICHARD), was a gallant ſea-officer in the time of James the ſecond. Although a proteſtant, he was alike the favourite of Rupert and of James. He died in 1714.

HADDON (Dr. WALTER), an eminent ſcholar, and great reſtorer of the learned languages in England, was deſcended from a good family in Buckinghamſhire, and born in 1516. He was educated at Eton-ſchool, and thence elected to King's college in Cambridge: where he greatly diſtinguiſhed himſelf by his parts and learning, and particularly by writing latin in a fine Ciceronian ſtyle. He ſtudied alſo the civil law, of which he became doctor; and read public lectures in it. In 1550, he was made profeſſor of it; he was alſo for ſome time profeſſor of rhetoric, and orator of the univerſity. During king Edward's reign, he was one of the moſt illuſtrious promoters of the reformation; and therefore, upon the deprivation of Gardiner, was thought a proper perſon to ſucceed him in the maſterſhip of Trinity-hall. September 1552, through the earneſt recommendation of the court, though not qualified according to the ſtatutes, he was choſen preſident of Magdalen-college in Oxford; but, October 1553, upon the acceſſion of queen Mary, he quitted the preſident's place for fear of being expelled, or perhaps worſe uſed, at Gardiner's viſitation of the ſaid college. He is ſuppoſed to have lain concealed in England all this reign; but, on the acceſſion of Elizabeth, was ordered by the privy council to repair to her majeſty at Hatfield in Herefordſhire, and ſoon after conſtituted by her one of the maſters of the court of requeſts. Biſhop Parker alſo made him judge of his prerogative-court. In the royal viſitation of the univerſity of Cambridge, performed in the beginning of Elizabeth's reign, he was one of her majeſty's commiſſioners, as appears by the ſpeech he then made, printed among his works. In 1566, he was one of the three agents ſent to Bruges to reſtore commerce between England and the Netherlands upon ancient terms. He died Jan. 1571-2, and was buried in Chriſt-church, London. He was engaged, with Sir John Cheke, in turning into latin and drawing up that uſeful code of eccleſiaſtical law, publiſhed in 1571,

1571, by the learned John Fox, under this title, "Reformatio Legum Ecclefiasticarum," in 4to. He publifhed, in 1653, a letter, or anfwer to an epiftle, directed to queen Elizabeth, by Jerom Oforio, bifhop of Silva in Portugal, and intituled, "Admonitio ad Elizabetham reginam Angliæ:" wherein the Englifh nation, and the reformation of the church, were taeated in a falfe, abufive, and fcurrilous, manner. His other works were collected and publifhed in 1567, 4to. under the title of "Lucubrationes." This collection contains ten latin orations; fourteen letters, befides the above-mentioned to Oforio; and alfo poems. Many of our writers fpeak in high terms of Haddon, and not without reafon; for, through every part of his writings, his piety appears equal to his learning and politnefs.

HADRIAN VI. Pope of Rome, was born at Utrecht, 1459. His father, whofe name was Florent Boyens, was in a low condition of life; fome fay a barge-maker, others a brewer, and others a weaver. Be this as it will, he was certainly fo poor, that his fon Hadrian, who, according to the cuftom of the country, took the furname of Florent, being defirous of a learned education, was forced to beg a place in the Pope's college at Louvain, where poor fcholars are brought up gratis. We are told, that he ufed to read at night by the light of the lamps that were hung up in the churches, or the corners of ftreets; which may ferve as a proof both of his poverty and his ftudious temper. As he had a genius for learning, he made great progrefs in all kinds of fciences, and became in a few years an able divine. The princefs Margaret, daughter of Maximilian the emperor, being informed of his learning and piety, (for, his manners were alfo exemplary,) gave him a cure in Holland, and furnifhed him with all neceffary charges to take his degree of D. D. which he did at Louvain in 1491. A little after he was made canon of St. Peter and divinity-profeffor in the fame city: and afterwards dean of St. Peter, and vice-chancellor of the univerfity. Being now in good circumftances, and willing to teftify his gratitude to the univerfity which had raifed him, he built a college at Louvain, of his own name, to receive poor fcholars. His reputation in a little time gained him many benefices, as the deanery of Antwerp, the treafury of the chapter of St. Mary the Greater at Utrecht, and the provoftfhip of our Saviour in the fame city.

In 1507, he was removed from a collegiate life to court; for, the emperor Maximilian, wanting a preceptor for his grandfon the archduke Charles, then about feven years old, thought he could not find a fitter perfon for that place than Dr. Hadrian Florent. The young prince made no great progrefs in latin under him; and it is faid that his governor
Chievres

Chievres was the cause of it; who, desiring to have the sole possession of his pupil, and all the glory of his progress, cultivated his inclination and bias, which lay for politics and arms, and made him quite indifferent about his improvement by the lessons of the Louvain professor. Hadrian, not able to stand it out against Chievres, contented himself with forewarning his young scholar, that he would repent of his negligence hereafter. He did so; and Jovius speaks of it as a thing that happened in his presence, how, upon hearing a speech made to him in latin, after he was emperor, and not understanding it, he cried out with a sigh, "Hadrian told me how it would be." However, the preceptor had as noble recompences for his pains, how ineffectual soever they might prove to his pupil, as any man of that employ ever had; for, it was Charles V's interest which raised him to the papacy.

Maximilian was so pleased with the service of Hadrian, that he sent him ambassador to Ferdinand of Spain, whose daughter he had married, to obtain the favour of that prince for the archduke Charles; and, it is said, he managed things with much greater address than could be expected from a man who had so long breathed the air of an university. Ferdinand honoured Hadrian with the bishopric of Tortosa; who still continued ambassador, and discharged all the functions of that office, till the death of Ferdinand. Charles, then becoming heir of his dominions, left the bishop of Tortosa in Spain, that he might have part of the government with Ximenes cardinal of Toledo. He was soon after made a cardinal by Leo X. at the recommendation of Maximilian, in a promotion made by that Pope, July 1517. Charles, going into Spain after Ximenes was sent home, was so pleased with the negotiations of Hadrian, that, when he went to receive the imperial crown, he appointed him governor of Spain in his absence.

The holy see becoming vacant by the death of Leo X. cardinal Julius de Medicis, who had a powerful faction in the conclave, not being able to carry it for himself, agreed at last with the other cardinals to give their votes for the cardinal of Tortosa, who was absent; judging him fit to be raised to the papacy, as one learned in theological matters to oppose Luther, and, in political, to quiet the troubles of Italy. These two qualifications, rarely to be found in the same man, met together in Hadrian; who had given proofs of the one by his lectures and writings, and of the other by his government of Spain. He received the news of his election at Victoria in Biscay, and assumed the next day the pontifical habit, in the presence of some bishops, whom he assembled in haste, without waiting for the legates, whom the sacred college should send.

fend. He departed a little after to Rome; and, having paſſed through Barcelona, and thence to Terragon, he embarked for Italy; where arriving, he made his entry at Rome in Auguſt, and was crowned the next day by the name of Hadrian VI.

Hadrian found no little buſineſs at his arrival. Italy was in a combuſtion, by reaſon of a war between the emperor and the king of France. The holy ſee was at variance with the dukes of Ferrara and Urbin. The city of Rome afflicted with ſickneſs: Rimini newly ſeized by the houſe of Malateſta: the cardinals divided, and defying one another: the Iſle of Rhodes beſieged by the Turks: the treaſury exhauſted: the goods of the church engaged by his predeceſſor: the whole eccleſiaſtical ſtate fallen into diſorder through an anarchy of eight months: and, what affected him the moſt, the reformation by Luther, which gained ground, and grew ſtronger every day in Germany. He applied himſelf as faſt as he could to remedy theſe diſorders, but the ſhortneſs of his pontificate permitted him to do but little; for he died October 24, 1523, in his 64th year, without being able to make any great progreſs in removing the evils which diſturbed the eccleſiaſtical ſtate, within or without. He had very little ſatisfaction in his triple crown, as we may learn from the inſcription he ordered to be engraved upon his tomb: " Adrianus VI. hic ſitus eſt, qui nihil ſibi infelicius in vita duxit, quam quod imperaret;" that is, " Here lies Hadrian VI. who eſteemed no misfortune, which happened to him in life, ſo great as that of being called to govern."

It has been thought ſtrange, that a Pope, who owed his advancement to his learning, and who was himſelf an author, ſhould give ſo little countenance to men of letters. One of the things which made him decried by the Italians was his ſlighting of poetry and delicacy of ſtyle: two accompliſhments, by which many under Leo X. had made their fortunes, and upon which they had valued themſelves principally in that country for fifty or ſixty years. He was ſo little diſpoſed to favour poets, that one of the reaſons, Jovius gives for experiencing his kindneſs, was, becauſe he had not joined poetry to the ſtudy of the liberal arts. The paganiſm which the poets ſcattered in their works contributed, it ſeems, not a little to this pope's coldneſs for them; for he did not underſtand raillery in this point, nor could he be prevailed on to be complaiſant in theſe matters. He was no admirer either of fine painting or of antique ſtatues; ſo that, when Vianeſius, the ambaſſador from Bologna, was commending the ſtatue of Laocoon, which pope Julius had bought at an immenſe price, and ſet up in the gardens of the Belvidere, he turned

away

away his eyes, to shew his dislike of the images of that idolatrous people. This contempt of poetry and the fine arts may easily be conceived to have rendered him very ungracious in the eyes of the Italians: it was however more pardonable than sinking the funds, as he did, which had been employed for the maintenance of learned men, who came from Greece into Italy, and to whom the West is indebted for the resurrection of letters. Cardinal Bessarion maintained at Rome part of those great genii, and established an academy for them in the Vatican. The greatest number subsisted upon the bounties of pope Nicolas V. of all whose successors, says a certain writer, there was none but Hadrian VI. who suppressed these gratifications by an œconomy, which doth no honour to his memory.

He was nevertheless a great and good man in many respects. He did not dissemble the abuses he observed in the church: he publicly acknowleged them, and that in a strong manner, in his instructions to the nuncio, who was to speak in his name at the diet of Nuremberg. He had long wished to introduce among the clergy a reformation of manners, and had laboured to effect this while he was dean of St. Peter's at Louvain: but the fruitlessness of his pains had obliged him to desist from the attempt.

We have said he was an author. He published a piece or two of school-divinity before his advancement to the pontificate, and "Regulæ Cancellariæ Apostolicæ" after. He wrote many letters to the princes of Germany, which were printed with the councils, and elsewhere.

HAEN (ANTONY DE), privy counsellor and physician to the empress Maria Theresa. He wrote several books, and with great ability. His principal performances are his "Ratio medendi," in 17 vols. 8vo. and a "Treatise on Magic," in which he vindicates the possibility and real existence of that art. He died in 1776.

HAGEDORN, a german poet of the present century, deserving of much praise for the spirit and delicacy of his sentiments. He was a great imitator of Fontaine; but wrote also many original works.

HAGGAI, is usually reckoned the tenth in order among the prophets. He appears to have been inspired by God to exhort Zerubbabel and Joshua the high priest to resume the work of the temple. He prophesied in the second year of Darius Hystaspes. Lowth and Newcome are at variance about the style of Haggai; the former calling it prosaic, the latter affirming that a great part of it admits of a metrical division. Haggai was probably of the sacerdotal race; and

Epiphanius relates that he was buried among the priests at Jerusalem.

HAQUENIER (JOHN), a french poet, born in Burgundy, of great facetiousness and convivial accomplishments. He wrote many light pieces of poetry, and died in 1738.

HAHN (SIMON FREDERIC), a young man of extraordinary abilities. At the age of ten years he knew many languages, and at twenty-four was professor of history at Holmstadt. He was made historiographer to the king of Great Britain, at Hanover; but died in 1729, at the early age of thirty-seven. He wrote a " History of the Empire," and a work entitled, " Collectio Monumentorum veterum et recentiorum ineditorum," 2 vols. 8vo.

HAILLAN (BERNARD DE GIRARD, lord of), a french historian, of an ancient family, was born at Bordeaux about 1535. He went to court at twenty years of age, and set up early for an author. His first appearance in the republic of letters was in the quality of a poet and translator. In 1559, he published a poem, intituled, " The Union of the Princes, by the Marriages of Philip King of Spain and the Lady Elizabeth of France, and of Philbert Emanuel Duke of Savoy, and the Lady Margaret of France;" and another intituled, " The Tomb of the most Christian King Henry II." In 1560 he published an abridged translation of " Tully's Offices," and of " Eutropius's Roman History:" and, in 1568, of " The Life of Æmilius Probus." He applied himself afterwards to the writing of history, and succeeded so well, that, by his first performances of this nature, he obtained of Charles IX. the title of historiographer of France 1571. He had published the year before at Paris a book intituled, " Of the State and Success of the Affairs of France;" which was reckoned very curious, and was often reprinted. He augmented it in several successive editions, and dedicated it to Henry IV. in 1694: the best editions of it are those of Paris 1609 and 1613, in 8vo. He had published also the same year a work intituled, " Of the Fortune and Power of France, with a Summary Discourse on the Design of a History of France:" though Niceron suspects that this may be the same with " The Promise and Design of the History of France," which he published in 1571, in order to let Charles IX. see what he might expect from him in support of the great honour he had conferred on him of historiographer of France. In 1576, he published a history, which reaches from Pharamond to the death of Charles VII. and was the first who composed a body of the french history in french. Henry III. was very well pleased with this, and shewed his satisfaction by the advantageous and honourable gratifications he made the author.

The

The reasons which induced de Haillan to conclude his work with Charles VIIth's death are very good, and shew that he understood the duties of an historian. He considered the alternative to which a man is exposed, who writes the history of monarchs lately dead; viz. that he must either dissemble the truth, or provoke persons who are most to be feared. However, he afterwards promised Henry IV. to continue this history to his time; as may be seen in his dedication to him of this work in 1594. As for the promises he made of continuing the history of France, they came to nothing. Nothing of this kind was found among his papers after his death: the booksellers, who added a continuation to his work as far as to 1615; and afterwards as far as to 1627, took it from Paulus Æmilius, de Comines, Arnoul Ferron, du Bellay, &c.

Du Haillan died at Paris, November 23, 1610. Dupleix remarks, that he was originally a calvinist, but changed his religion, in order to ingratiate himself at court. It must not be forgotten, that he attended, in quality of secretary, Francis de Noailles, bishop of Acqs, in his embassies to England and Venice, in 1556 and 1557. His dedications and prefaces shew, that he was not disinterested enough, either as to glory or fortune. He displays too much his labours, and the success of his books, their several editions, translations, &c. and he too palpably manifests desires of reward. It was with du Haillan, as it always will be with men who make no other use of letters than to serve the purposes of avarice and ambition: for, learning, if it be not applied to correct the depravity of the human heart, is but too apt to increase it, and so is often found to inflame the passions, instead of appeasing them.

HAINES (JOSEPH), commonly called Count Haines, was a very eminent low comedian, and a person of great facetiousness of temper and readiness of wit. When, or where, or of what parents, he was born, are particulars about which the historians of his life are totally silent. It is certain, however, that the earlier parts of his education were communicated to him at the school of St. Martin's in the Fields, where he made so rapid a progress as to become the admiration of all who knew him. From this place he was sent by the voluntary subscription of a number of gentlemen, to whose notice his quickness of parts had strongly recommended him to Queen's college, Oxford, where his learning and great fund of humour gained him the esteem and regard of Sir Joseph Williamson, who was afterwards secretary of state, and minister plenipotentiary at the concluding of the peace of Ryswick. When Sir Joseph was appointed to the first of those high offices, he took our author as his latin secretary. But taciturnity not being one of those qualities for which Haines was eminent, Sir Joseph found that, through his means, affairs of great importance frequently transpired, even before they came to the knowlege of those who were more immediately concerned in them. He was, therefore, obliged to remove him from an employment for which he seemed so ill calculated; but recommended him, however, to one of the heads of the university of Cambridge, where he was very kindly received; but, a company of comedians coming to perform at Stirbridge fair, Mr. Haines took so sudden an inclination for their employment and way of living, that he threw away his cap and band, and immediately joined their company. It was not long, however, before the reputation of his theatrical abilities procured him an invitation to the Theatre-royal, in Drury-lane, where his inimitable performance, together with his vivacity and pleasantry in private conversation, introduced him not only to the acquaintance, but even the familiarity, of persons of the most exalted abilities, and of the first rank in the kingdom. Insomuch, that a certain noble duke, being appointed ambassador to the french court, thought it no disgrace to take Joe Haines with him as a companion, who being, besides his knowledge of the dead languages, as perfect master of the french and italian, as if he had been a native of the respective capitals of Paris and Rome, was greatly caressed by many of the french nobility. On his return from France, where he had assumed the title of count, he again applied himself to the stage, on which he continued till 1701, on the 4th of April in which year he died of a fever, after a very short illness, at his lodgings in Hart-street, Long-acre, and was buried in the church-yard of St. Paul's, Covent-garden.

HAKEM, the third of the Fatamite Caliphs, a frantic youth, alike remarkable for his impiety and despotism. At first he pretended to be a zealous mussulman; but his vanity became finally so extravagant, that he styled himself the visible image of the most high God, who, after nine apparitions, was at length manifested in Hakem's person. At his very name every knee was bent in religious adoration, and at the present hour the Druses are persuaded of his divinity. He persecuted the jews and christians, he destroyed the sacred edifices at Jerusalem; and, after many barbarous and frantic acts, was assassinated by the emissaries of his sister.

HAKEWILL (GEORGE), a learned divine, was the son of a merchant in Exeter, and born there in 1579. After a proper education in classical literature, he was admitted of St. Alban's hall, in Oxford; where he became so noted a disputant and orator, that it seems he was unanimously elected

ed fellow of Exeter-college at two years standing. He was afterwards made chaplain to prince Charles, and archdeacon of Surrey, in 1616; but never raised to any higher dignity, on account of the zealous opposition he made to the match of the Infanta of Spain with the prince-his master. Wood relates the story thus. After Hakewill had written a small tract against that match, not without reflecting on the spaniard, he caused it to be transcribed in a fair hand, and then presented it to the prince. The prince perused it, and shewed it to the king; who, being highly offended at it, caused the author to be imprisoned. This was in August 1621; soon after which, being released, he was dismissed from his attendance on the prince. He was afterwards elected rector of Exeter-college, but resided very little there; for, the civil war breaking out, he retired to his rectory of Heanton near Barnstable in Devonshire, and there continued to the time of his death; which happened in 1649. He wrote several things; but his principal work, and that for which he is most known, is, "An Apology or Declaration of the Power and Providence of God in the Government of the World, proving that it doth not decay, &c." in four books, 1627. To which were added two more in the third edition, 1635, folio.

He had a brother John, who was mayor of Exeter in 1632; and an elder brother William, who was of Exeter-college, and removed thence to Lincoln's inn, where he arrived at eminence in the study of the common law. He was always a puritan, and therefore had great interest with the prevailing party in the civil war. He published some pieces in his own way; and, among the rest, "The Liberty of the Subject against the pretended Power of Impositions, &c. 1641," 4to,

HAKLUYT (RICHARD), famous for his skill in the naval history of England, was descended from an ancient family at Eyton, in Herefordshire, and born about 1553. He was trained up at Westminster-school; and, in 1570, removed to Christ-church college in Oxford. While he was at school, he used to visit his cousin Richard Hakluyt, of Eyton, Esq; at his chambers in the Middle-temple: which Richard Hakluyt was well known and esteemed, not only by some principal ministers of state, but also by the most noted persons among the mercantile and maritime part of the kingdom, as a great encourager of navigation, and the improvement of trade, arts, and manufactures. At this gentleman's chambers young Hakluyt met with books of cosmography, voyages, travels, and maps; and he was so infinitely pleased with them, that he resolved henceforth to direct his studies that way, to which he was not a little encouraged by his cousin. For this

this purpose, as soon as he got to Oxford, he made himself a master in the modern as well as ancient languages; and then read over whatever printed or written discourses of voyages and discoveries, naval enterprizes, and adventures of all kinds, he found either extant in greek, latin, italian, spanish, portuguese, french, or english. By these means he became so conspicuous in these branches of science, that he was chosen to read public lectures in them at Oxford, and was the first man there who introduced maps, globes, spheres, and other instruments of this art, into the common schools. In process of time, he became known and respected by the principal sea-commanders, merchants, and mariners of our nation; and, though it was but a few years after that he went to reside a long time beyond sea, yet his fame travelled thither long before him. He held a correspondence with the learned in these matters abroad, as with Ortelius, the king of Spain's cosmographer, Mercator, &c.

In 1582, he published a small "Collection of Voyages and Discoveries;" in the epistle dedicatory of which to Mr. Philip Sidney it appears, that his lecture upon navigation above mentioned was so well approved of by Sir Francis Drake, that the latter made some proposals to continue and establish it in Oxford. The same year, he received particular encouragements from secretary Walsingham to pursue the study of cosmography, and to persevere in the same commendable collections and communications. The secretary also gave him a commission to confer with the mayor and merchants of Bristol, upon the naval expedition they were undertaking to Newfoundland; and incited him to impart to them such intelligence and advertisements as he should think useful. Hakluyt did so; and in acknowledgement of the services he had done them, the secretary sent him the following letter, to be found in the third volume of his voyages in folio. "Sir Francis Walsingham to Mr. Richard Hakluyt, of Christ-church in Oxford. I understand, as well by a letter I long received from the mayor of Bristol, as by conference with Sir George Peckham, that you have endeavoured and given much light for the discovery of the Western parts yet unknown. As your studie in these things is very commendable, so I thanke you for the same; wishing you to continue your travel in these and like matters, which are like to turne, not only to your owne good in private, but to the public benefite of this realm. And so I bid you farewell. From the court, the 11th of March 1582. Your loving friend, Francis Walsingham."

About 1584, he attended Sir Edward Stafford as his chaplain, when that gentleman went over ambassador to France; and continued there some years with him. He was made a
prebendary

prebendary of Briſtol in his abſence. During his reſidence at Paris, he contracted an acquaintance with all the eminent mathematicians, coſmographers, and other literati in his own ſphere of ſtudy. He enquired after every thing that had any relation to our engliſh diſcoveries; and prevailed with ſome to ſearch their libraries for the ſame. At laſt, having met with a choice narrative in MS. containing "The notable Hiſtory of Florida," which had been diſcovered about twenty years before by captain Loudonniere and other french adventurers, he procured the publication thereof at Paris at his own expence. This was in 1586; and, May 1587, he publiſhed an engliſh tranſlation of it, which he dedicated, after the example of the french editor, to Sir Walter Raleigh. The ſame year he publiſhed a new edition of Peter Martyr's book, intituled "De Orbe Novo," illuſtrated with marginal notes, a commodious index, a map of New England and America, and a copious dedication, alſo, to Sir Walter Raleigh; and this book he afterwards cauſed to be tranſlated into engliſh.

Hakluyt returned to England in the memorable year 1588, and applied himſelf to ſet forth the naval hiſtory of England more expreſsly and more extenſively than it had ever yet appeared; and in this he was encouraged by Sir Walter Raleigh in particular. He applied himſelf ſo cloſely to collect, tranſlate, and digeſt, all voyages, journals, narratives, patents, letters, inſtructions, &c. relating to the engliſh navigations, which he could procure either in print or MS. that, towards the end of 1589, he publiſhed his ſaid collections in one volume folio, with a dedication to Sir Francis Walſingham, who was a principal patron and promoter of the work. About 1564, he entered into the ſtate of matrimony; yet it did not divert him from going on with his collections of engliſh voyages, till he had increaſed them into three volumes folio: and, as he was perpetually employed himſelf, ſo he did not ceaſe to invite others to the ſame uſeful labours. Thus Mr. John Pory, whom he calls his honeſt, induſtrious, and learned friend, undertook, at his inſtigation, and probably under his inſpection, to tranſlate from the ſpaniſh "Leo's Geographical Hiſtory of Africa," which was publiſhed at London, 1600, in folio. Hakluyt himſelf appeared in 1601, with the tranſlation of another hiſtory, written by Antonio Galvano in the portugueſe tongue, and corrected and amended by himſelf. This hiſtory was printed in 4to, and contains a compendious relation of the moſt conſiderable diſcoveries in various parts of the univerſe from the earlieſt to the later times.

In 1605, he was made a prebendary of Weſtminſter; which, with the rectory of Wetheringſet in Suffolk, is all the eccleſiaſtical fiaftical promotion we find he obtained. About this time the tranflation of Peter Martyr's "Hiftory of the Weft-Indies" was undertaken, and firft publifhed by Mr. Lock, at the requeft and encouragement of our author: for, befides his own publications of naval hiftory, far fuperior to any thing of the like kind that had ever appeared in this kingdom, he was no lefs active in encouraging others to tranflate and familiarize among us the conquefts and difcoveries of foreign adventurers. This, and the fpirit with which he alfo animated thofe of his countrymen, who were engaged in naval enterprizes, by his ufeful communications, gained the higheft efteem and honour to his name and memory, from mariners of all ranks, in the moft diftant nations no lefs than his own. Of this there are feveral inftances; and particularly in thofe Northern difcoveries, that were made at the charges of the Mufcovy merchants in 1608, under captain W. Hudfon: when among other places there denominated, on the continent of Greenland, which were formerly difcovered, they diftinguifhed an eminent promontory, lying in 80 degrees northward, by the name of Hakluyt's Headland. In 1609, he publifhed a tranflation from the Portuguefe of an hiftory of Virginia, intitled, "Virginia richly valued, by the Defcription of the maine Land of Florida, her next Neighbour, &c." and dedicated to the right worfhipful counfellors, and others the chearful adventurers for the advancement of that chriftian and noble plantation of Virginia. Upon the revival of our plantation in that country, which afterwards enfued, Drayton the poet thus apoftrophifes our author, in his "Ode to the Virginian Voyage:"

> "Thy voyages attend,
> Induftrious Hakluyt;
> Whofe reading fhall inflame
> Men to feek fame,
> And much to commend
> To after-times thy wit."

In 1611, we find Edmund Hakluyt, the fon of our author, entered a ftudent of Trinity-college, Cambridge. In the fame year, the Northern difcoverers, in a voyage to Peckora in Ruffia, called a full and active current, they arrived at, by the name of Hakluyt's River; and, in 1614, it appears that the banner and arms of the king of England were erected at Hakluyt's Headland above-mentioned. Our hiftorian died November 23, 1616, and was buried in Weftminfter-abbey. His MS. remains, which might have made another volume, falling into the hands of Mr. Purchas, were difperfed by him throughout his four volumes of voyages.

HALDE (JOHN BAPTIST DU), a learned frenchman, was born at Paris in 1674; and entered into the fociety of the Jefuits, among whom he died in 1743. He was extremely well verfed in all which regarded the Afiatic geography; and we have of his a work, intitled, "Grande Defcription de la Chine & de la Tartarie," which he compofed from original memoirs of the jefuitical miffionaries. This great and learned work, on which he fpent much time and pains, was publifhed after his death in four volumes folio; and contains many curious and interefting particulars. He was concerned in a collection of letters, called, "Des Lettres Edifiantes," in 18 volumes, begun by father Gobien. He publifhed alfo fome latin poems 'and orations.

HALE (Sir MATTHEW), a moft learned lawyer, and chief juftice of the King's-bench, was born at Alderfly in Gloucefterfhire, November 1, 1600. His father was a barrifter of Lincoln's inn; and, being puritanically inclined, caufed him to be inftructed in grammar-learning by Mr. Staunton, vicar of Wotton-under-Edge, a noted puritan. In 1626, he was admitted of Magdalen-hall in Oxford, where he laid the foundation of that learning and knowledge, on which he afterwards raifed fo vaft a fuperftructure. Here however he fell into many levities and extravagances, and was preparing to go along with his tutor, who went chaplain to lord Vere into the Low-countries, with a refolution of entering himfelf into the prince of Orange's army: from which mad fcheme he was diverted, by being engaged in a law-fuit with Sir William Whitmore, who laid claim to part of his eftate. Afterwards, by the perfuafions of ferjeant Glanville, he refolved upon the ftudy of the law, and was admitted of Lincoln's-inn, November 1629. And now he became as grave as before he had been gay; ftudied at the rate of fixteen hours a day; and threw afide all appearance of vanity in his apparel. He is faid indeed to have neglected the point of drefs fo much, that, being a ftrong and well-built man, he was once taken by a prefs gang, as a perfon very fit for fea-fervice: which pleafant miftake made him regard more decency in his cloaths for the future, though never to any fuperfluity or vanity in them. What confirmed him ftill more in a ferious and regular way of life, was an accident, which is related to have befallen one of his companions. Hale, with other young ftudents of the Inn, being invited out of town, one of the company called for fo much wine, that, notwithftanding all Hale could do to prevent it, he went on in his excefs, till he fell down as dead before them: fo that all prefent were not a little affrighted at it, and did what they could to bring him to himfelf again. This particularly affected Hale, being naturally of a religious make:

who thereupon went into another room, and, falling down upon his knees, prayed earnestly to God, both for his friend, that he might be restored to life again, and for himself, that he might be forgiven the being present and countenancing so much excess: and he vowed to God, that he would never again keep company in that manner, nor drink a health while he lived. His friend recovered; and henceforward he forsook all his gay acquaintance, and divided his whole time between the duties of religion and the studies of his profession.

Not satisfied with the law-books then published, but resolved to take things from the fountain-head, he was very diligent in searching records; and with collections out of the books he read, together with his own learned observations, he made a most valuable common-place book. He was early taken notice of by the attorney-general Noy, who directed him to his studies, and admitted him to such an intimacy with him, that he came to be called young Noy. Selden also soon found him out, and took such a liking to him, that he not only lived in great friendship with him, but left him at his death one of his executors. Selden put him upon a more enlarged pursuit of learning, which he had before confined to his own profession; so that he arrived in time to a considerable knowledge in the civil law, in arithmetic, algebra, and other mathematical sciences, as well as in physic, anatomy, and surgery. He was also very conversant in experimental philosophy, and other branches of philosophical learning; and in ancient history and chronology. But above all, he seemed to have made divinity his chief study, so that those who read what he has written, might be inclined to think, that he had studied nothing else.

Some time before the civil wars broke out, he was called to the bar, and began to make a figure in the world; but, observing how difficult it was to preserve his integrity, and yet live securely, he resolved to follow those two maxims of Atticus, whom he proposed to himself as a pattern, viz. "To engage in no faction nor meddle in public business, and constantly to favour and relieve those that were lowest." He often relieved the royalists in their necessities, which so ingratiated him with them, that he came generally to be employed by them in his profession. He was one of the counsel to the earl of Strafford, archbishop Laud, and king Charles himself; as also the duke of Hamilton, the earl of Holland, the lord Capel, and the lord Craven: but being esteemed a plain honest man, and of great knowledge in the law, he was entertained by both parties, the presbyterians as well as the loyalists. In 1643, he took the covenant, and appeared several times with other lay persons among the assembly of divines. He was then in great esteem with the parliament, and employed by them in several affairs,

affairs, particularly in the reduction of the garrison at Oxford; being, as a lawyer, added to the commissioners named by the parliament, to treat with those appointed by the king. In that capacity he did good service, by advising them, especially the general Fairfax, to preserve the seat of learning from ruin. Afterwards, though he was greatly grieved at the murder of Charles I. yet he took the oath called 'The Engagement;' and, January 1651-2, was one of those appointed to consider of the reformation of the law. Cromwell, who well knew the advantage it would be to have the countenance of such a man as Hale to his courts, never left importuning him, till he accepted the place of one of the justices of the common bench, as it was called; for which purpose he was by writ made serjeant at law, January 25, 1653-4. In that station he acted with great integrity and courage. He had at first great scruples concerning the authority under which he was to act; and, after having gone two or three circuits, he refused to sit any more on the crown side; that is, to try any more criminals. He had indeed so carried himself in some trials, that the powers then in being were not unwilling he should withdraw himself from meddling any farther in them: of which Burnet gives the following instance. Soon after he was made a judge, a trial was brought before him, upon the circuit at Lincoln, concerning the murder of one of the townsmen, who had been of the king's army, and was killed by a soldier of the garrison there. He was in the field with a fowling-piece on his shoulder, which the soldier seeing, he came to him and said, he was acting against an order the protector had made, viz. "That none who had been of the king's party should carry arms;" and so would have forced the piece from him. But the other not regarding the order, and being the stronger man, threw down the soldier; and, having beat him, left him. The soldier went to the town, and telling a comrade how he had been used, got him to go with him, and help him to be revenged on his adversary. They both watched his coming to town, and one of them went to him to demand his gun; which he refusing, the soldier struck at him; as they were struggling, the other came behind, and ran his sword into his body, of which he presently died. It was in the time of the assizes, so they were both tried. Against the one there was no evidence of malice prepense, so he was only found guilty of manslaughter, and burnt in the hand; but the other was found guilty of murder: and though colonel Whaley, who commanded the garrison, came into the court, and urged, that the man was killed only for disobeying the protector's order, and that the soldier was but doing his duty; yet the judge regarded both his reasonings and threatenings very little, and therefore not only gave sentence against him, but ordered the execution to be so suddenly done, that it might not be possible to procure a reprieve.

When Cromwell died, he not only excused himself from accepting the mourning that was sent him, but also refused the new commission offered him by Richard; alleging, that "he could act no longer under such authority." He did not sit in Cromwell's second parliament in 1565; but in Richard's, which met in January 1658-9, he was one of the burgesses for the university of Oxford. In the healing parliament in 1660, which recalled Charles II. he was elected one of the knights for the county of Gloucester; and moved, that a committee might be appointed to look into the propositions that had been made, and the concessions that had been offered by Charles I. during the late war, that thence such propositions might be digested as they should think fit to be sent over to the king at Breda. The king upon his return recalled him in June, by writ, to the degree of serjeant at law: and, upon settling the courts in Westminster-hall, constituted him in November chief baron of the Exchequer. When chancellor Clarendon delivered him his commission, he told him, that, "if the king could have found out an honester and fitter man for that employment, he would not have advanced him to it; and that he had therefore preferred him, because he knew none that deserved it so well." He continued eleven years in that place, and very much raised the reputation and practice of the court by his impartial administration of justice, as also by his generosity, vast diligence, and great exactness in trials. According to his rule of favouring and relieving those that were lowest, he was now very charitable to the Nonconformists, and took care to cover them as much as possible from the severities of the law. He thought many of them had merited highly in the affair of the king's restoration, and at least deserved that the terms of conformity should not have been made stricter than they were before the war. In 1671, he was promoted to the place of lord chief justice of England, and behaved in that high station with his usual strictness, regularity, and diligence; but, about four years and a half after this advancement, he was on a sudden brought very low by an inflammation in his midriff, which in two days time broke his constitution to that degree, that he never recovered; for, his illness turned to an asthma, which terminated in a dropsy. Finding himself unable to discharge the duties of his function, he petitioned, in January 1675-6, for a writ of ease; which being delayed, he surrendered his office in February. He died December 25th following, and was interred in the church-yard of Alderley among his ancestors: for, he did not approve

approve of burying in churches, but used to say, "That churches were for the living, and church-yards for the dead." He was knighted soon after the Restoration; and twice married, having by his first wife ten children.

He was the author of several things which were published by himself: namely, 1. "An Essay touching the Gravitation or Non-gravitation of Fluid Bodies, and the Reasons thereof." 2. "Difficiles Nugæ, or Observations touching the Torricellian Experiment, and the various Solutions of the same, especially touching the Weight and Elasticity of the Air." 3. "Observations touching the Principles of natural Motion, and especially touching Rarefaction and Condensation." 4. "Contemplations moral and divine." 5. "An English Translation of the Life of Pomponius Atticus, written by Corn. Nepos; together with Observations political and moral." 6. "The primitive Origination of Mankind considered and explained according to the Light of Nature, &c." He left also at his decease other works, which were published; namely, 1. "Pleas of the Crown; or a methodical Summary of the principal Matters relating to that Subject." 2. "Discourse touching Provisions for the Poor." 3. "A Treatise touching the Sheriffs Accounts:" to which is joined his "Trial of Witches at the Assizes held at Bury St. Edmund's on March 1, 1664." 4. "His Judgement of the Nature of true Religion, the Causes of its Corruption, and the Church's Calamity by Men's Addition and Violences, with the desired Cure." 5. "Several Tracts; as, "A Discourse of Religion under three Heads, &c." His "Treatise concerning Provision for the Poor" already mentioned. "A Letter to his Children, advising them how to behave in their Speech." "A Letter to one of his Sons after his Recovery from the Smallpox." 6. "Discourse of the Knowledge of God and of ourselves, first by the Light of Nature; secondly, by the sacred Scriptures." 7. "The original Institution, Power, and Jurisdiction, of Parliaments." 8. "The History of the Pleas of the Crown;" first published in 1736 from his original MS. and the several references to the records examined by the originals, with large notes, by Sollom Emlyn of Lincoln's inn, Esq; 2 vols. folio. The House of Commons had made an order, November 29, 1680, that it should be printed then; but it never was printed till 1736. By his will he bequeathed to the Society of Lincoln's inn his MS. books, of inestimable value, which he had been near forty years in gathering with great industry and expence. "He desired they should be kept safe and all together, bound in leather, and chained; not lent out or disposed of: only, if any of his posterity of that society should desire to transcribe any book, and give good
caution

caution to reſtore it again in a prefixed time, they ſhould be lent to him, and but one volume at a time: They are," ſays he, "a treaſure not fit for every man's view; nor is every man capable of making uſe of them."

HALES (JOHN), uſually called the Ever Memorable, was born at Bath in Somerſetſhire, in 1584, and educated in garmmar-learning there. At thirteen years of age, he was ſent to Corpus-Chriſti college in Oxford: and, in 1605, choſen fellow of Merton by the intereſt and contrivance of Sir Henry Saville, warden of that college; who, obſerving the prodigious pregnancy of his parts, reſolved to bring him in, and employed him, though young, in his edition of the works of St. Chryſoſtom. His knowledge of the greek tongue was ſo conſummate, that he was not only appointed to read the greek lecture in his college, but alſo made in 1612 greek profeſſor to the univerſity. Sir Thomas Bodley, founder of the Bodleian library, dying in 1613, Hales was choſen by the univerſity to make his funeral oration, and the ſame year admitted a fellow of Eton-college. Five years after, in 1618, he accompanied Sir Dudley Carleton, king James's ambaſſador to the Hague, in quality of chaplain; and by theſe means procured admiſſion to the ſynod of Dort, held at that time. He had the advantage of being preſent at the ſeſſions or meetings of that ſynod, and was witneſs to all their proceedings and tranſactions; of which he gave Sir Dudley an account in a ſeries of letters, printed afterwards among his "Golden Remains." His friend Farindon tells us, in a letter prefixed to this collection, that Hales "in his younger days was a calviniſt, and even then when he was employed at that ſynod; and that at the well preſſing of St. John iii. 16. by Epiſcopius there, 'I bid John Calvin good night,' as he hath often told me." He grew very fond of the remonſtrants method of theologizing; and after his return to England, being of a frank and open diſpoſition, wrote and talked in ſuch a manner as brought him under the ſuſpicion of being inclined to ſocinianiſm, ſo far, in ſhort, that books actually written by ſocinians were attributed to him.

In the mean time, he had a moſt ardent thirſt after truth, and a deſire to have religion freed from whatever did not belong to it, and reduced to its primitive purity and ſimplicity; which temper of his was ſufficiently made known by a ſmall tract, he wrote for the uſe of his friend Chillingworth, concerning ſchiſm and ſchiſmatics; in which he traced the original cauſe of all ſchiſm, and, delivered with much freedom, his principles about eccleſiaſtical peace and concord. This tract being handed about in MS. a copy of it fell into the hands of Laud; who, being diſpleaſed with ſome things in it, occaſioned Hales to draw up a vindication of himſelf in a remarkable

remarkable letter, which was first printed in the seventh edition of a pamphlet entituled ".Difficulties and Discouragements, &c." He also sent for him, in 1638, to Lambeth, and, after a conference of several hours, appears not only to have been reconciled to him, but even to have admitted him into his friendship. Some are of opinion, that the archbishop used Hales's assistance in composing the second edition, in 1639, of his "Answer to the Jesuit Fisher," where the objections of A. C. against the first addition are so fully and so learnedly confuted; and it is certain that Hales was, the same year, preferred to a canonry of Windsor, which could not be done without the approbation and favour of the archbishop. This, however, he did not enjoy longer than to the beginning of the civil wars in 1642. About the time of Laud's death, he retired from the college at Eton to private lodgings in that town, where he remained for a quarter of a year unknown to any one, living only upon bread and beer; and, when he heard of the archbishop's death, wished his own head had been taken off instead of his. He continued in his fellowship at Eton, though refusing the covenant, nor complying in any thing with the times; but was ejected upon his refusal to take the Engagement. After this, he underwent incredible hardships, and was obliged to sell one of the most valuable libraries that ever was in the possession of a private man for the support of himself and his friends.

Nothing shews the unfortunate condition, he was and had been in, better than the conversation he had one day with his intimate friend Farindon. This worthy person coming to see Hales some few months before his death, found him in very mean lodgings at Eton, but in a temper gravely chearful, and well becoming a good man under such circumstances. After a slight and homely dinner, suitable to their situation, some discourse passed between them concerning their old friends, and the black and dismal aspect of the times; and at last Hales asked Farindon to walk out with him into the church-yard. There this unhappy man's necessities pressed him to tell his friend, that he had been forced to sell his whole library, save a few books, which he had given away, and six or eight little books of devotion, which lay in his chamber; and that, for money, he had no more than what he then shewed him, which was about seven or eight shillings; and "besides," says he, " I doubt I am indebted for my lodging." Farindon, it seems, did not imagine that it had been so very low with him, and therefore was much surprised to hear it; but said, that " he had at present money to command, and to-morrow would pay him fifty pounds, in part of the many sums he and his wife had received of him in their great necessities, and would pay him more as he should want it." But Hales replied,

plied, "No, you don't owe me a penny; or, if you do, I here forgive you; for, you shall never pay me a penny. I know you and yours will have occasion for much more than what you have lately gotten: but if you know any other friend that hath too full a purse, and will spare me some of it, I will not refuse that." To this Hales added, "When I die, which I hope is not far off, for I am weary of this uncharitable world, I desire you to see me buried in that place in the church-yard," pointing to the place. "But why not in the church," said Farrindon, "with the provost (Sir Henry Savile), Sir Henry Wotton, and the rest of your friends and predecessors?" "Because," says he, "I am neither the founder of it, nor have I been a benefactor to it, nor shall I ever now be able to be so." He died May 19, 1656, aged 72; and the day after was buried in Eton-college church-yard. He is reported to have said in his former days, that he "thought he should never die a martyr;" but he suffered more than many martyrs have suffered, and certainly died little less than a martyr to the establishment in church and state.

All writers and parties have agreed in giving to him the character of one of the greatest as well as best of men that any age has produced. "He was," says Wood, "highly esteemed by learned men beyond and within the seas; from whom he seldom failed to receive letters every week, wherein his judgement was desired as to several points of learning." And as, with the profound learning of a scholar, he had all the politeness of a man of wit, so the same historian tells us, that "when the king and court resided at Windsor, he was frequented by noblemen and courtiers, who delighted much in his company; not for his severe or retired walks of learning, but for his polite discourses, stories, and poetry, in which last, it is supposed, he was excellent. That he had a talent for poetry, appears from Sir John Suckling's mentioning him in his "Session of Poets:"

"Hales, set by himself, most gravely did smile
To see them about nothing keep such a coil.
Apollo had spied him, but knowing his mind
Past by, and called Falkland that sat just behind."

And it is well known, that he was intimately acquainted with the most eminent wits and poets of his time, such as Falkland, Suckling, Davenant, Jonson, &c. But his talent for poetry, how excellent soever, was far from being the most considerable of his accomplishments.

We do not find that Hales ever suffered any thing to be published in his life-time, except his oration at the funeral of Sir Thomas Bodley, in 1613: this was printed at Oxford that year.

year, and again in the "Vitæ felectorum aliquot virorum, &c," by Bates, in 1681. Bifhop Pearfon fays, that "while he lived, none was ever more folicited and urged to write, and thereby truly teach the world, than he; but that none was ever fo refolved, pardon the expreffion, fo obftinate againft it." However, two or three years after his death, namely in 1659, there came out a collection of his works with this title, "Golden Remains of the ever-memorable Mr. John Hales of Eton-college, &c." which was enlarged with additional pieces in a fecond edition of 1673. This collection confifts of fermons, mifcellanies, and letters; all of them written upon particular occafions. In 1677, there appeared another collection of his works, intituled, "Several Tracts by the ever-memorable Mr. John Hales, &c." The 1ft of which is, "Concerning the Sin againft the Holy Ghoft;" 2. "Concerning the Sacrament of the Lord's Supper, and whether the Church may err in Fundamentals;" 3. "A Paraphrafe on the 12th chapter of the Gofpel according to St. Matthew;" 4. "Concerning the Power of the Keys, and auricular Confeffion;" 5. "Concerning Schifm and Schifmatics;" and fome fhort pieces intituled, "Mifcellanies." There is no preface nor advertifement to this volume, which feems to have been put out by the unknown editor with caution; but it is finely and correctly printed, with Mr. Hales's picture before it. To thefe volumes of pofthumous works we muft add the letter to archbifhop Laud, mentioned before, which was printed in 1716.

HALES (STEPHEN), was born in 1677, of a good family in Kent; his grandfather having been created a baronet by Charles II. In 1696, he was entered a penfioner at Benet-college in Cambridge; and was admitted a fellow in 1703. The bent of his genius to natural philofophy began foon to fhew itfelf. Botany was his firft ftudy; in which he took infinite pains, when he was a very young man. With Ray's "Catalogue of Cambridge Plants" in his pocket, we are told, he took many a painful walk among Gogmagog hills, and the bogs of Cherryhunt Moor. In thefe expeditions likewife he ufed to collect fof..s, and fometimes infects, and contrived a curious inftrument for taking fuch of them as could fly; and in chemiftry is faid, even when very young, to have made a confiderable progrefs. He not only conftantly attended the lectures, which Vigain read in the cloifters of Queen's college; but himfelf went through the procefs of moft of Mr. Boyle's experiments. But what made him moft remarkable at the univerfity was the invention of a machine of brafs, to demonftrate the motions of the planets. This machine was conftructed with great ingenuity, and was nearly the fame with that which was afterwards invented by Rowley, under the name of the Orrery.

Our philosopher, who had now been admitted to a doctor's degreee, began to be much taken notice of in the philosophical world; and was elected a fellow of the Royal Society. He soon after received the thanks of that learned body, for some experiments he communicated to them, on the nature of vegetation. In 1741, he published his invention of ventilators, which he continued to improve as long as he lived. About six or seven years afterwards, one of these machines was put up in the prison of the Savoy; the benefit of which was soon, acknowledged. In general between fifty and one hundred had died every year of the gaol-distemper in that place; but, after his machine was erected, four persons only died in two years, though the number of prisoners often exceeded two hundred. The use of ventilators afterwards became general, in the king's ships and other places. In the last war, after long solicitations, he procured an order from the French king to erect ventilators in the prisons where the English were kept; and the writer of this memoir has heard him merrily say, " he hoped nobody would inform against him for corresponding with the enemy." It would be endless to mention his various researches into nature, and his various schemes for the benefit of mankind; most of which are to be found in the Transactions of the Royal Society, which he chose as his vehicle for the communication of them to the public. They all discover great knowledge of the secrets of nature, which he was able to apply to agriculture, physic, and various other arts in life. His " Statical Essays," in two volumes 8vo, have been often printed, and are well known.

He spent most of the latter part of his life at his parsonage at Teddington, near Hampton-Court. Here he was honoured with the friendship of some of the greatest persons in the nation, whom, without any of the fashionable modes of polite breeding, he visited and received with patriarchal simplicity. Among those who honoured him with a particular esteem was the late prince Frederic, father of the present king; who would often take great pleasure in surprising him in his laboratory. After the death of that prince, when the household of the princess was settled, he was appointed her almoner; and soon afterwards nominated to a canonry of Windsor. When he first heard of the honour that was designed him, he immediately waited upon the princess, and engaged her to put a stop to the affair. His circumstances, he said, were such as entirely satisfied him; and a better income would only be a greater incumbrance,

Hales

Hales deferved, as much as any man ever did, the title of a Chriftian Philofopher. All his ftudies, and all his refearches into nature, tended only to one point, that of doing good to mankind. In 'this employment, bleft with ferenity of mind, and an excellent conftitution of body, he attained the age of eighty-four years; and died, after a fhort illnefs, January 4, 1761.

HALI-BEIGH, a polander, whofe original name was Bobowfki, was born a chriftian; but, being taken by the tartars while a child, was fold to the turks, who educated him in their religion. He acquired the knowledge of feventeen languages, among the reft, of the french, englifh, and german, having had part of his education in thefe countries; and became interpreter to the Grand Signior. He tranflated into the turkifh language the catechifm of the church of England, and all the bible. He compofed a turkifh grammar and dctionary, and other things which were never printed. His principal work is, "A Treatife upon the Liturgy of the Turks, their Pilgrimages to Mecca, their Circumcifion, and Manner of vifiting the Sick;" which was publifhed by Thomas Smith in latin, in the appendix of the "Itinera Mundi ab Abrahamo Peritfol," printed at Oxford in 1691. His death, which happened in 1675, prevented the execution of a defign which he had formed of returning to the chriftian religion. He is fuppofed to have furnifhed Ricaut, the conful of Symrna, with fome materials for his book, intituled, "The State of the Ottoman Empire."

HALL (JOSEPH), an eminent and learned divine, and fucceffively bifhop of Exeter and Norwich, was born July 1, 1574, in Briftow Park, within the parifh of Afhby de la Zouch in Leicefterfhire, of honeft parentage. His fchooleducation was at his native place; and, at the age of fifteen, he was fent to Emanuel-college in Cambridge; of which in due time, after taking his degrees, he became fellow. He often difputed and preached before the univerfity; and he read alfo the rhetoric-lecture in the public fchools for two years with great applaufe. He diftinguifhed himfelf as a wit and poet in this early feafon of his life; for he publifhed, in 1597, "Virgidemiarum; Satires in Six Books." The three firft are called toothlefs fatires, poetical, academical, moral: the three laft, biting fatires. They were reprinted at Oxford in 1753, 8vo. He calls himfelf in the prologue the firft fatyrift in the Englifh language:

"I firft adventure, follow me who lift,
And be the fecond englifh fatyrift."

After

After six or seven years stay in college, he was presented to the rectory of Halstead, in Suffolk, by Sir Robert Drury; and, being thus settled, married a wife, with whom he lived happily forty-nine years. In 1605, he accompanied Sir Edmund Bacon to the Spa, where he composed his second "Century of Meditations." He had an opportunity, in this journey, of informing himself of the state and practice of the romish church; and at Brussels he entered into a conference with Coster the jesuit. After his return, having some misunderstanding with his patron about the rights of his living, he resolved to quit it, as soon as he could conveniently; and, while he was meditating on this, Edward lord Denny, afterwards earl of Norwich, gave him the donative of Waltham Holy-Cross in Essex. About the same time, which was in 1612, he took the degree of D. D. He had been made chaplain a little before to Prince Henry, who was much taken with his meditations, and with two sermons he had preached before him; and on that account conferred this honour upon him. In the second year of his monthly attendance, when he solicited a dismission, the prince ordered him to stay longer, promising him suitable preferments: but, being loth to forsake his noble patron, who had placed his heart much upon him, he waved the offer, and remained twenty-two years at Waltham. In the mean time he was made prebendary of the collegiate church of Wolverhampton; and, in 1616, dean of Worcester, though he was then absent, attending the embassy of lord Hay into France. The year after, he attended his majesty into Scotland as one of his chaplains; and the year after that, viz. in 1688, was sent to the synod of Dort, with others of our English divines. Indisposition obliged him to return home very soon; however, before his departure, he preached a latin sermon to that famous assembly, which by their president and assistants took a solemn leave of him; and the deputies of the States dismissed him with an honourable retribution, and sent after him a rich gold medal, having on it the portraiture of the synod.

Having refused in 1624 the bishopric of Gloucester, he accepted in 1627 that of Exeter. Though he was reckoned a favourer of puritanism, yet he wrote, in the beginning of the troubles, with great strength in defence of episcopacy. November 1641, he was translated to the see of Norwich; but on December 30 following, having joined with other bishops in the protestation against the validity of all laws made during their forced absence from the parliament, he was voted amongst the rest to the tower, and committed thither January 30, in all the extremity of frost, at eight o'clock in a dark evening. About June 1642, he was released upon giving 5000l. bail,

and

and withdrew to Norwich, where he lived in tolerable quiet till April 1643. But then, the order for sequestering notorious delinquents being passed, in which he was included by name, all his rents were stopped, and he had nothing but what the parliament allowed him; all the while suffering the greatest inconveniences, which he has given an account of in a piece, intituled his "Hard Measure." In 1647, he retired to a little estate, which he rented at Heigham, near Norwich; and in this retirement he ended his life September 8, 1656, in his 82d year. He was buried in the church-yard of that parish without any memorial: for in his will he has this passage, "I do not hold God's house a meet repository for the dead bodies of the greatest saints."

He is universally allowed to have been a man of great wit and learning, and of as great meekness, modesty, and piety. He was so great a lover of study, that he earnestly wished his health would have allowed him to do it even to excess. His work, besides the "Satires" above mentioned, make in all five volumes in folio and 4to; and "are filled," says Mr. Bayle, "with fine thoughts, excellent morality, and a great deal of piety." His writings shew, that he was very zealous against popery; neither was he more favourable to those who separated from the mother-church without an extreme necessity. He lamented the divisions of protestants, and wrote something with a view of putting an end to them.

Two of his pieces were published in 1662, with Dury's "Irenicorum Tractatuum Prodromus." His "Miscellaneous Letters" are, in the judgement of Mr. Bayle, very good: they are without date; but, being dedicated to prince Henry, we may conclude they were written before 1613, because that prince died November 6, 1612. He observes, in his epistle dedicatory, that it was not as yet usual in England to publish discourses in forms of letters, as was done in other nations. In the catalogue of his works is a satyrical piece, intituled, "Mundus idem, & alter, &c." that is, "The World different, yet the same." This is, as Mr. Bayle says, a learned and ingenious fiction, wherein he describes the vicious manners of several nations; the drunkenness of one, the lewdness of another, &c. and does not spare the court of Rome. We cannot find out in what year it was first published; but it was reprinted at Utrecht, 1643, in 12mo. to which edition, adorned with maps, is joined, because of the conformity of the matter, Campanella's "City of the Sun," and the "New Atalantis" of Chancellor Bacon. Gabriel Naude says of his work, that "it is calculated less to divert the readers than to inflame their minds with the love of virtue." Our author did not approve of English gentlemen travelling into foreign countries;

and compofed a book on that fubject, which he dedicated to lord Denny his patron. It is intituled, "Quo vadis? or a juft Cenfure of Travel, as it is commonly undertaken by the Gentlemen of our Nation."

HALL (JOHN), born at Durham, in Auguft 1627, after one year fpent at St. John's college, Cambridge, removed to Gray's inn, London, where he was called to the bar; but entering into the politics of the times, and writing on fubjects of that fort, he attracted the notice of parliament, who fent him into Scotland to attend Oliver Cromwell, and afterwards diftinguifhed him by other marks of favour: but, being too much addicted to pleafure, he fell a facrifice to its indulgence; and returning to his native city of Durham, died there, Auguft 1, 1656. In 1646 (during his fhort refidence at Cambridge), being then but nineteen years of age, he publifhed "Horæ Vacivæ, or Effayes," a fufficient proof of his abilities. His poems came out the fame year. He publifhed the firft englifh verfion of Longinus, which he intituled "The Height of Eloquence, Lond. 1652." 8vo. This he tranflated from the greek, as he alfo did "Hierocles upon the Golden Verfes of Pythagoras;" before which is an account ot the ingenious tranflator and his works, by John Davis of Kidwelly, by whom it was publifhed in 1657, 8vo. More of him and his writings may be feen in Wood's Athen. Oxon, 2d Ed. Vol. I. p. 534. Several of his poems are preferved in the "Select Collection," reprinted from a little volume (intituled, "Poems by John Hall, Cambridge, printed by Roger Daniel, Printer to the Univerfitie, 1646, for J. Rothwell at the Sun in St. Paul's Church-Yard, to which in 1647 was added 'The Second Booke of Divine Poems by J. H.'") which is now become exceedingly fcarce. Recommendatory verfes are prefixed to it by Jo. Pawfon (his tutor), H. More, W. Dillingham, W. Harrington, Ja. Windet, R. Marfhall, T. Smithfby, and Edw. Holland.

HALL (HENRY), M. A. born in London in 1716, was fent early to Eton, admitted on the foundation in 1729; and elected to King's college, Cambridge, in 1735, where of courfe he became a fellow in 1738, and took the degrees in arts. Being recommended by Dr. Chapman to archbifhop Potter, his grace appointed him his librarian at Lambeth in 1748, on the refignation of Mr. Jones. In that ftation he continued till the death of his patron in 1747; when archbifhop Herring, who fucceeded to the primacy, being fenfible of his merit [A], not only continued him in that office, but,

---

[A] His Grace, in one of his letters to Mr. Duncombe, faid, "I have an excellent young man for my librarian, who never did and never can offend me."

on his taking orders, appointed him one of his chaplains; and, in April 1750, collated him to the rectory of Harbledown (vacant by the promotion of Mr. Thomas Herring to the rectory of Chevening); in November 1752, the archbishop collated him also to the vicarage of Herne, which he held by dispensation; to which his grace afterwards added the sinecure rectory of Orpington, in the deanery of Shoreham, one of his peculiars. In 1756, Mr. Hall vacated Herne, on being presented to the vicarage of East Peckham by the dean and chapter of Canterbury, by whom he was much esteemed, having greatly assisted their auditor in digesting many of the records, charters, &c. preserved in their registry. In return, the late Dr. Walwyn (one of the prebendaries, who vacated that vicarage) was called by the archbishop to the rectory of Great Mongeham, void by the death of Mr. Byrch. On the death of archbishop Herring in 1757, he resigned the librarianship of Lambeth, and from that time resided chiefly at Harbledown, in a large house, which he hired, now the seat of Robert Mead Wilmott, Esq. only son of Sir Edward. Soon after the death of archbishop Herring, Mr. Hall was presented by his executors to the treasurership of the cathedral of Wells, one of his grace's options. He was also at first a competitor for the precentorship of Lincoln, an option of archbishop Potter (which Dr. Richardson gained in 1760 by a decree of the House of Lords); but soon withdrew his claim, well-grounded as it seemed. His learning and abilities were great, but not superior to his modesty; and by his singular affability he obtained the love and esteem of all who knew him. His charitable attention to his poor parishioners, especially when they were ill, was constant and exemplary. At archbishop Secker's primary visitation at Canterbury, in 1758, Mr Hall was " pitched upon" (his Grace's official expression) to preach before him at St. Margaret's church, which he did from Acts xvii. 21. " For all the Athenians and strangers which were there spent their time in nothing else, but to tell or hear some new thing." He died a bachelor, at Harbledown, Nov. 2, 1763, in the 47th year of his age, after a short illness, occasioned by a violent swelling in the neck, which could not be accounted for by the eminent physicians who attended him. He was buried under the communion table, at Harbledown churchurch, without any epitaph.

HALL (JOHN), a surgeon in the reign of Elizabeth. He resided at Maidstone in Kent, and translated several chirurgical treatises, of which an account may be found in bishop Tanner. Hall was also author of a book of hymns, with musical notes.

HALL (JACOB), a celebrated rope-dancer in the reign of Charles the second. His eloquence and symmetry of person were

were so remarkable, and were united with so much strength and agility, that he captivated many of the females belonging to that licentious Court, and in particular the dutchess of Cleveland, from whom he received a pension.

HALL (RICHARD), an English divine of the roman communion, who left England, it is said, in consequence of the penal laws then exacted against the papists by queen Elizabeth. He went to the spanish Netherlands, and was professor of divinity at Douay. He published several books, and died in the year 1604.

HALLE (PETER), professor of canon law in the university of Paris, was born at Bayeux in Normandy, September 8, 1611. He studied philosophy, the law, and divinity, for five years in the university of Caen; and also applied himself to poetry, under the direction of his uncle Anthony Hallé, who was an eminent poet, with such success, that he gained the prizes in the poetical exercises that are performed every year in these two cities, "to the honour of the immaculate conception of the Virgin Mary." This procured him so much reputation, that, though he was still very young, he was chosen teacher of rhetoric in the university of Caen. Some time after, being rector of the university, he made an oration to M. Seguier, chancellor of France, then in Normandy, to suppress some popular insurrections; which was so much approved by that head of the law, that he received a doctor of law's cap from his hands in 1640. He attended M. Seguier to Paris, and gained such reputation by some pieces he published, that they offered him the mastership of five different colleges; and he was incorporated in his absence (a very unusual thing) into the body of the university, 1641. He was made king's poet, and reader of the latin and greek tongues in the royal college, 1646. His strong application to study having ruined his health, he was obliged to rest for two years, in order to recover it. He afterwards resolved to raise the glory of the faculty of the law, which was miserably sunk; and, in 1655, he obtained the post of regius professor of the canon law, when he vigorously began, and, though he met with great difficulties, successfully executed what he had resolved.

Besides "Canonical Institutions," which he published in 1685, he wrote also for the use of his pupils several treatises upon the civil and canon laws; as, concerning councils, the Pope's authority, the regale, simony, usury, censures, regular persons, ecclesiastical benefices, matrimony, last wills and testaments, &c. He had published in 1655, 8vo. "A Collection of Latin Poems and Orations." He died December 27, 1689.

HALLE (Antony), profeffor of eloquence at the univerfity of Caen, and one of the beft latin poets of his age. He publifhed fome treatifes, a "Latin Grammar," and various pieces of poetry. He died at Paris in 1675, at the age of eighty-threee.

HALLE (Claude Guy), a french painter of no mean eminence, and director of the Academy of Painters, at Paris. He adorned many of the public edifices, in and near Paris, with his works, particularly the church of Notre Dame. He was remarkable for his fweetnefs of manners, and died univerfally lamented in 1736, aged eighty-five.

HALLER (Albert), an illuftrious phyfician, who died at Bern, in Switzerland, December 12, 1777, in his 75th year. While profeffor of medicine at Gottingen, he filled fucceffively the botanical, chemical, and anatomical, chairs; and raifed the reputation of that univerfity to a very high pitch. He is fuppofed to have been the moft acute, various, and original, genius, that has appeared in the medical world fince Boerhaave. His ftudies, however, were not confined to medicine: he wrote many ingenious moral effays, fome theological tracts, and a few odes, which, for elegance of diction and harmony of numbers, are not reckoned inferior to any poetical productions in the german language. In 1760, he retired to Bern, where he was elected a fenator, and enjoyed the firft authority in the adminiftration of public affairs till the time of his death.

HALLEY (Edmund), a moft eminent Englifh philofopher and aftronomer, was born in the parifh of St. Leonard, Shoreditch, near London, October 29, 1656. His father, a wealthy citizen in Winchefter-ftreet, put him to St. Paul's fchool under the learned Dr. Thomas Gale; where he not only excelled in all parts of claffical learning, but made an uncommon advance in mathematics; fo much that, as Wood fays, he had perfectly learnt the ufe of the celeftial globe, and could make a complete dial; and we are informed by Halley himfelf, that he obferved the change of the variation of the magnetic needle at London in 1672, that is, one year before he left fchool. In 1673, he was entered a commoner of Queen's college in Oxford, where he applied himfelf to practical and geometrical aftronomy, in which he was greatly affifted by a curious apparatus of inftruments, which his father, willing to encourage his fon's genius, had purchafed for him. At nineteen, he began to oblige the public with new obfervations and difcoveries, and continued to do fo to the end of a very long life. It would greatly exceed the bounds, propofed in thefe memoirs, to enter into a detail of all Halley's productions; and the reader will be able to form as clear a notion of the man from a relation of some of the most considerable. Besides particular observations, made from time to time, upon the celestial phænomena, he had, from his first admission into college, pursued a general scheme for ascertaining the true places of the fixed stars, and thereby correcting the errors of Tycho Brahe. His original view was to carry on the design of that first restorer of astronomy, by completing the catalogue of those stars from his own observations; but, upon farther enquiry, finding this province taken up by Hevelius and Flamstead, he dropped that pursuit, and formed another; which was, to perfect the whole scheme of the heavens by the addition of the stars which lie so near the south pole that they could not be observed by those astronomers, as never rising above the horizon either at Dantzick or Greenwich. With this view he left the university, before he had taken a degree, and applied himself to Sir Joseph Williamson, then secretary of state, and to Sir Jonas Moore, surveyor, both encouragers of these studies; who applauding his purpose, mentioned it to Charles II. The king was much pleased with the thing, and immediately recommended him to the East-India company, who thereupon promised to supply him with all the accommodations and conveniences they could, and to carry him to St. Helena, then in their possession by a grant from the crown, which he pitched upon as a proper situation for his design. Accordingly he embarked for that island, November 1676; and, arriving there safely in three months, stuck close to his telescope, till he finished his task, and completed his catalogue. This done, he returned to England, November 1678; and, having delineated a planisphere, wherein he laid down the exact places of all the stars near the south pole, from his own observations, he presented it, with a short description, to his majesty. Among these stars there appeared (such was his address) the "Constellation of the Royal Oak," with this description: "Robur Carolinum in perpetuam sub illius latebris servati Caroli Secundi Magnæ Britanniæ Regis memoriam, in cœlum merito translatum." The king was greatly satisfied with Halley, and gave him, at his own request, a letter of mandamus to the university of Oxford for the degree of M. A. the words of which are, that "his majesty has received a good account of his learning as to the mathematics and astronomy, whereof he has gotten a good testimony by the observations he has made during his abode in the island of St. Helena." This letter was dated November 18, and the same month he was also chosen fellow of the Royal Society. Indeed his catalogue of these southern stars merited particular honour: it was an entirely-new acquisition to the astronomical world, and

might

might not unaptly be called "Cœlum Auſtrale eo uſque incognitum;" and thence he acquired a juſt claim to the title, which, by Flamſtead, was not long after given him, the Southern Tycho.

In 1697, he was pitched upon by the Royal Society to go to Dantzick, for the ſatisfaction of Hevelius the conſul, to adjuſt a diſpute between him and our Hooke, about the preference of plain or glaſs ſights in aſtroſcopical inſtruments. He ſet out May 14 of this year, with a letter recommendatory from that ſociety, and arrived at that city on the 26th. He waited on the conſul immediately, and, after ſome converſation, agreed to enter upon the buſineſs of his viſit that ſame night; on which, and every night afterwards, when the ſky permitted, the two aſtronomers made their obſervations together, till July 18, when Halley left Dantzick, and returned to England. Here he continued till the latter end of the following year 1680; when he ſet out upon what is uſually called the grand tour, accompanied by the celebrated Mr. Nelſon, who had been his ſchool-fellow, and was his friend. They croſſed the water in December to Calais; and, in the mid-way thence to Paris, Halley had, firſt of any one, a ſight of the remarkable comet, as it then appeared a ſecond time that year in its return from the ſun. He had, the November before, ſeen it in its deſcent, and now haſtened to complete his obſervations upon it, in viewing it from the Royal Obſervatory of France. That building had been finiſhed not many years before; and Halley's deſign in this part of his tour was to ſettle a friendly correſpondence between the two royal aſtronomers of Greenwich and Paris; watching, in the mean time, all occaſions of improving himſelf under ſo great a maſter as Caſſini, as he had done before under Hevelius. From Paris he went with his fellow-traveller by the way of Lyons to Italy, where he ſpent a great part of the year 1681; but his affairs then calling him home, he left Mr. Nelſon at Rome, and returned to England, after making ſome ſtay a ſecond time at Paris.

Soon after his return to England, he married the daughter of Mr. Tooke, auditor of the Exchequer; and took a houſe at Iſlington, near London, where he immediately ſet up his tube and ſextant, and eagerly purſued his favourite ſtudy. In 1683, he publiſhed his "Theory of the Variation of the Magnetical Compaſs," wherein he ſuppoſes, "the whole globe of the earth to be one great magnet, having four magnetical poles or points of attraction, &c." The ſame year alſo, he entered early upon a new method of finding out the longitude by a moſt accurate obſervation of the moon's motion. His purſuits are ſaid to have been interrupted about this

this time by the death of his father, who, having suffered greatly by the fire of London, as well as by a second marriage into which he had imprudently entered, was found to have wasted his fortunes. He soon, however, resumed his usual occupations; for, January 1614, he turned his thoughts upon the subject of Kepler's sesquialterate proportion, and, after some meditation, concluded from it, that the centripetal force must decrease in proportion to the squares of the distances reciprocally. He found himself, however, unable to make it out in any geometrical way, and therefore first applied to Mr. Hooke and Sir Christopher Wren; who not affording him any assistance, he went to Cambridge to Mr. Newton, who supplied him fully with what he had so ardently sought. But Halley, having now found an immense treasure, could not rest, till he had prevailed with the owner to enrich the public therewith; and to this interview the world is in some measure indebted for the "Principia Mathematica Philosophiæ Naturalis." The "Principia" were published in 1686; and Halley, who had the whole care of the impression by the direction of the Royal Society, presented it to James II. with a discourse of his own, giving a general account of the astronomical part of that book. He also wrote a very elegant copy of verses in latin, which are prefixed to the "Principia."

The same year he undertook to explain the cause of a natural phænomenon, which had, till then, baffled the researches of the ablest geographers. The Mediterranean Sea is observed not to swell in the least, although there is no visible discharge of the prodigious quantity of water which runs into it from nine large rivers, besides several small ones, and the constant setting in of the current at the mouth of the Streights. His solution of this difficulty gave so much satisfaction to the society, that he received orders to prosecute these enquiries. He did so; and having shewn, by the most accurate experiments, how that great increase of water was actually carried off in vapours raised by the action of the sun and wind upon the surface, he proceeded with the like success to point out the method used by nature to return the said vapours into the sea. This circulation he supposes to be carried on by the winds driving these vapours to the mountains; where, being collected, they form springs, which uniting become rivulets or brooks, and many of these again meeting in the valleys grow into large rivers, emptying themselves at last into the sea: thus demonstrating, in the most beautiful manner, the way in which the equilibrium of receipt and expence is continually preserved in the universal ocean. Mr. Halley still continued to give his labours to the world by the canal of the
" Philo-

"Philolofophical Tranfactions," of which, for many years, his pieces were the chief ornament and fupport.

Halley publifhed his "Theory of the Variation of the Magnetical Compafs," as we have already obferved, in 1683; which, though it was well received both at home and abroad, he found upon a review liable to great and infuperable objections. Yet the phænomena of the variation of the needle, upon which it is raifed, being fo many certain and indifputed facts, he fpared no pains to poffefs himfelf of all the obfervations, relating to it, he could poffibly come at. To this end he procured an application to be made to king William, who appointed him commander of the Paramour Pink, Auguft 19, 1698: with exprefs orders to feek by obfervations the difcovery of the rule of the variations, and, as the words of his commiffion run, "to call at his majefty's fettlements in America, and make fuch farther obfervations as are neceffary for the better laying down the longitude and latitude of thofe places, and to attempt the difcovery of what land lies to the fouth of the Weftern ocean." He fet out on this attempt November 24th following, and proceeded fo far as to crofs the line; but his men growing fickly and untractable, and his firft lieutenant mutinying, he returned home in June 1699 After getting his lieutenant tried and cafhired, he fet off, September following, a fecond time, having the fame fhip with another of lefs bulk, of which he had alfo the command. He traverfed the vaft Atlantic ocean from one hemifphere to another, as far as the ice would permit him to go; and, in his way back, touched at St. Helena, the coaft of Brazil, Cape Verd, Barbadoes, Madeiras, the Canaries, the coaft of Barbary, and many other latitudes, arriving in England in September 1700. Having thus furnifhed himfelf with a competent number of obfervations, he publifhed in 1701 "A General Chart, fhewing at one View the Variation of the Compafs in all thofe Seas, where the Englifh Navigators were acquainted;" and hereby, firft of any one, laid a fure foundation for the difcovery of the law or rule whereby the faid variation changes all over the world.

Halley had been at home little more than half a year, when he went in the fame fhip, with another exprefs commiffion from the king, to obferve the courfe of the tides in every part of the Britifh channel at home, and to take the longitude and latitude of the principal head-lands, in order to lay down the coaft truly. Thefe orders were executed with his ufual expedition and accuracy; and foon after his return he publifhed, in 1702, a large map of the Britifh channel. The emperor of Germany having refolved to make a convenient and fafe harbour for fhipping in that part of his dominions

which

which borders upon the Adriatic, Halley was sent this year by queen Anne to view the two ports on the Dalmatian coast, lying to that sea. He embarked November 27, went over to Holland, and passing thence through Germany to Vienna, proceeded to Istria, with a view of entering upon the execution of the emperor's design; but, some opposition being given to it by the dutch, it was laid aside: nevertheless, the emperor presented him with a rich diamond ring from his own finger, and gave him a letter of high commendation, written with his own hand to queen Anne. He was likewise received with great respect by the king of the Romans, by prince Eugene, and the principal officers of that court. Presently after his arrival in England, he was dispatched again upon the same business; and, passing through Osnaburgh and Hanover, arrived at Vienna, and was presented the same evening to the emperor, who directly sent his chief engineer to attend him to Istria.

He returned to England November 1703; and, Wallis being deceased a few weeks before, Halley was appointed Savilian professor of geometry at Oxford in his room, and had the degree of LL. D. conferred upon him by that university. He was scarcely settled at Oxford, when Aldrich, dean of Christ-church, engaged him to translate into latin from the arabic "Apollonius de Sectione Rationis." At the same time, from the account given of them by Pappus, he restored the two books, which are lost, of the same author, "De Sectione Spatii;" and the whole was published by him in one volume 8vo, at Oxford, 1706. Afterwards he took a share with his colleague, Dr. David Gregory, in preparing for the press the same Apollonius's "Conics;" and ventured to supply the whole 8th book, which is lost, of the original. He likewise added Serenus on the "Section of the Cylinder and Cone," printed from the original greek, with a latin translation, and published the whole, 1710, in folio; not to mention, that in the midst of all these publications the "Miscellanea Curiosa," in 3 vols. 8vo. had come out under his direction in 1708. In 1713, he succeeded Dr. afterwards Sir, Hans Sloan, in the post of secretary to the Royal Society; and, upon the death of Flamstead, in 1719, was appointed to succeed him at Greenwich by George I. which made Halley, that he might be more at liberty for the proper business of his situation, resign the post of secretary to the Royal Society in 1721.

Upon the accession of the late king, his consort queen Caroline thought proper to make a visit at the Royal Observatory; and, being pleased with every thing she saw, took notice that Dr. Halley had formerly served the crown as a

captain

captain in the navy; and she soon after obtained a grant of his half-pay for that commission, which he enjoyed from that time during his life. An offer was also made him of being appointed mathematical preceptor to the duke of Cumberland; but he declined that honour, by reason of his advanced age, and because he deemed the ordinary attendance upon that employ not consistent with the performance of his duty at Greenwich. August 1729, he was admitted as a foreign member of the Academy of Sciences at Paris. About 1737, he was seized with a paralytic disorder in his right hand, which, it is said, was the first attack he ever felt upon his constitution: however, he came as usual once a week till within a little while before his death, to see his friends in town on Thursday, before the meeting of the Royal Society. His paralytic disorder increasing, his strength gradually wore away, and he came at length to be wholly supported by such cordials as were ordered by his physician Dr. Mead. He expired as he sat in his chair, without a groan, January 14, 1741-2, in his 86th year.

HALLIFAX (SAMUEL), bishop of St. Asaph, was a man of great learning and abilities. He was the eldest son of an apothecary at Chesterfield, and educated at Jesus College, Cambridge. He was regius professor of Civil Law in that University, and acquired great reputation by his "Analysis of the Civil Law." In 1781, he was made bishop of Gloucester, and, in 1787, bishop of St. Asaph. His sermons, at bishop Warburton's lectures, have been deservedly admired; he was also an incomparable civilian, and remarkable for his acuteness as a public speaker. Dr. Hallifax also published an analysis of Butler's Analogy, which is written with great eloloquence, and evinces much profound thinking. He died at the age of sixty in 1790.

HAMBERGER (GEORGE ALBERT), a native of Franconia, and an eminent mathematician. He published many valuable pieces on philosophical subjects, and particularly on Optics and Hydraulics. He died at Jena in 1726.

HAMBERGER (GEORGE CHRISTOPHER), member of the University of Gottingen, published a number of books on various subjects; and seems to have been a man of considerable talents and erudition. He is best known in the literary world by an edition of the works of Orpheus, in which he was materially assisted by Gesner. He was born in 1726, and died in 1773.

HAMEL (JOHN BAPTISTE DU), a French philosopher and divine, was born at Vire in Lower Normandy, 1614. He passed through his first studies at Caen, and his course of rhetoric and philosophy at Paris. At eighteen, he wrote
a trea-

a treatise, in which he explained, in a very simple manner, and by one or two figures, Theodosius's three books upon Spherics; to which he added a tract upon Trigonometry extremely short yet precious, and designed as an introduction to astronomy. In one of his latter works he observes, that he was prompted by the vanity natural to a young man to publish this book: but, as Fontenelle remarks, there are few persons of that age capable of such an instance of vanity. At nineteen, he entered himself in the congregation of the oratory, where he continued ten years, and left it in order to be curate of Neuilli upon the Marne. He applied in the mean time intensely to study, and distinguished himself greatly by publishing works upon astronomy and philosophy. In 1666, Colbert proposed to Lewis XIV. a scheme, which was approved by his majesty, for establishing a Royal Academy of Sciences; and appointed our author secretary of it. In 1668, he attended M. Colbert de Croissy, plenipotentiary for the peace at Aix la Chapelle; and, upon the conclusion of it, accompanied him in his embassy to England, where he formed an acquaintance with the most eminent persons of this nation, particularly with Boyle, Ray, and Willis. Thence he went over to Holland, and so returned to France, having made a great number of useful observations in his Travels. In 1678, his " Philosophia Vetus & Nova, ad Usum Scholæ accommodata in Regia Burgundia pertractata," was printed at Paris in 4 vols. 12mo; and, in 1681, enlarged and reprinted there in six. This work, which was done by the order of M. Colbert, contains a judicious collection of the ancient and modern opinions in philosophy. Several years after its publication, the Jesuits carried it to the East-Indies, and taught it with success; and father Bovet, a missionary in China, wrote to Europe, that when his brethren and himself engaged in drawing up a system of philosophy in the Tartarian language for the emperor, one of their chief aids was Du Hamel's " Philosophia & Astronomica:" and they were then highly valued, though the improvements in philosophy since his time have brought them into discredit, by rendering them of little use. In 1697, he resigned his place of secretary of the Royal Academy of Sciences, which by his recommendation he procured for M. de Fontenelle. He had some years before this devoted himself to divinity, and published large works in this way. However, he did not lose all care of his former studies, but published at Paris, in 1698, " Regiæ Scientiarum Academiæ Historia," 4to, in four books; which, being greatly liked, he afterwards augmented with two books more. It contains an account of the foundation of the Royal Academy of Sciences and its Transactions, from 1666

1666 to 1700, and is now the moſt uſeful of any of his works relating to philoſophy; as perhaps the moſt uſeful which he publiſhed in theology, is his laſt work printed at Paris, 1706, in folio, and intituled, "Biblia Sacra Vulgatæ Editionis, una cum ſelectis ex optimis quibuſque interpretibus notis, prolegomenis, novis tabulis chronologicis & geographicis."

He died at Paris Auguſt 6, 1796, without any ſickneſs, and of mere old age, being almoſt eighty-three. Though he had quitted his cure at Neuilli in 1663, yet he went every year to viſit his old flock; and the day he ſpent there was kept as an holy-day by the whole village. He was highly eſteemed by the moſt eminent prelates of France, though he enjoyed but very ſmall preferments. He was a man of great modeſty, affability, piety, and integrity; he was diſintereſted, averſe to all conteſts, and exempt from jealouſy and affectation. He wrote Latin with remarkable purity and elegance.

HAMILTON (ANTONY COUNT), of an ancient Scotch family, but born in Ireland, whence with his family he paſſed over to France followers of the fate of Charles the Second. At the Reſtoration he again returned to England, but was a ſecond time compelled to baniſhment at the Revolution. He was an elegant and accompliſhed character; and was for many years the delight and ornament of the moſt ſplendid circles of ſociety, by his wit, his taſte, and, above all, his writings. His works have been often publiſhed, and conſiſt of pieces of Poetry, Fairy Tales, and "Memoirs of the Count de Grammont," all of which are excellent in their kind. The Fairy Tales were intended as a refined piece of ridicule on the paſſion for the marvellous, which made the Arabian Nights Entertainments ſo eagerly read at their firſt appearance. The Memoirs of Grammont will always excite curioſity, as giving a ſtriking and too faithful detail of the diſſolute manners of Charles the Second's Court. Count Hamilton died at St. Germains in 1720, aged ſeventy-four.

HAMILTON (GEORGE), earl of Orkney, a brave general, and fifth ſon of the earl of Selkirk. He greatly diſtinguiſhed himſelf at the battle of the Boyne, and at many ſieges and battles. William the Third made him a peer of Scotland. On the acceſſion of Queen Anne, he ſerved under the duke of Marlborough, and greatly contributed to the victories of Blenheim and Malplaquet. After paſſing through various honourable employments in theſe different reigns, he died in 1737.

HAMLET, the name of a prince of Denmark, whoſe hiſtory is related in Saxo Grammaticus the Daniſh hiſtorian, and whose name deserves a place in our volumes, as having furnished Shakspeare with the ground-work of one of the finest of his plays.

HAMMOND (Dr. HENRY), a learned English divine, was born at Chertsey in Surrey, August 18, 1605; and was the youngest son of Dr. John Hammond, physician to Henry prince of Wales, who was his godfather, and gave him his own name. He was educated at Eton-school, and sent to Magdalen-college, Oxford, in 1618; of which, after taking his degrees in a regular way, he was elected fellow in 1625. Some time after, he applied himself to divinity; which however he did not pursue in the ordinary way, by having recourse to modern systems and voluminous compilations of men who perhaps knew as little of the matter as himself, but, as Fell says, " by beginning that science at the upper end, as conceiving it most reasonable to search for primitive truth in the primitive writers, and not to suffer his understanding to be prepossessed by the contrived and interested schemes of modern, and withal obnoxious, authors." In 1633, he was presented to the rectory of Penshurst in Kent, by Robert Sidney earl of Leicester. That nobleman happening to be one of his auditors while he was supplying a turn at court for Dr. Frewen, the president of his college, and one of his majesty's chaplains, was so deeply affected with the sermon, and formed so just a measure of the preacher's merit, that he conferred on him this living, then void, and in his gift. Upon this he quitted his college, and went to his cure, where he resided as long as the times permitted him, punctually performing every branch of the ministerial function in the most diligent and exemplary manner. In 1640, he was chosen one of the members of the convocation, called with the long parliament, which began that year; and, in 1643, made archdeacon of Chichester by the unsolicited favour of Dr. Brian Duppa, then bishop of Chichester, and afterwards of Winchester. The same year also he was named one of the assembly of divines, but never sat amongst them.

In the beginning of the national troubles he continued undisturbed at his living, till the middle of July 1643; but joining in the fruitless attempt then made at Tunbridge in favour of the king, and a reward of 100 l. being soon after promised to the person that should produce him, he was forced to retire privily and in disguise to Oxford. Having procured an apartment in his own college, he sought that peace in retirement and study, which was no where else to be found. Among the few friends he conversed with, was Dr. Christopher Potter, provost of Queen's college; by whose persuasion it was, that he published his " Practical Cate-
" chism,

"chifm, in 1644." This was one of the moſt valuable books publiſhed at that time; yet, becauſe it did not ſuit the nonſenſe then prevailing, nor the principles of thoſe who cried up Faith to the ſkies, but condemned Works as fit for little elſe but to make a man's damnation more ſure, great objections were raiſed againſt it by 52 miniſters within the province of London; and eſpecially by the famous Francis Cheynell, who has contrived to perpetuate his good name by his extraordinary treatment of the excellent Chillingworth. Hammond however defended his book, and the ſame year, and the following, put out ſeveral uſeful pieces, adapted to the times. December, 1645, he attended as chaplain the duke of Richmond and earl of Southampton; who were ſent to London by Charles 1. with terms of peace and accommodation, to the parliament; and when a treaty was appointed at Uxbridge, he appeared there as one of the divines on the king's ſide, where he managed, greatly to his honour, a diſpute with Richard Vines, one of the Preſbyterian miniſters ſent by the parliament.

A few days after the breaking of this treaty, a canonry of Chriſt-church in Oxford becoming vacant, the king beſtowed it upon him about March, 1645; and the univerſity choſe him their public orator. His majeſty alſo, coming to reſide in that city, made him one of his chaplains in ordinary: notwithſtanding all which employments, he did not remit from his ſtudies, or ceaſe to publiſh books, principally contrived to do ſervice in the times when they were written. When Oxford ſurrendered, his attendance as chaplain was ſuperſeded; but when the king came into the power of the army, he was permitted to attend him again, in his ſeveral confinements and removes of Wooburn, Caverſham, Hampton-court, and the Iſle of Wight: at which laſt place he continued till Chriſtmas, 1647, the time that all his majeſty's ſervants were put away from him. He then returned again to Oxford, where he was choſen ſub-dean of Chriſt-church; in which office he continued till March 30, 1648, when he was forcibly turned out of it by the parliamentary viſitors. Inſtead of being commanded immediately to quit Oxford, as others were, a committee of parliament voted him and Dr. Sheldon to be priſoners in that place, where they continued in reſtraint for about ten weeks. During this confinement he began his "Paraphraſe and Annotations on the New Teſtament;" the ground-work of which is ſaid to be this. Having written in Latin two large volumes of the way of interpreting the New Teſtament, with reference to the cuſtoms of the Jews, and of the firſt Heretics in the Chriſtian church, and alſo of the Heathens, eſpecially in the Grecian games; and, above all, of the importance of the Hellenistical dialect; he began to consider, that it might be more useful to the English reader, to write in our vulgar language, and set every observation in its natural order, according to the direction of the text. And having some years before collated several Greek copies of the New Testament, and observed the variation of our English from the original, and made an entire translation of the whole for his own private use, he cast his work into that form, in which it now appears. It came out first in 1653; in 1656, with additions and alterations; and, in 1698, Le Clerc put out a Latin translation of it, viz. of the "Paraphrase and Annotations," with the text of the Vulgate, in which he has intermixed many of his own animadversions, explained those points which Dr. Hammond had but slightly touched, and corrected many of his mistakes This is the most useful of all his works; which however let us quit for the present, and look a little after its author.

We left him under confinement at Oxford; whence he was afterwards removed to the house of Sir Philip Warwick at Clapham in Bedfordshire. The trial of king Charles drawing on, and Dr. Hammond being in no other capacity to interpose than by writing, he drew up an address to the general and council of officers, which he published under this title : " To the Right Honourable the Lord Fairfax, and his Council of War, the humble Address of Henry Hammond." His grief for the death of his royal master was extreme; but after having indulged it for a while, he resumed his studies, and published several pieces. The rigour of his restraint being taken off in the beginning of 1649, he removed to Westwood in Worcestershire, the seat of the loyal Sir John Packington, from whom he received a kind invitation ; and here spent the remainder of his days. In 1651, when Charles II. came into those parts, he waited upon him, and received a letter from his own hand of great importance, to satisfy his loyal subjects concerning his adherence to the religion of the church of England. In 1653, he published, as we have already observed, his great work on the New Testament, and went on applying antidotes to the distempers of the church and state, and opposing those monstrous ill-grounded and absurd tenets, which were daily broached under the name of religion; particularly those of the Anabaptists and other enthusiasts. Afterwards he undertook a " Paraphrase and Commentary on all the Books of the Old Testament;" of which he published the Psalms, and went through a third part of the book of Proverbs. His want of health, only, hindered him from proceeding farther: for that strength of body, which had hitherto attended his indefatigable mind,

begin-

beginning to fail him about 1654, he was seized by those four tormenting diftempers, each of which has been judged a competent trial of human patience, namely, the ftone, the gout, the colic, and the cramp; but the ftone put an end to his life. For, while Charles II. was defigning him for the bifhopric of Worcefter, and he was preparing to go to London, whither he had been invited by the moft eminent divines, he was feized with a fharp fit of the ftone the 4th of April, of which he died the 25th of the fame month, 1660.

HAMMOND (ANTHONY, Efq.), defcended from a family long fituated at Somerfham-place, in Huntingdonfhire, was born in 1668, and educated at St. John's college, Cambridge. He was a commiffioner of the navy, a good fpeaker in parliament, had the name of "filver-tongued Hammond" given him by lord Bolingbroke, and was a man of note among the wits, poets, and parliamentary writers, in the beginning of this century. A volume of "Mifcellany Poems" was infcribed to him, in 1694, by his friend Mr. Hopkins; and in 1720 he was himfelf the editor of "A new Mifcellany of Original Poems," in which he had himfelf no fmall fhare. His own pieces, he obferves in his preface, "were written at very different times, and were owned by him, left in a future day they fhould be afcribed to other perfons to their prejudice, as the 'Ode on Solitude' has been, in wrong, to the earl of Rofcommon, and as fome of the reft have been to others." He was the intimate friend of Mr. Moyle, and wrote the "Account of his Life and Writings," prefixed to his works in 1727. Their acquaitance began, through Sir Robert Marfham, in the latter end of 1690, foon after Hammond's return from a fhort tour into Holland and fome parts of Flanders. The places of refort for wits at that period were Maynwaring's coffee-houfe in Fleet-ftreet, and the Grecian near the Temple; where Moyle, having taken a difguft againft the clergy, had feveral friendly difputes with Hammond, and at the fame place had a fhare with Trenchard in writing the argument againft a ftanding army. In Moyle's works are three valuable letters to Hammond; a copy of verfes, by Hammond, to Moyle; another, by Hopkins, to the fame; and a third, by Hopkins, to Hammond. In the latter, in 1694, we have the following intimation of what Dr. Johnfon calls "the moft arduous work of its kind:"

> With joy I learn'd Dryden's defign to crown
> All the great things he has already done:
> No lefs, no change of vigour can he feel,
> Who dares attempt the facred Mantuan ftill.

Thefe

These lines are a remarkable confirmation of our excellent Biographer's obfervation, that "the expectation of this work was undoubtedly great: the nation confidered its honour as interefted in the event. One gave him the different editions of his author, and another helped him in the fubordinate parts. The arguments of the feveral books were given him by Addifon. The hopes of the public were not difappointed." "He produced," fays Pope, "the moft noble and fpirited tranflation that I know in any language. It certainly excelled whatever had appeared in Englifh, and appears to have fatisfied his friends; and, for the moft part, to have filenced his enemies."

HAMMOND (JAMES), well remembered as a man efteemed and careffed by the elegant and great, was the fecond fon of Anthony Hammond mentioned above. He was born about 1710, and educated at Weftminfter-fchool; but it does not appear that he was of any univerfity. He was equerry to the prince of Wales, and feems to have come very early into public notice, and to have been diftinguifhed by thofe whofe patronage and friendfhip prejudiced mankind at that time in favour of thofe on whom they were beftowed; for he was the companion of Cobham, Lyttelton, and Chefterfield. He is faid to have divided his life between pleafure and books; in his retirement forgetting the town, and in his gaiety lofing the ftudent. Of his literary hours all the effects are exhibited in his memorable "Love Elegies," which were written very early, and his Prologue not long before his death. In 1733, he obtained an income of 400l. a year by the will of Nicholas Hammond, efq. a near relation. In 1741, he was chofen into parliament for Truro in Cornwall, probably one of thofe who were elected by the prince's influence; and died next year in June at Stowe, the famous feat of the lord Cobham. His miftrefs long outlived him, and, in 1779, died unmarried bed-chamber woman to the queen. The character which her lover bequeathed her was, indeed, not likely to attract courtfhip. Yet it was her own fault that fhe remained fingle, having had another very honourable offer. The "Elegies" were publifhed after his death; and while the writer's name was remembered with fondnefs, they were read with a refolution to admire them. The recommendatory preface of the editor, who was then believed, and is now affirmed by Dr. Maty, to be the earl of Chefterfield, raifed ftrong prejudices in their favour.

HAMON (JOHN), a French phyfician, born at Cherbourg. He publifhed ferious works, remarkable both for their folidity of argument and elegance of ftyle. His works were principally on religious fubjects, for he was a good and pious

pious man, and lived a life of folitude and devotion. He is commended in very animated terms by Boileau, who calls him

" Tout brillant de favoir, d'efprit, et d'éloquence."

He died in 1687, aged fixty-nine.

HAMPDEN (JOHN, Efq.), of Hamden, in Buckinghamfhire, famous for fuftaining, fingly, the weight of a royal profecution, on his refufing to pay the fhip-money in the reign of Charles I. was born at London in 1594. He was of as ancient, Whitlocke fays, the ancienteft, extraction as any gentleman in his county; and coufin-german to Oliver Cromwell, his father having married the protector's aunt. In 1609, he was fent to Magdalen-college in Oxford; whence, without taking any degree, he removed to the inns of court, where he made a confiderable progrefs in the ftudy of the law. Sir Philip Warwick obferves, that " he had great knowledge both in fcholarfhip and the law." In his entrance into the world, he is faid to have indulged himfelf in all the licence of fports, and exercifes, and company, fuch as were ufed by men of the moft jovial converfation; but afterwards to have retired to a more referved and auftere fociety, preferving, however, his natural chearfulnefs and vivacity. In the fecond parliament of king Charles, which met at Weftminfter, February, 1625-6, he obtained a feat in the Houfe of Commons, as he alfo did in two fucceeding parliaments; but made no figure till 1636, when he became univerfally known, by a folemn trial at the King's bench, on his refufing to pay the fhip-money. He carried himfelf, as Clarendon tell us, through this whole fuit with fuch fingular temper and modefty, that he actually obtained more credit and advantage by lofing it, than the king did fervice by gaining it. From this time he foon grew to be one of the moft popular men in the nation, and a principal leading member in the long parliament. " The eyes of all men," fays the fame writer, " were fixed upon him as their *pater patriæ*, and the pilot that muft fteer the veffel through the tempefts and rocks which threatened it." After he had held the chief direction of his party in the Houfe of Commons againft the king, he took up arms in the fame caufe, and was one of the firft who opened the war by an action at a place called Brill, a garrifon of the king's, upon the edge of Buckinghamfhire, about five miles from Oxford. He took the command of a regiment of foot under the earl of Effex, and fhewed fuch fkill and bravery, that, had he lived, he would, probably, foon have been raifed to the poft of a general. But he was cut off early by a mortal wound, which he received in a fkirmifh with

with prince Rupert, at Chalgrove-field, in Oxfordshire: for he was there shot in the shoulder with a brace of bullets, which broke the bone, June 18, 1643; and, after suffering much pain and misery, he died the 24th, to as great a consternation of all his party as if their whole army had been defeated. Many men observed, says Clarendon, that the field in which this skirmish was, and upon which Hampden received his death-wound, namely, Chalgrove-field, was the same place in which he had first executed the ordinance of the militia, and engaged that county, in which his reputation was very great, in this rebellion: and it was confessed by the prisoners that were taken that day, and acknowledged by all, that upon the alarm that morning, after their quarters were beaten up, he was exceeding solicitous to draw forces together to pursue the enemy; and, being a colonel of foot, put himself amongst those horse as a volunteer, who were first ready, and that, when the prince made a stand, all the officers were of opinion to stay till their body came up, and he alone persuaded and prevailed with them to advance: so violently did his fate carry him to pay the mulct in the place where he had committed the transgression about a year before. This, says Clarendon, was an observation made at that time; but his lordship does not adopt it as an opinion of his own.

Hampden, if we form our judgement of him only from the account of those who were engaged in the opposite party to him, was, perhaps, one of the most extraordinary men that ever lived; and it must certainly be very amusing to contemplate the portrait of him, as it is thus delineated by the earl of Clarendon. "He was," says the noble historian, "a man of much greater cunning, and it may be of the most discerning spirit, and of the greatest address and insinuation to bring any thing to pass which he desired, of any man of that time, and who laid the design deepest—He was not a man of many words, and rarely began the discourse, or made the first entrance upon any business that was assumed, but a very weighty speaker; and after he had heard a full debate, and observed how the house was like to be inclin'd, took up the argument, and shortly, and clearly, and craftily, so stated it, that he commonly conducted it to the conclusion he desired. —He was of that rare affability and temper in debate, and of that seeming humility and submission of judgement, as if he brought no opinion of his own with him, but a desire of information and instruction: yet he had so subtle a way, and under the notion of doubts insinuating his objections, that he infused his own opinions into those from whom he pretended to learn and receive them. And even with them who were able to preserve themselves from his infusions, and discerned

those

those opinions to be fixed in him with which they could not comply, he always left the character of an ingenuous and conscientious person. He was, indeed, a very wise man, and of great parts, and possessed with the most absolute spirit of popularity, and the most absolute faculties to govern the people, of any man I ever knew. For the first year of the parliament, he seemed rather to moderate and soften the violent and distempered humours than to inflame them. But wise and dispassionate men plainly discerned, that that moderation proceeded from prudence, and observation that the season was not ripe, rather than that he approved of the moderation; and that he begot many opinions and notions the education whereof he committed to other men; so far disguising his own designs, that he seemed seldom to wish more than was concluded. And in many gross conclusions, which would hereafter contribute to designs not yet set on foot, when he found them sufficiently backed by a majority of voices, he would withdraw himself before the question, that he might seem not to consent to so much visible unreasonableness; which produced as great a doubt in some as it did approbation in others of his integrity.—After he was among those members accused by the king of high treason, he was much altered; his nature and carriage seeming much fiercer than it did before: and without question, when he first drew his sword, he threw away the scabbard.—He was very temperate in diet, and a supreme governor over all his passions and affections; and had thereby a great power over other men's. He was of an industry and vigilance not to be tired out or wearied by the most laborious; and of parts not to be imposed upon by the most subtle and sharp; and of a personal courage equal to his best parts: so that he was an enemy not to be wished, wherever he might have been made a friend; and as much to be apprehended, where he was so, as any man could deserve to be. And therefore his death was no less pleasing to the one party than it was condoled in the other. In a word, what was said of Cinna might well be applied to him: he had a head to contrive, a tongue to persuade, and a hand to execute, any mischief, or," as the historian says elsewhere, " any good." Thus is Hampden described by Clarendon, agreeably to the notions usually formed of his character after the Restoration; which, we see, was that of a great rather than a good man. But as the characters of statesmen, commanders, or men acting in a public capacity, always vary with the times and fashions of politics, so at the Revolution, when passive obedience and non-resistance were disgraced by law, he came to be esteemed a good man as well as a great; and, bating a small interval in the days of
Sacheverell,

Sacheverell, has continued to be thought so from that time to this. Thus a poet of our own days, in an elegant piece, intituled, "An Elegy in a Country Church yard," has painted him in the glorious colours of a warm and active patriot:

> "Perhaps in this neglected spot is laid
> "Some heart once pregnant with celestial fire;
> "Hands that the rod of empire might have sway'd,
> "Or wak'd to extasy the living lyre."
> * * * * * * * * * * * *
> "Some village Hampden, that with dauntless breast
> "The little tyrant of his fields withstood,
> "Some mute inglorious Milton here may rest;
> "Some Cromwell, guiltless of his country's blood."

HAMSA, a mahometan doctor, remarkable for having undertaken the arduous task of extirpating Mahometanism, and establishing a new religion in its stead. His motive does no great honour to his principles, for it originated in political discontent. He composed a book in opposition to the Alcoran, which, in point of purity and elegance, is thought by many equal, and by some even superior, to that celebrated production. But his zeal and his talents were of no avail. Hamsa's book was translated into French: it is called, "Evidences of the Mysteries of the Unity." He lived about the year 1020.

HANDEL (GEORGE-FREDERIC), an illustrious master in music, was born at Hall, a city of Upper Saxony, February 24, 1684, by a second wife of his father, who was an eminent physician and surgeon of the same place, and then above 60 years of age. From his very childhood he discovered such a propensity to music, that his father, who always intended him for the civil law, was alarmed at it; and took every method to oppose this inclination, by keeping him out of the way of, and strictly forbidding him to meddle with, musical instruments of any kind. Nevertheless, the son found means to get a little clavicord privately conveyed to a room at the top of the house; and with this he used to amuse himself when the family was asleep. While he was yet under seven years of age, he went with his father to the duke of Saxe Weisenfels, where it was impossible to keep him from harpsichords, and other musical instruments. It happened one morning, that, while he was playing on the organ, after the service was over, the duke was in the church; and something there was in his manner of playing, which affected his highness so strongly, that he asked his valet de chambre (who, by the way, was Handel's brother-in-law) who it was that he heard at the organ? The valet replied, that it was his brother. The duke demanded to see him; and, after making proper enquiries about him, expostulated very seriously with the old doctor, who still retained his prepossessions in favour of the civil law. He told him, at length, that every father had certainly a right to dispose of his children as he should think most expedient; but that, for his own part, he could not but consider it as a sort of crime against the public and posterity to rob the world of such a rising genius. The issue of this debate was, not only a toleration for music, but consent also that a master should be called in to forward and assist him.

The first thing his father did, at his return to Hall, was to place him under one Zackaw, organist to the cathedral church; who was a person of great abilities in his profession, and not more qualified than inclined to do justice to any pupil of promising hopes. Handel pleased him so much, that he never thought he could do enough for him. He was proud of a pupil, who already began to attract the attention of the public; and also glad of an assistant, who, by his prodigious talents, was capable of supplying his place, whenever he had a mind to be absent. It may seem strange to talk of an assistant at seven years of age; but it is stranger, that at nine he began to compose the church-service for voices and instruments, and from that time actually did compose a service every week for three years successively. Having far surpassed his master, the master himself confessing it, and made all the improvements he could at Hall, it was agreed he should go to Berlin; and to Berlin he went in 1698, where the opera was in a flourishing condition under the encouragement of the king of Prussia, grandfather of the present. Handel had not been long at court, before his abilities became known to the king, who frequently sent for him, and made him large presents. He farther offered to send him to Italy, where he might be formed under the best masters, and have opportunities of hearing and seeing all that was excellent in the kind: but there were reasons for refusing this offer, and also for leaving Berlin, which he did soon after. During his stay there, he became acquainted with two Italian composers, Buononcini and Attilio; the same who afterwards came to England while Handel was here, and were at the head of a formidable opposition against him.

Next to the opera of Berlin, that of Hamburg was in the highest request; and thither it was resolved to send him on his own bottom; and chiefly with a view to improvement: but his father's death happening soon after, and his mother being left in narrow circumstances, he thought it necessary to procure scholars, and obtain some employment in the orcheſtra ; and by this means, inſtead of a burden, he proved a great relief to her. He had a diſpute at Hamburg with one of the maſters, in oppoſition to whom he laid claim to the firſt harpſichord ; and he had the luck to have it determined in his favour. The honour however had like to have coſt him dear ; for his antagoniſt ſo reſented his being conſtrained to yield to ſuch a ſtripling competitor, that, as they were coming out of the orcheſtra, he made a puſh at him with a ſword, which had infallibly pierced his heart, but for the friendly Score, which he carried accidentally in his boſom. " Had this happened," ſays his hiſtorian, " in the early ages, not a mortal but would have been perſuaded that Apollo himſelf interpoſed to preſerve him in the form of a muſic-book."

From conducting the performance he became compoſer to the houſe ; and " Almeria," his firſt opera, was made here, when he was not much above 14 years of age. The ſucceſs of it was ſo great, that it ran f r 30 nights without interruption ; and this encouraged him to make others, as he did alſo a conſiderable number of ſonatas not extant, during his ſtay at Hamburg, which was about four or five years. He contracted an acquaintance at this place with many perſons of note, among whom was the prince of Tuſcany, brother to the grand duke. The prince, who was a great lover of the art for which his country was famous, would often lament Handel's not being acquainted with the Italian muſic ; ſhewed him a large collection of it ; and was very deſirous he ſhould return with him to Florence. Handel plainly anſwered, that he could ſee nothing in the muſic anſwerable to the prince's character of it ; but, on the contrary, thought it ſo very indifferent, that the ſingers, he ſaid, muſt be angels to recommend it. The prince ſmiled at the ſeverity of his cenſure ; yet preſſed him to return with him, and intimated, that no convenience ſhould be wanting. Handel thanked him for the offer of a favour which he did not chuſe to accept ; for he reſolved to go to Italy, on his own bottom, as ſoon as he could make a purſe ſufficient for the purpoſe. He had in him, from his childhood, a ſtrong ſpirit of independence, which was never known to forſake him in the moſt diſtreſsful ſeaſons of his life : and it is remarkable, that he refuſed the greateſt offers from perſons of the firſt diſtinction ; nay, and even the higheſt favours from the faireſt of the fair ſex, only becauſe he would not be cramped or confined by particular attachments.

" Soon after he went to Italy, and Florence was his firſt deſtination ; where, at the age of 18, he made the opera of " Rodrigo,"

"Rodrigo," for which he was presented with 100 sequins, and a service of plate. This may serve to shew, what a reception he met with at a place, where the highest notions were conceived of him before he arrived. Vittoria, a celebrated actress and singer, bore a principal part in this opera. She was a fine woman, and had been some time in the good graces of his serene highness; yet Handel's youth and comeliness, joined with his fame and abilities in music, had raised emotions in her heart, which, however, we do not find that Handel in the least encouraged. After about a year's stay at Florence, he went to Venice; where he was first discovered at a masquerade, while he was playing on a harpsichord in his vizor. Scarlatti happened to be there, and affirmed it could be no one but the famous Saxon or the devil. Being earnestly importuned to compose an opera, he finished his "Agrippina" in three weeks; which was performed 27 nights successively, and with which the audience were so enchanted, that they seemed to be all distracted. From Venice he went to Rome, where his arrival was no sooner known, than he received polite messages from persons of the first distinction. Among his greatest admirers was the cardinal Ottoboni, a person of refined taste and princely magnificence; at whose court he met with the famous Corelli, with whom he became well acquainted. Attempts were made at Rome to convert him to Popery; but he declared himself resolved to die a member of that communion, whether true or false, in which he had been born and bred. From Rome he went to Naples; and, after he quitted Naples, made a second visit to Florence, Rome, and Venice. The whole time of his abode in Italy was six years; during which he had made abundance of music, and some in almost every species of composition. These early fruits of his studies would doubtless be great curiosities, could they be met with.

He was now returned to his native country, but yet had not done travelling, nor was likely to have done, while there was any musical court which he had not seen. Hanover was the first he stopped at, where he met with Steffani, with whom he had been acquainted at Venice; and who was then master of the chapel to George I. when he was only elector of Hanover. At Hanover, also, there was a nobleman who had taken great notice of him in Italy, and who afterwards did him great service, when he came to England for the second time. This person was baron Kilmanseck. He introduced him at court, and so well recommended him to his electoral highness, that he immediately offered him a pension of 1500 crowns per ann. as an inducement to stay. Handel
excused

excused his not accepting this high favour, because he had promised the court of the elector palatine, and also resolved to pass over into England, whither it seems he had received strong invitations from the duke of Manchester: upon which he had leave to be absent for a twelvemonth or more, and to go whithersoever he pleased, and on these conditions he thankfully accepted the pension.

After paying a visit to his mother, who was now extremely old and blind, and to his old master Zackaw, he set out for Dusseldorp. The elector was highly pleased with him, and at parting made him a present of a fine set of wrought plate for a desert. From Dusseldorp he made the best of his way through Holland; and, embarking for England, he arrived at London in the winter of 1710. He was soon introduced at court, and honoured with marks of the queen's favour. Many of the nobility were impatient for an opera from him; whereupon he composed "Rinaldo," in which the famous Nicolini sang. Its success was great, and his engagements at Hanover the subject of much concern. He returned thither in about a twelvemonth; for, besides his pension, Steffani had resigned to him the mastership of the chapel; but in 1712, he obtained leave of the elector to make a second visit to England, on condition that he returned within a reasonable time. The poor state of music here, and the wretched proceedings at the Haymarket, made the nobility desirous that he might be employed in composing for the theatre. To their applications the queen added her own authority; and, as an encouragement, settled on him for life a pension of 200l. per annum. All this made Handel forget his obligations to return to Hanover; so that when his late majesty came over, at the death of the queen, in 1714, conscious how ill he had deserved at his hands, he durst not appear at court. It happened, however, that his noble friend baron Kilmanseck was here; and he, with others of the nobility, contrived the following scheme for reinstating him in his majesty's favour. The king was persuaded to form a party on the water; and Handel was bid to prepare some music for that occasion. It was performed and conducted by himself, unknown to his majesty, whose pleasure, on hearing it, was equal to his surprize. Upon his enquiring whose it was, the baron produced the delinquent, and presented him to his majesty, as one that was too conscious of his fault to attempt an excuse for it. Thus Handel was restored to favour, and his music honoured with the highest approbation; and as a token of it, the king was pleased to add a pension for life of 200l. a year to that which queen Anne had before given him. Some years after, when he was employed to

teach

teach the young princesses, another pension was added to the former by her late majesty.

Handel was now settled in England, and well provided for. The three first years he was chiefly, if not constantly, at the earl of Burlington's; where he frequently met Pope. The poet one day asked his friend Arbuthnot, of whose knowledge in music he had an high idea, what was his real opinion of Handel, as a master of that science? who replied, "Conceive the highest you can of his abilities, and they "are much beyond any thing that you can conceive."—Pope neverthelefs declared, that Handel's finest things, so untoward were his ears, gave him no more pleasure than the airs of a common ballad. The two next years he spent at Cannons, then in its glory, and composed music for the chapel there. While he was here, a project was formed by the nobility, for erecting an academy in the Haymarket; the intention of which was to secure a constant supply of operas, to be composed by Handel, and to be performed under his direction. For this purpose a large sum was subscribed, the king subscribing 1000l. the nobility 4000l. and Handel went to Dresden in quest of singers, whence he brought Senesino and Duristanti. At this time Buononcini and Attilio, whom we have mentioned before, composed for the opera, and had a strong party in their favour, and by whom a violent opposition was maintained; but at last the parties were all united, and each was to have his particular part.

The academy being now firmly established, and Handel appointed composer to it, all things went on prosperously for a course of ten years. Handel maintained an absolute authority over the singers and the band, or rather kept them in total subjection. Having one day a dispute with Cuzzoni on her refusing to sing something or other, "Oh, madam," said he, "I know very well that you are a true devil; but "I will make you know, that I am Beelzebub the chief of "the devils." With this he took her up by the waist, and swore, that if she made any words, he would fling her out of the window. This may serve to shew what a spirit he possessed, and how well the company were governed. What, however, they regarded hitherto as legal government, at length appeared to be downright tyranny; upon which a rebellion commenced, with Senesino at the head of it, and all became tumult and civil war. Handel, perceiving that Senesino was grown less tractable and obsequious, resolved to subdue him. To manage him by gentle means he disdained; yet to controul him by force he could not, Senesino's interest and party being too powerful. The one, therefore, was quite refractory, the other quite outrageous. The merits of the quarrel are not known; but, whatever they were, the nobility would not confent to his defign of his parting with Senefino, and Handel had refolved to have no farther concerns with him. And thus the academy, after it had gone on in a flourifhing ftate for above nine years, was at once diffolved.

Handel ftill continued at the Haymarket, but his audience gradually funk away. New fingers muft be fought, and could not be had any nearer than Italy. Difcouraging this! yet to Italy he went, and, returning with feveral fingers, he embarked on a new bottom. He carried it on for three or four years, but it did not do. Many of the nobility raifed a new fubfcription for another opera at Lincoln's inn fields, and fent for Farinelli and others; and, in fhort, the oppofition was fo ftrong, that, in fpite of his great abilities, his affairs declined; all for want of a little prudence, and a fpirit that knew how to yield on proper occafions. His fortune was not more impaired than his health and his underftanding. His right arm was become ufelefs to him from a ftroke of a palfy; and his fenfes were greatly difordered at intervals for a long time. In this unhappy ftate it was thought neceffary, that he fhould go to the vapour-baths at Aix-la-Chapelle; and thence he received a cure, which, from the manner, as well as quicknefs of it, paffed with the nuns for a miracle.

Soon after his return to London in 1736, his "Alexander's Feaft" was performed at Covent Garden, and applauded; and feveral other attempts of the like nature were made to reinftate him, but they did not prevail: the Italian party were too powerful; fo that, in 1741, he went to Dublin, where he was well received. Pope has recorded this paffage of his hiftory. A poor phantom, which is made to reprefent the genius of the modern Italian opera, expreffes her apprehenfions, and gives her inftructions to Dullnefs, already alarmed for her own fafety, in the following lines:

"But foon, ah! foon, rebellion will commence,
If mufic meanly borrows aid from fenfe:
Strong in new arms, lo! giant Handel ftands,
Like bold Briareus with his hundred hands;
To ftir, to roufe, to fhake the foul he comes,
And Jove's own thunders follow Mars's drums.
Arreft him, emprefs; or you fleep no more—
She heard,—and drove him to th' Hibernian fhore."

DUNCIAD, Book iv. 63.

At his return to London in 1741-2, the minds of moſt men were difpoſed in his favour, and the æra of his proſperity returned, He immediately began his oratorios in Covent-Garden, which he continued, with uninterrupted ſucceſs and unrivalled glory, till within eight days of his death. The laſt was performed on the 6th, and he expired on the 14th of April, 1759. He was buried in Weſtminſter abbey, where, by his own order, and at his own expence, a monument is erected to his memory.

HANKIUS (MARTIN), born at Breſlaw in 1633, where he was profeſſor of hiſtory, politics, and eloquence. He wrote many works which eſtabliſhed his reputation among his countrymen as an acute critic and profound ſcholar. His principal performance, and that for which he is moſt eſteemed among ſcholars, is his book " De Romanarum rerum Scriptoribus," to which was added another book " De Byzantinarum rerum Scriptoribus Græcis. His other publications, alſo on Hiſtory and Antiquities are in conſiderable repute. He died in 1709.

HANMER (SIR THOMAS. Bart.), a diſtinguiſhed ſtateſman and polite writer, was born about 1676; and had his education at Weſtminſter-ſchool, and Chriſt-church, Oxford. When he arrived at years of maturity, he was choſen knight of the ſhire for the county of Suffolk, and he ſat in parliament near 30 years, either as a repreſentative for that county, or for Flintſhire, or for the borough of Thetford. In this venerable aſſembly he was ſoon diſtinguiſhed: and his powerful elocution and unbiaſſed integrity drew the attention of all parties. In 1713, he was choſen ſpeaker of the houſe of commons; which office, difficult at all times, but at that time more particularly ſo, he diſcharged with becoming dignity. All other honours and emoluments he declined. Having withdrawn himſelf by degrees from public buſineſs, he ſpent the remainder of his life in an honourable retirement amongſt his books and friends: and there prepared an elegant and correct edition of the works of Shakſpeare. This he made a preſent of to the univerſity of Oxford; and it was printed there 1744, in ſix volumes 4to. with elegant engravings, by Gravelot, at the expence of Sir Thomas. He died at his ſeat in Suffolk, April 5, 1746.

HANNEKEN (MENNON), a Lutheran clergyman, born at Blaxen in Oldenbourg in 1595. He was profeſſor of Morals, Theology, and the Oriental languages, at Marpurg. He wrote an Hebrew Grammar, and an expoſition of St. Paul's Epiſtle to the Epheſians. He died at Lubeck in 1671.

HANNEKEN (Philip Louis), eldeſt ſon of the preceding, was profeſſor of Eloquence and Hebrew at Gieſſen in 1663. He publiſhed a great variety of works principally on ſubjects of theological controverſy. He died at Wittenberg in 1706.

HANNIBALIANUS (Flavius Claudius), nephew of Conſtantine, and the only one, of the whole ſeries of Roman princes in any age of the Empire, who was diſtinguiſhed by the title of king. The emperor aſſigned him the city of Cæſarea as his reſidence; and the provinces of Pontus, Cappadocia, and the leſs Armenia, as his kingdom. He was cruelly murdered by Conſtantius, the ſon and ſucceſſor of Conſtantine, notwithſtanding he had married the ſiſter of the emperor.

HANNO, king of the Ammonites. Ambaſſadors were ſent by David to compliment him on his acceſſion to the throne. Hanno's courtiers told him, that their men were ſpies, in conſequence of which he ſhaved their beards, and treated them otherwiſe with great indignity. But this coſt him his life and his crown; for, David deprived him of both.

HANNO, a Carthaginian general, employed by his countrymen to make the circuit of Africa, in conſequence of which he explored various regions, and made great geographical diſcoveries. Pliny and Plutarch ſeverally relate a curious anecdote of this Hanno. He had by much perſeverance ſo tamed a lion, that it followed him as a dog and carried his baggage. The Carthaginians thought that the man, who could accompliſh a thing ſeemingly ſo impoſſible as this, might ſucceed in whatever he undertook. They therefore baniſhed him, that he might not carry into execution any deſigns againſt the liberties of his country. Some ſuppoſed voyages of this man are publiſhed in the Oxford geographers.

HANNSACHS, a german poet, who publiſhed his works in five large folio volumes, among which ſome few and thinly-ſcattered rays of genius are to be found. He was a native of Nuremberg.

HANNEMAN (Adrian), a native of the Hague, and an eminent painter. He ſtudied the works of Vandyke; and was a favourite painter with Mary princeſs of Orange, daughter of Charles the Firſt. Many of his works are in England, but the moſt conſiderable abroad. Some of his performances are in the chamber of ſtate at the Hague.

HANWAY (Jonas), a benevolent and amiable character, born at Portſmouth in 1712. He was at a very early age bound apprentice to a merchant at Liſbon, and afterwards connected
himſelf

himself with a mercantile house at Petersburgh, in consequence of which he was induced to make a journey into Persia. On leaving Russia with an independent fortune, he returned to his own country, and passed the remainder of his life as a private gentleman, honourably to himself and useful to the world. In 1753, he published an account of his travels through Russia into Persia, and back again through Russia, Germany, and Holland. To this work also was added an account of the Revolutions of Persia during the present century. His other publications are very numerous, most of them were well received, and all of them calculated to prove him an excellent citizen and liberal-minded man. The institution of the Marine Society is to be attributed to his activity and benevolence; the usefulness of which requires no panegyric, its truest praise is its extraordinary success. This was the favourite object of Mr. Hanway's care; but, in 1758, he was also particularly instrumental in the establishment of the Magdalen charity. His public spirit, and, above all, his disinterestedness were so conspicuous, that a deputation of the principal merchants in London waited upon the earl of Bute, when prime minister, and represented to him that an individual like Mr. Hanway, who had done so much public good to the injury of his private fortune, was deserving of some signal mark of the public esteem. He was accordingly made a commissioner of the navy, a situation which he held more than twenty years, and, when he resigned, he was allowed to retain the salary for life, on account of his known exertions in the cause of universal charity. To enumerate the various instances in which the benevolent character of his heart were successfully exerted, would be no easy task. Sunday-schools in a great measure may look upon Mr. Hanway as their father; the chimney-sweepers' boys are much indebted to his humanity; and perhaps there never was any public calamity in any part of the British empire which he did not endeavour to alleviate. So greatly and so universally was he respected, that when he died, in 1786, a subscription of many hundred pounds was raised to erect a monument to his memory. Some may think so whimsical a circumstance not worth recording; but Mr. Hanway was the first person who ventured to walk in the streets of London with an umbrella; he, however, lived to see them brought into general use. The great character of his numerous works is a strong masculine spirit of good sense, and a very chaste simplicity. In his private life he was remarkable for the strictest integrity of conduct, and for a frankness and candour which naturally inspired confidence. The number of his publications amounted to almost seventy, which are enumerated by Mr. Pugh, a gentleman who wrote his life.

HARDING (THOMAS), a famous divine, and the antagonist of bishop Jewel, was born at Comb-Martin in Devonshire, 1512. His school education was at Winchester, whence he was removed to New-college, Oxford, and chosen fellow there in 1536. He was afterwards chosen Hebrew professor of the university by Henry VIII. and, as his religion probably kept pace with the king's, so being consequently half reformed at Henry's death, Edward no sooner ascended the throne, than Harding became a very good Protestant. He was afterwards chaplain to the duke of Suffolk, father of Jane Grey: he had the honour to instruct this young lady in the then true religion; but, on the accession of queen Mary, he immediately saw his error, and became a confirmed Papist. There is a curious epistle preserved by Fox, said to be written by lady Jane to Harding on his apostacy; but many are of opinion, and not without reason, that the violent flaming zeal, with the coarse indelicate language of it, can never be the genuine effusion of a mild and amiable young lady of seventeen. He had taken his degrees in arts: in 1554, he proceeded D. D. at Oxford, and was the year after made treasurer of the cathedral of Salisbury, as he had been a little before prebendary of Winchester. When Elizabeth came to the crown, being deprived of his preferment, he left the kingdom; and, having fixed his abode at Louvain in Flanders, he became, says Wood, " the target of Popery," in a warm controversy with bishop Jewel, against whom, between 1554 and 1567, he wrote seven pieces.

He was a man of parts and learning, and not an inelegant writer. Humphrey, in his " Life of Jewel," comparing him with his adversary, says,—" in multis pares sunt, & ambo doctrinæ & eloquentiæ gloria præcellentes," p. 142.

HARDINGE (NICHOLAS,) of Canbury, near Kingston in Surrey, (brother of Caleb Hardinge, M. D. grandson of Sir Robert Hardinge of King's Newton, in the county of Derby, Knt. and father of George Hardinge, esq. of the Middle-Temple, barrister, an eminent counsel, and of Henry, vicar of Kingston) fellow of King's college, Cambridge, many years clerk of the house of commons, and at last member of parliament for Eye in Suffolk, and one of the secretaries of the treasury. In December, 1732, he was appointed law reader to the duke of Cumberland, with a salary of 100l. He married in December, 1738, Jane second daughter of Sir John Pratt, of Wilderness in Kent (chief justice of the Common Pleas), and sister to the present lord
Cam-

Camden; and died April 9, 1758. His library was fold by auction in 1759. His "Dialogue in the Senate-houfe at Cambridge," is preferved in the "Poetical Calendar," Vol. IX. p. 92, and his "Denhill Iliad," a poem occafioned by the hounds running through lady Gray's gardens at Denhill in Eaft Kent, 1747, in the fixth volume of the "Select Collection, 1780," p. 82. His Latin poems (in every meafure and ftyle) are much admired. Two of them are in the "Mufæ Anglicanæ," and another in the "Select Collection," Vol. VI. p. 87. He was a very diligent and able officer in both his departments; and thought one of the beft claffical fcholars of his age, deeply verfed in the hiftory, laws, and conftitution, of England, on which he could exprefs himfelf with the greateft precifion. He obliged his friends with an engraving, by Mr. Vertue, of two views of the chapel of St. Mary, adjoining to the fouth fide of the parochial church of Kingfton upon Thames, in the county of Surrey, in which feveral Englifh Saxon kings are faid to have been crowned, which was ruined in 1730 by the falling down of one of the pillars and arch next the church.

HARDION (JAMES), a native of Tours, and member of the Academy of Infcriptions, was a very accomplifhed fcholar and critic. Many of his differtations are publifhed in the "Memoirs of the Academy," and do the higheft credit to his tafte, acutenefs, and learning. He publifhed alfo a "Treatife on French Poetry and Rhetoric," in three fmall volumes, and a "Univerfal Hiftory" in eighteen volumes. His works are much admired for their eloquence and ftyle, and for erudition untinctured by pedantry. He died at Paris in 1766, aged eighty.

HARDOUIN (JOHN), a French Jefuit, eminent for his great parts, learning, and fingularities of opinion, was born of obfcure parents, at Kimper in Bretagne, in 1647. He entered young in the fociety of Jefuits and devoted himfelf to the ftudy of the belles lettres, the learned languages, hiftory, philofophy, and divinity. In 1684, he publifhed, in 4to, a work, intituled, "Nummi antiqui populorum & urbium illuftrati:" in which he often gave explications very fingular, and as contrary to truth as to good fenfe. The fame year, in conjunction with Petavius, "Themiftii Orationes xxxiii. cum notis," fol. The year following, in 5 vols. 4to, for the ufe of the Dauphin, "Plinii Hiftoriæ naturalis libris xxxvii, interpretatione & notis illuftrati." Hitherto he confined himfelf to profane learning, where his whimfies were not fuppofed capable of doing much harm; but now, to the great uneafinefs of many good perfons, he was going to tamper with religious fubjects; and, in 1687, he publifhed his book intituled, "De

"De Baptismo quæstio triplex." Two years after appeared his "Antirrheticus de nummis antiquis coloniarum & municipiorum," in 4to; and also "S. Joannis Chrysostomi Epistola ad Cæsarium Monachum, notis ac dissertatione de sacramento altaris," in 4to. Le Clerc having made some reflections upon "St. Chrysostom's Letter to Cæsarius," Hardouin replied, in a piece printed in 1690, and intituled, "Défence de la Lettre de S. Jean Chrysostome, addressée à l'Auteur de la Bibliotheque Universelle:" to which Le Clerc returned an answer in the 19th volume of that work.

In 1693, he printed at Paris, in 2 vols. 4to, Chronologiæ ex nummis antiquis restitutæ prolusio, de nummis Herodiadum;" in which he opened more fully that strange paradoxical system, of which he had yet done little more than hint. He undertakes to prove from medals, that the greater part of those authors, which have passed upon the moderns for ancient, were forged by some monks of the thirteenth century, who gave to them the several names of Homer, Plato, Aristotle, Plutarch, &c. Tertullian, Origen, Basil, Augustin, &c. He only excepts out of this monkish manufacture the works of Cicero, Pliny's "Natural History," Virgil's "Georgics," and Horace's "Satires and Epistles." These he supposes the only genuine monuments of antiquity remaining, except some few Inscriptions and Fasti: and with the assistance of these, he thinks, that these monks drew up and fashioned all the other ancient writings, as Terence's "Plays," Livy's and Tacitus's "Histories," Virgil's "Eneid," Horace's "Odes," &c. Nay, he pushed this chimera so far, that he fancied he could see plainly enough that Æneas in Virgil was designed for Jesus Christ, and Horace's mistress Lalage for the Christian religion. An absurder system never came out of the brain of man: however, he appears to have seriously believed it himself, and was persuaded that his reasons for it were clear and evident; though he would not publish them to the world, nor explain his system, though he was frequently called upon so to do. This work was suppressed by public authority at Paris. He afterwards published "A Letter upon three Samaritan Medals;" "An Essay towards the restoring Chronology by Medals of Constantine's Age," and "A Chronology of the Old Testament, conformable to the vulgar Translation, illustrated by ancient Medals:" all which books were likewise suppressed, on account of the parodoxes contained in them.

However, he continued still in his opinion; for, in his letters, written to Monf. Ballanfaux, and printed at Luxemburg in 1700, he speaks of "an impious faction begun a
long

long while ago, which still subsists, and which by forging an infinite number of writings, that seem to breathe nothing but piety, appears to have no other design than to remove God out of the hearts of mankind, and to overturn all religion." Mr. La Croze refuted his notion concerning the forgery of the antient writings, in " Differtations historiques sur divers sujets, Rot. 1707;" and in " Vindiciæ veterum Scriptorum contra J. Harduinum." La Croze imagined, that Hardouin advanced his notions in concert with the society of Jesuits, or at least with his superiors, in order to set aside the ancient Greek and Latin sacred and profane writers, and so leave all clear to infallibility and tradition only; but Le Clerc was of opinion, that there was no ground for this supposition. In 1700, there was published at Amsterstam a volume in folio, intituled, " Joannis Harduini opera selecta, tum quæ jampridem Parisiis edita nunc emendatiora & multo auctiora prodeunt, tam quæ nunc primum edita." These select works consist of his " Nummi antiqui populorum & urbium illustrati;" " De Baptismo quæstio triplex; edition of " St. Chrysostom's Letter to Cæsarius," with the dissertation " De Sacramento Altaris;" " De nummis Herodiadum;" his " Discourse on the Last Supper," which had been printed in 1693; a treatise in which he explains the medals of the age of Constantine; " Chronology of the old Testament, adjusted by the Vulgate translation, and illustrated by Medals;" " Letters to M. de Ballanfaux;" and other pieces. This volume made a great deal of noise, before it was published. The author had corrected what he thought proper in the works he had already published; and then put them into the hands of a bookseller, who undertook to print them faithfully from the copy he had received. He began the impression with the author's consent, and was considerably advanced in it; when the clamour raised against the paradoxes in those works obliged Hardouin to send an order to the bookseller, to retrench the obnoxious passages. But the bookseller refused to do it, and wrote an answer to him, alleging the reasons of his refusal; upon which was issued " A Declaration of the Father Provincial of the Jesuits, and of the Superiors of their houses at Paris, concerning a new Edition of some works of Father John Hardouin of the same Society, which has been actually made contrary to their will by the Sieur de Lorme, Bookseller at Amsterdam, &c." At the bottom of this was Hardouin's recantation, which runs in these curious terms: " I subscribe sincerely to every thing contained in the preceding declaration; I heartily condemn in my writings what it condemns in them, and particularly what I have said concerning an impious faction, which had forged some ages ago the greatest part of the ecclesiastical or profane writings, which have hitherto been considered as ancient. I am extremely sorry that I did not open my eyes before in this point. I think myself greatly obliged to my superiors in the society, who have assisted me in divesting myself of my prejudices. I promise never to advance in word or writing any thing directly or indirectly contrary to my present recantation. And if hereafter I shall call in question the antiquity of any writing either ecclesiastical or profane, which no person before shall have charged as suppositious, I will only do it by proposing my reasons in a writing published under my name, with the permission of my superiors, and the approbation of the public censors. In testimony of which I have signed, this 27th of December, 1708, J. Hardouin, of the society of Jesus."

Here we have a notable proof of the glorious latitude which Jesuitical morality allowed its professors; for, notwithstanding this solemn protestation, nothing can be more certain, than that Hardouin never departed a tittle from his opinions; but, on the contrary, industriously cherished and propagated them to the last moment of his life. Thus in 1723, when he reprinted his edition of Pliny in three volumes folio, he greatly augmented it with notes, in which were dispersed many paradoxical conceits, tending to support his general system: insomuch, that Mr. Crevier and father Desmolets of the Oratory thought themselves obliged to point them out to the public, and to refute them. Notwithstanding the clamour raised against this Jesuit and his writings, he yet maintained his credit so well with the clergy of France, that they engaged him to undertake a new edition of "The Councils," and gave him a pension for that purpose. It was printed, 1715, in 12 vols. folio, at the royal printing-house; but the sale of it was prohibited by the parliament, who commissioned some doctors, among whom was the celebrated Dupin, to examine it. These doctors gave in their report, that the edition should either be suppressed, or at least corrected in a great number of places; because it contained many maxims injurious to the doctrines and discipline of the church in general, and to those of the Gallican church in particular; and because some very essential things were omitted, while others that were spurious were inserted.

Father Hardouin died at Paris, September 3, 1729, in his eighty-third year; and after his death, a volume of his "Opuscula" in folio was published by an anonymous friend. The largest and most singular of these is intituled, "Athei detecti;" among whom are to be found Jansenius, Malbranche, Thomasin, Descartes, Regis, Arnaud, Nicole, Paschal,

Paschal, Quesnel; whose irreligion, no doubt, consisted chiefly in their being enemies to the Jesuits. The society, however, thought proper, in their " Mémoires de Trevoux," to disown any concern in the publication of these " Opuscula;" and affected to censure freely the errors contained in them.

A posthumous work was published in 1766, under the title of ' Joannis Harduini, Jesuitæ, ad Censuram Scriptorum Veterum Prolegomena," with a valuable preface by Mr. Bowyer, to whom a curious Latin pamphlet was addressed on that occasion by his friend the Rev. Cæsar De Missy.

We will conclude our account of this famous Jesuit with a characteristic epitaph by M. de Boze.

" In expectatione judicii,
Hic jacet
Hominum paradoxotatos,
Natione Gallus, Religione Romanus:
Orbis litterati portentum ;
Venerandæ antiquitatis cultor & destructor.
Docte febricitans,
Somnia & inaudita commenta
Vigilans edidit.
Scepticum pie egit,
Credulitate puer, audacia juvenis, deliriis senex."

HARDWICKE (PHILIP YORKE, earl of), was born at Dover in Kent, December 1, 1690; and educated under Mr. Samuel Morland, of Bethnal-Green, in classical and general learning, which he ever cultivated amidst his highest employments. He studied the law in the Middle Temple; and, being called to the bar in 1714, he soon became very eminent in his profession. In 1718, he sat in parliament as member for Lewes in Sussex; and, in the two successive parliaments, for Seaford. March 1719-20, he was promoted to the office of solicitor-general, by the recommendation of the lord chancellor Parker: an obligation he never forgot, returning it by all possible marks of personal regard and affection. The trial of Mr. Layer at the king's bench for high treason, November, 1722, gave him an opportunity of shewing his abilities: his reply, in which he summoned up late at night the evidence against the prisoner, and answered all the topics of defence, being justly admired as one of the ablest performances of that kind extant. About the same time, he gained much reputation in parliament, by opening the bill against Kelly, who had been principally concerned in bishop Atterbury's plot, as his secretary. February, 1723-4, he was appointed attorney-general; in the execution of which important office, he was remarkable for his candour and lenity.

nity. As an advocate for the crown, he spoke with the veracity of a witness and a judge: and, though his zeal for justice and the due course of law was strong, yet his tenderness to the subject, in the court of exchequer, was so distinguished, that upon a particular occasion, in 1733, the house of commons assented to it with a general applause. He was unmoved, by fear or favour, in what he thought right and legal; and often debated and voted against the court, in matters relating to the South-Sea company, when he was solicitor; and, in the affair of lord Derwentwater's estate, when he was attorney-general. Upon the resignation of the great seal by Peter lord King, in October, 1733, Sir Philip Yorke was appointed lord chief-justice of the king's bench. He was soon after raised to the dignity of a baron of this kingdom, with the title of lord Hardwicke, baron of Hardwicke, in the county of Gloucester, and called to the cabinet council. The salary of chief-justice of the king's bench, being thought not adequate to the weight and dignity of that high office, was raised, on the advancement of lord Hardwicke to it, from 2000 l. to 4000 l. per ann. to the chief-justice and his successors; his lordship refusing to accept the augmentation of it: and the adjustment of the two vacancies of the chancery and king's bench (which happened at the same time) between his lordship and lord Talbot, upon terms honourable and satisfactory to both, was thought to do as much credit to the wisdom of the crown, in those days, as the harmony and friendship, with which they co-operated in the public service, did honour to themselves. In the midst of the general approbation, with which he discharged his office there, he was called to that of lord high chancellor, on the decease of lord Talbot, February 17, 1736-7.

The integrity and abilities with which he presided in the court of chancery, during the space of almost twenty years, appears from this remarkable circumstance, that only three of his decrees were appealed from, and even those were afterwards affirmed by the house of lords. After he had executed that high office about seventeen years, in times and circumstances of accumulated difficulty and danger, and had twice been called to the exercise of the office of lord high steward, on the trials of peers concerned in the rebellion; he was, April 1754, advanced to the rank of an earl of Great Britain, with the titles of viscount Royston, and earl of Hardwicke. This favour was conferred unasked, by his sovereign, who treated him through the whole of his reign with particular esteem and confidence, and always spoke of him in a manner which shewed, that he set as high a value on the man as on the minister. His resignation of the great seal, in November,

vember, 1756, gave an univerſal concern to the nation, however divided at that time in other reſpects. But he ſtill continued to ſerve the public in a more private ſtation; at council, at the houſe of lords, and upon every occaſion where the courſe of public buſineſs required it, with the ſame aſſiduity as when he filled one of the higheſt offices in the kingdom. He always felt and expreſſed the trueſt affection and reverence for the laws and conſtitution of his country: this rendered him as tender of the juſt prerogatives inveſted in the crown, for the benefit of the whole, as watchful to prevent the leaſt incroachment upon the liberty of the ſubject. The part which he acted in planning, introducing, and ſupporting, the "Bill for aboliſhing the heretable Juriſdictions in Scotland," and the ſhare which he took, beyond what his department required of him, in framing and promoting the other bills relating to that country, aroſe from his zeal to the Proteſtant ſucceſſion, his concern for the general happineſs and improvement of the kingdom, and for the preſervation of this equal and limited monarchy; which were the governing principles of his public conduct through life. And theſe, and other bills which might be mentioned, were ſtrong proofs of his talents as a legiſlator. In judicature, his firmneſs and dignity were evidently derived from his conſummate knowledge and talents; and the mildneſs and humanity, with which he tempered it, from the beſt heart. He was wonderfully happy in his manner of debating cauſes upon the bench. His extraordinary diſpatch of the buſineſs of the court of chancery, increaſed as it was in his time, beyond what had been known in any former, was an advantage to the ſuitor, inferior only to that ariſing from the acknowledged equity, perſpicuity, and preciſion, of his decrees. The manner in which he preſided in the houſe of lords added order and dignity to that aſſembly, and expedition to the buſineſs tranſacted there. His talents, as a ſpeaker in the ſenate as well as on the bench, were univerſally admired: he ſpoke with a natural and manly eloquence, without falſe ornaments or perſonal invectives; and, when he argued, his reaſons were ſupported and ſtrengthened by the moſt appoſite caſes and examples which the ſubject would allow. His manner was graceful and affecting; modeſt, yet commanding; his voice peculiarly clear and harmonious, and even loud and ſtrong, for the greater part of his time. With theſe talents for public ſpeaking, the integrity of his character gave a luſtre to his eloquence, which thoſe who oppoſed him felt in the debate, and which operated moſt powerfully on the minds of thoſe who heard him with a view to information and conviction.

'Convinced of the great principles of religion, and steady in his practice of the duties of it, he maintained a reputation of virtue, which added dignity to the stations which he filled, and authority to the laws which he administered. His attachment to the national church was accompanied with a full conviction, that a tender regard to the Rights of conscience, and a temper of lenity and moderation, are not only right in themselves, but most conducive in their consequences to the honour and interest of the church. The strongest recommendation to him of the clergy, to the ecclesiastical preferments in his disposal, was their fitness for the discharge of the duties of their profession. And that respectable body owes a particular obligation to his lordship, and his predecessor lord Talbot, for the opposition which they gave in the house of lords to the "Act for the more easy recovery of Tithes, Church-rates, and other ecclesiastical Dues, from the People called Quakers," which might have proved of dangerous consequences to the rights and property of the clergy; though it had passed the other house, and was known to be powerfully supported. Many facts and anecdotes which do him honour may be recollected and set down, when resentments, partialities, and contests, are forgot.

The amiableness of his manners, and his engaging address, rendered him as much beloved by those who had access to him as he was admired for his great talents by the whole nation. His constitution, in the earlier part of his life, did not seem to promise so much health and vigour as he afterwards enjoyed, for a longer period than usually falls to the share of men of more robust habit of body. But his care to guard against any excesses secured to him an almost uninterrupted tenour of health: and his habitual mastery of his passions gave him a firmness and tranquillity of mind unabated by the fatigues and anxieties of business; from the daily circle of which, he rose, to the enjoyment of the conversation of his family and friends, with the spirits of a person entirely vacant and disengaged. Till the latter end of his seventy-third year, he preserved the appearance and vivacity of youth in his countenance, in which the characters of dignity and amiableness were remarkably united: and he supported the tedious disorder which proved fatal to him, and which was of the dysenteric kind, with an uncommon resignation, and even chearfulness, till the close of life. He died, in his seventy-fourth year, at his house in Grosvenor-square, March 6, 1764. His body lies interred at Wimple in Cambridgeshire, by that of his lady, Margaret, daughter of Charles Cocks, Esq. of Worcestershire, and niece of lord-chancellor Sommers:

HARDY

HARDY (ALEXANDER), a French dramatist, wrote an incredible number of pieces for the theatre, some say so many as six hundred, and some even more. Of these however no more remain than forty-one, which were published by himself in six volumes octavo. He had a remarkable facility in writing; and it was said that he would write two thousand lines in twenty-four hours: in three days his play was composed, learned, and acted. He certainly had. considerable talents, but, as he was very necessitous and compelled to write against time, his abilities had not fair scope. He was the first French dramatist who introduced the custom of being paid for his pieces. He died at Paris in 1630.

HARDY (CHARLES), was the grandson of a distinguished naval commander in the reign of Queen Anne. He was a gallant and able officer; and, passing through the different ranks of his profession with the highest reputation, was, in 1779, appointed commander in chief of the grand Western squadron. He died of an inflammation in his bowels in the same year at Spithead.

HARE (Dr. FRANCIS), an English bishop, of whose birth we have no particulars, was bred at Eton school, and from that foundation became a fellow of King's college, Cambridge; where he had the tuition of the marquis of Blandford, only son of the illustrious duke of Marlborough, who appointed him chaplain general to the army. He afterwards obtained the deanery of Worcester, and thence was promoted to the bishopric of Chichester, which he held with the deanery of St. Paul's to his death, which happened in 1740. He was dismissed from being chaplain to George I. in 1718, by the strength of party prejudices, in company with Dr. Moss and Dr. Sherlock, persons of distinguished rank for parts and learning.

About the latter end of queen Anne's reign he published a remarkable pamphlet, intituled, "The Difficulties and Discouragements which attend the Study of the Scriptures, in the Way of private Judgement:" in order to shew, that, since such a study of the Scriptures is an indispensable duty, it concerns all Christian societies to remove, as much as possible, those discouragements. In this work, his manner appeared to be so ludicrous, that the convocation fell upon him, as if he were really against the study of the holy Scriptures: and Whiston says, that, finding this piece likely to hinder that preferment he was seeking, he aimed to conceal his being the author. The same writer charges him with being strongly inclined to Scepticism; that he talked ludicrously of sacred matters; and that he would offer to lay wagers, about the fulfilling of Scripture prophecies. But the principal

principal ground for thefe invidious infinuations feems to be, that, though he never *denied* the genuinenefs of the apoftolical conftitutions (of which by the bve he procured for Whifton the collation of two Vienna MSS.), yet " he was not firm believer enough, nor ferious enough in Chriftianity, to hazard any thing in this world for their reception." He publifhed many pieces againft bifhop Hoadly, in the Bangorian controverfy, as it is called; and alfo other learned works, which were collected after his death, and publifhed in four volumes, 8vo. 2. An edition of " Terence," with notes, in 4to. 3. " The Book of Pfalms in the Hebrew, put into the original poetical Metre," 4to. In this laft work, he pretends to have difcovered the Hebrew metre, which was fuppofed to be irretrievably loft. But his hypothefis, though defended by fome, yet has been confuted by feveral learned men, particularly by Dr. Lowth in his " Metricæ Hareanæ brevis confutatio," annexed to his lectures " De Sacrâ Poefi Hebræorum."

HARIOT (THOMAS), an eminent mathematician, was born at Oxford, or, as Anthony Wood expreffes it, " tumbled out of his mother's womb in the lap of the Oxonian Mufes," in 1560. Having been inftructed in grammar-learning in that city, he became a commoner of St. Mary-hall, where he took the degree of B. A. in 1579. He had then fo diftinguifhed himfelf, by his uncommon fkill in mathematics, as to be recommended foon after to Sir Walter Raleigh as a proper preceptor to him in that fcience. Accordingly, that noble knight became his firft patron, took him into his family, and allowed him a handfome penfion. In 1585, he was fent over by Sir Walter with his firft colony to Virginia; where, being fettled, he was employed in difcovering and furveying that country, in obferving what commodities it produced, together with the manners and cuftoms of its inhabitants. He publifhed an account of it under this title, " A brief and true Report of the Newfoundland of Virginia;" which was reprinted in the third voyage of Hakluyt's " Voyages." Upon his return to England, he was introduced by his patron to the acquaintance of Henry earl of Northumberland; who " finding him," fays Wood, " to be a gentleman of an affable and peaceable nature, and well read in the obfcure parts of learning," allowed him a yearly penfion of 120l. About the fame time, Robert Hues, well known by his " Treatife upon the Globes," and Walter Warner, who is faid to have communicated to the famous Harvey the firft hint concerning the circulation of the blood, being both of them mathematicians, received penfions from him of

lefs

less value. So that in 1606, when the earl was committed to the Tower for life, Hariot, Hues, and Warner, were his constant companions, and were usually called the earl of Northumberland's Magi. They had a table at the earl's charge, who did constantly converse with them, to divert the melancholy of his confinement; as did also Sir Walter Raleigh, who was then in the Tower. Hariot lived for some time at Sion-college, and died in London, July 2, 1621, of a cancer in his lip. He was universally esteemed on account of his learning. When he was but a young man, he was styled by Mr. Hakluyt "Juvenis in disciplinis mathematicis excellens;" and by Camden, "Mathematicus insignis." A MS. of his, intituled "Ephemeris Chryrometrica," is preserved in Sion-college library; and his "Artis Analyticæ Praxis" was printed after his death, in a thin folio, and dedicated to Henry earl of Northumberland. Des Cartes is said to have been obliged to this book for a great many of his improvements in algebra.

As to his religion, Wood says, that, "notwithstanding his great skill in mathematics, he had strange thoughts of the Scripture, always undervalued the old story of the Creation of the World, and could never believe that trite position, 'Ex nihilo nihil fit.' He made a Philosophical Theology, wherein he cast off the Old Testament, so that consequently the New would have no foundation. He was a Deist, and his doctrine he did impart to the earl, and to Sir Walter Raleigh, when he was compiling the 'History of the World,' and would controvert the matter with eminent divines of those times: who therefore, having no good opinion of him, did look on the manner of his death, as a judgement upon him for those matters, and for nullifying the Scripture." Wood mentions no authority for this assertion: and we may observe, that Hariot assures us himself, that when he was with the first colony settled in Virginia, in every town where he came, "he explained to them the contents of the Bible, &c. And though I told them," says he, "the book materially and of itself was not of such virtue as I thought they did conceive, but only the doctrine therein contained; yet would many be glad to touch it, to embrace it, to kiss it, to hold it to their breasts and heads, and stroke over all their bodies with it, to shew their hungry desires of that knowledge which was spoken of." To which we may add, that, if Hariot was reputed a Deist, it is by no means probable that Dr. Corbet, an orthodox divine, and successively bishop of Oxford and Norwich, sending a poem, dated December 9, 1618, to Sir Thomas Aylesbury, when the comet appeared, should speak of

"——— Deep Hariot's mine,
"In which there is no dross, but all refine."

Lastly,

Laſtly, it is very unlikely that his noble executors, Sir Thomas Ayleſbury and Robert Sidney, viſcount Liſle, would have ſuffered an inſcription to be engraved upon his monument in St. Chriſtopher's church, which might have been contradicted by all the town, if it had been falſe, and which, upon the ſuppoſition of his being an infidel, would have been ridiculous:

" Qui omnes ſcientias calluit, & in omnibus excelluit ;
" Mathematicis, Philoſophicis, Theologicis,
" Veritatis indagator ſtudioſiſſimus,
" Dei Triniunius cultor piiſſimus."

HARLEY (ROBERT), afterwards earl of Oxford and earl Mortimer, and lord high treaſurer in the reign of queen Anne, was eldeſt ſon of Sir Edward Harley, and born at London, in Bow-ſtreet, Covent Garden, December 5, 1661. He was educated under the Rev. Mr. Birch, at Shilton, near Burford, Oxfordſhire, which, though a private ſchool, was remarkable for producing at the ſame time a lord high treaſurer, viz. lord Oxford; a lord high chancellor, viz. lord Harcourt; a lord chief juſtice of the Common pleas, viz. lord Trevor; and ten members of the Houſe of Commons, who were all contemporaries, as well at ſchool as in parliament. Here he laid the foundation of that extenſive knowledge and learning, which rendered him afterwards ſo conſpicuous in the world. At the Revolution, Sir Edward Harley, and this his eldeſt ſon, raiſed a troop of horſe at their own expence; and, after the acceſſion of king William and queen Mary, he was firſt choſen member of parliament for Tregony in Cornwall, and afterwards ſerved for the town of Radnor, till he was called to the Houſe of Lords. In 1690, he was choſen by ballot one of the nine members of the Houſe of Commons, commiſſioners for ſtating the public accounts; and alſo one of the arbitrators for uniting the two India companies. In 1694, the Houſe of Commons ordered Mr. Harley, November 19, to prepare and bring in a bill " For the frequent meeting and calling of parliaments ;" which he accordingly did upon the 22d, and it was received and agreed to by both houſes, without any alteration or amendment. On February 11, 1701-2, he was choſen ſpeaker of the Houſe of Commons; and that parliament being diſſolved the ſame year by king William, and a new one called, he was again choſen ſpeaker December 31 following, as he was in the firſt parliament called by queen Anne.

April 17, 1704, he was ſworn of her majeſty's privy council; and, May 18th following, ſworn in council one of the principal

principal secretaries of state, being also speaker of the House of Commons at the same time. In 1706, he was appointed one of the commissioners for the treaty of union with Scotland, which took effect; and resigned his place of principal secretary of state in February 1707-8. August 10, 1710, he was constituted one of the commissioners of the treasury, also chancellor and under-treasurer of the exchequer. On the 8th of March following, he was in great danger of his life; the marquis of Guiscard, a french papist, then under examination of a committee of the privy council at Whitehall, stabbing him with a penknife, which he took up in the clerk's room, where he waited before he was examined. Guiscard was imprisoned, and died in Newgate the 17th of the same month: whereupon an act of parliament passed, making it felony, without benefit of clergy, to attempt the life of a privy counsellor in the execution of his office; and a clause was inserted "To justify and indemnify all persons, who in assisting in defence of Mr. Harley, chancellor of the exchequer, when he was stabbed by the sieur de Guiscard, and in securing him, did give any wound or bruise to the said sieur de Guiscard, whereby he received his death." The wound Mr. Harley had received confined him some weeks; but the house being informed that it was almost healed, and that he would in a few days come abroad, resolved to congratulate his escape and recovery; and accordingly, upon his attending the house on the 26th of April, the speaker addressed him in a very respectful speech, to which Mr. Harley returned as respectful an answer. They had before addressed the queen on this alarming occasion.

In 1711, queen Anne, to reward his many eminent services, was pleased to advance him to the peerage of Great Britain, by the style and titles of baron Harley of Wigmore, in the county of Hereford, earl of Oxford, and earl Mortimer, with remainder, for want of issue male of his own body, to the heirs male of Sir Robert Harley, knight of the Bath, his grandfather. May 29, 1711, he was appointed lord high-treasurer of Great Britain; and, August 15th following, at a general court of the South-Sea company, he was chosen their governor, as he had been their founder and chief regulator. October 26, 1712, he was elected a knight companion of the most noble order of the garter. July 27, 1714, he resigned his staff of lord high-treasurer of Great Britain, at Kensington, into the queen's hand, she dying upon the 1st of August following. June 10, 1715, he was impeached by the House of Commons of high-treason, and high crimes and misdemeanors; and, on July the 16th, was committed to the tower by the House of Lords, where he suffered confinement

finement till July 1, 1717, and then, after a public trial, was acquitted by his peers. He died in the 64th year of his age, May 21, 1724, after having been twice married. Pope has celebrated his memory in the following lines:

> "A foul fupreme, in each hard inftance tried,
> Above all pain, all anger, and all pride,
> The rage of power, the blaft of public breath,
> The luft of lucre, and the dread of death."

From our account of this noble lord, he muft naturally pafs for a very great as well as good man; yet he has been reprefented by others as very remote from either greatnefs or goodnefs; and particularly by the late lord Bolingbroke, in his curious "Letter to Sir William Windham," where the portrait given of him is not only mean, but odious. However, as it is but reafonable to fuppofe, that lord Oxford had his allay of infirmities, notwithftanding the fine things that were faid of him, and the honours that were done to him; fo, on the other hand, it is as reafonable not to believe all that contemporary minifters fay of each other, and efpecially when they have quarrelled. He was a great encourager of learning, and not only fo, but the greateft collector in his time of all curious books in print and manufcript, efpecially thofe concerning the hiftory of his own country, which were preferved and much augmented by the earl his fon. He was alfo a man of tafte and letters himfelf; and under this character we find a propofal addreffed to him by Dr. Swift, "for correcting, improving, and afcertaining the Englifh tongue."

HARMER (THOMAS), a diffenting minifter at Waterfield, in Suffolk; was much and defervedly efteemed in the literary world. His moft important and valuable work was, "Obfervations on Paffages of Scripture," in four volumes, octavo. This has gone through different editions, and, as the author had the advantage of Sir John Claudius's manufcripts, great light is thrown on his performances, not only on fcripture, but on the manners of the Eaft. Mr. Harmer alfo publifhed "Notes on Solomon's Song." He was eminently diftinguifhed for his accomplifhments in oriental learning, and for his fkill in the ftudy of antiquities. He died at an advanced age in 1788.

HARMODIUS, the friend of Ariftogiton, who in conjunction delivered their country from the tyranny of the Pififtratidæ. They received immortal honour from their fellow-citizens; and have been celebrated in every age and country where the value of liberty was known. See an account of thofe deliverers of their country in Herodotus—Beloc's Tranflation, Vol. II. p. 420.

HAROLD,

HAROLD, succeffor to the crown of England, at the death of Edward the Confeffor. It was in his reign that William the firft, actuated by courage, refentment, and ambition, invaded England with his Norman army. Harold gave him battle in perfon, and the English and Normans prepared for this important decifion at Haftings. This terminated in favour of William; for, Harold was flain by an arrow as he was combating with great bravery at the head of his troops. With this prince terminated the authority of the anglo-faxon monarchs, who had governed England for the fpace of fix hundred years.

HARPALUS, a great aftronomer, who flourifhed about 480 years before Chrift. He corrected the cycle of eight years, invented by Cleoftratus, and in its ftead propofed a new one of nine years, in which he fuppofed that the fun and moon returned to the fame point; but this cycle of Harpalus was afterwards altered by Meton, who added ten years to it.

HARPOCRATION (VALERIUS), an ancient rhetorician of Alexandria, has left us an excellent "Lexicon upon the ten Orators of Greece;" for that is the title ufually given to it, though Meurfius will have it, that the author infcribed it only λεξεις; and he is followed, in this opinion, by James Gronovius. Harpocration fpeaks in this work, with much feeming exactnefs, of magiftrates, pleadings at the bar, places in Attica, names of men who had the chief management of affairs in the republic, and of every thing, in fhort, which has been faid to the glory of this people by their orators. Aldus firft publifhed this Lexicon in greek at Venice. 1603, in folio; many learned men, as Meurfius, Mauffac, Valefius, have laboured upon it; and James Gronovius gave an edition of it at Leyden, 1696, in 4to.

HARRINGTON (Sir JOHN), an ingenious Englifh poet, was the fon of John Harrington, Efq. who was imprifoned in the tower, under queen Mary, for holding a correfpondence with the lady Elizabeth, with whom he continued in great favour to the time of his death. Sir John was born at Kelfton, near Bath, in Somerfetfhire, and had queen Elizabeth for his godmother. He was inftructed in claffical learning at Eton-fchool, and from removed to Cambridge, where he took the degree of M. A. Before he was thirty, he publifhed a tranflation of Ariofto's "Orlando Furiofo," by which he gained a confiderable reputation, and for which he is now principally known. He was knighted in the field by the earl of Effex, which gave much offence to the queen, who was fparing of fuch honours, and chofe to confer them herfelf. In the reign of James, he was created knight of the Bath; and, being a courtier, prefented a MS. to prince Henry, levelled chiefly againft the married bifhops, which was intended only for the private ufe of his royal highnefs; but, being publifhed afterwards, created great clamour, and made feveral of the clergy fay, that his conduct was of a piece with his doctrines; fince he, together with Robert earl of Leicefter, fupported Sir Walter Raleigh in his fuit to queen Elizabeth for the manor of Banwell, belonging to the bifhopric of Bath and Wells; on a prefumption, that the Right Rev. Incumbent had incurred a *præmunire*, by marrying a fecond wife. Wood's account of it is this: "That Sir John Harrington, being minded to obtain the favour of prince Henry, wrote a difcourfe for his private ufe, intituled, 'A brief View of the State of the Church of England, as it ftood in Queen Elizabeth's and King James's Reign, to the year 1608.' This book is no more than a character and hiftory of the bifhops of thofe times, and was written to the faid prince Henry, as an additional fupply to the Catalogue of Bifhops of Dr. Francis Godwin, upon occafion of that proverb."

'Henry the eighth pulled down monks and their cells,
Henry the ninth fhall pull down bifhops and their bells.'

"In the faid book the author Harrington doth, by imitating his godmother, queen Elizabeth, fhew himfelf a great enemy to married bifhops, efpecially to fuch as had been married twice; and many things therein are faid of them, that were by no means fit to be publifhed, being written only for private ufe. But fo it was, that the book coming into the the hands of one John Chetwind, grandfon by a daughter to the author, a perfon deeply principled in prefbyterian tenets, did, when the prefs was open, print it at London in 1653; and no fooner was it publifhed, and came into the hands of many, but it was exceedingly clamoured at by the loyal and orthodox clergy, condemning him that publifhed it."

We have not been able to fix the time of Sir John Harrington's birth, nor are we more certain about that of his death; but, as the former may be moft probably placed about the middle of queen Elizabeth's reign, fo we think the latter might happen towards the latter end of king James's. We will fubjoin an epigram, as a fpecimen of his poetry; fince his productions in this way are not every day to be met with.

"IN CORNUTUM.

What curl'd pale youth is he that fitteth there,
So near my wife, and whifpers in her ear,

And takes her hand in his, and foft doth wring her,
Sliding her ring ftill up and down her finger?
Sir, 'tis a proctor, feen in both the laws,
Retained by her in fome important caufe;
Prompt and difcreet both in his fpeech and action,
And doth her bufinefs with great fatisfaction.
And think'ft thou fo? a horn-plague on thy head!
Art thou fo like a fool, and wittol led,
To think he doth the bufinefs of thy wife?
He doth thy bufinefs, I dare lay my life."

A mifcellaneous collection of Harrington's works, in profe and verfe, was publifhed by the Rev. Henry Harrington, under the title of "Nugæ Antiquæ," which contains many curious things. Sir John had formed a plan for the hiftory of his own times, but he did not live to execute it. He died in 1612, at the age of fifty-one.

HARRINGTON (JAMES), an eminent political writer, was born in January 1611; being the eldeft fon of Sir Sapcote Harrington, and Jane the daughter of Sir William Samuel of Upton, in Northamptonfhire, the place of his nativity. When he had made a progrefs in claffical learning, he was admitted in 1629 a gentleman-commoner of Trinity-college, in Oxford, and placed under Mr. Chillingworth, who had lately been elected fellow of that college; from whom he might poffibly acquire fome portion of that fpirit of reafoning and thinking for himfelf, which afterwards fhone forth fo confpicuoufly in his writings. About three years after, his father died; upon which he left the univerfity, and began to think of travelling, having previoufly furnifhed himfelf with the knowledge of feveral foreign languages for that purpofe. His firft ftep was into Holland, then the principal fchool of martial difcipline; and, what may be fuppofed to have affected him more fenfibly, a country wonderfully flourifhing, under the aufpices of liberty, commerce, ftrength, and grandeur. Here, it is probable that he began to make government the fubject of his meditations; for, he was often heard to fay, that, "before he left England, he knew no more of anarchy, monarchy, ariftocracy, democracy, oligarchy, or the like, than as hard words, whofe fignification he found in his dictionary." On coming into the Netherlands, he entered a volunteer, and fo continued fome months, in lord Craven's regiment; during which time, being much at the Hague, he had the farther opportunity of accomplifhing himfelf in two courts; namely, thofe of the prince of Orange, and the queen of Bohemia, daughter of our James I. who was then a fugitive in Holland. He was taken into great favour by this princefs, and alfo by the prince elector, whom he attended to Copenhagen, when his highness paid a visit to the king of Denmark; and, after his return from travelling, was entrusted by him with the affairs of the palatinate, so far as they were transacted at the British court.

He stayed, however, but a short time in Holland; no temptations or offers could divert or restrain him from the resolution he had formed to travel, and therefore, taking Flanders in his way, he set out on a tour through part of Germany, France, and Italy. While he was at Rome, the pope performed the ceremony of consecrating wax-lights on Candlemas-day. When his holiness had sanctified these torches, they were distributed among the people, who fought for them very eagerly. Harrington was desirous to have one of them; but, perceiving that it was not to be obtained without kissing the pope's toe, he declined to accept it on such a condition. His companions were not so scrupulous, and when they came home spoke of his squeamishness to the king. The king told him, " he might have done it only as a piece of respect to a temporal prince;" but Harrington replied, that "since he had the honour to kiss his majesty's hand, he thought it beneath him to kiss any other prince's foot." He is said to have preferred Venice to all other places in Italy, as he did its government to that of the whole world; it being, in his opinion, immutable by any external or internal causes, and to finish only with mankind. Here he cultivated an acquaintance with all the men of letters, and furnished himself with the most valuable books in the italian tongue, such especially as were written upon politics and government.

After having thus seen Italy, France, the Low-countries, Denmark, and some parts of Germany, he returned home to England, perfectly accomplished. In the beginning of the civil war, 1642, he manifestly sided with the parliament, and endeavoured to get a seat in the house, but could not. His inclination to letters kept him from seeking public employments, so that we hear no more of him till 1646; when attending out of curiosity the commissioners, appointed by parliament to bring Charles I. from Newcastle nearer to London, he was by some of them named to wait on his majesty, as a person known to him before, and engaged to no party or faction. The king approved the proposal, and Harrington entered on the station of a domestic; but would never presume to come into his presence, except in public, till he was particularly commanded by the king, and made one of the grooms of the bed-chamber, as he was in May 1647. He had the good fortune to please the king much: "His Majesty loved his company," says Wood, "and finding him to be an ingenious

ous man, chose rather to converse with him, than with others of his chamber. They had often," says he, " discourses concerning government; but, when they happened to talk of a commonwealth, the king seemed not to endure it." Harrington conceived a high notion of the king, finding him to be a different person from what he had been represented, as to parts, morals, religion, &c. and therefore, after the king was removed out of the Isle of Wight to Hurst-castle, in Hampshire, was forcibly turned out of his service, because he vindicated some of his majesty's arguments against the parliament commissioners at Newport, and thought his concessions more satisfactory than they did. There is no ground to imagine that he saw the king any more, till the day he was brought to the scaffold; whither Harrington found means to accompany him, and where, or a little before, he received a token of his majesty's affection. The king's execution affected him extremely. He often said, " nothing ever went nearer him; and that his grief on that account was so great as to bring a disorder upon him."

After the king's death, he was observed to keep much in his library, and more retired than usual, which his friends attributed to discontent and melancholy. But, to convince them that this was not the cause of his retirement, he produced a copy of his " Oceana;" which " he had been writing," he said, " not only because it was agreeable to the studies which he pursued, but because, if ever it should be the fate of England to be, like Italy of old, overrun by a barbarous people, or to have its government and records destroyed by some merciless conqueror, they might not be then left to their own invention in framing a new government." This "Oceana" is a kind of political romance, in imitation of Plato's " Atlantic Story," where, by Oceana, Harrington means England; exhibiting a plan of republican government, which he would have had erected here, in case these kingdoms had formed themselves into a genuine commonwealth. This work, however, as it reflected severely upon Oliver's usurpation, met with many difficulties in the publishing; for, it being known to some of the courtiers that it was printing, they hunted it from one press to another, till at last they found it, and carried it to Whitehall. All the solicitations he could make were not able to retrieve his papers, till he bethought himself of applying to lady Claypole, who was a good-natured woman, and Oliver's favourite daughter; and who, upon his declaring that it contained nothing prejudicial to her father's government, got them restored to him. He printed it in 1656, and dedicated it, as he promised lady Claypole, to her father; who, it is said, perused it, but declared, agreeable to his principles of policy, that "the

"the gentleman muft not think to cheat him of his power and authority; for that what he had won by the fword, he would not fuffer himfelf to be fcribbled out of."

This work was no fooner publifhed, than many undertook a refutation of it. This occafioned him to reply, and to explain his fcheme, in feveral fucceffive pieces; which however we will not ftay to enumerate here, becaufe they are fo eafy to be feen in the collection of his works. In the mean time, he not only endeavoured to propagate his republican notions by writing, but, for the more effectually advancing a caufe, of which he was enthufiaftically enamoured, he formed a fociety of gentlemen, agreeing with him in principles, who met nightly at Miles's coffee-houfe, in New Palace-yard, Weftminfter, and were called the Rota. Wood has given a very particular account of this affociation, or gang, as he calls them. "Their difcourfes about government," fays he, "and of ordering a commonwealth, were the moft ingenious and fmart that ever were heard; for the arguments in the parliament-houfe were but flat to thofe. This gang had a balloting-box, and balloted how things fhould be carried by way of effay; which not being uf'd, or known in England before on this account, the room was every evening very full. The doctrine there inculcated was very taking; and the more, becaufe as to human forefight there was no poffibility of the king's return. The greateft part of the parliament-men hated this rotation and balloting, as being againft their power: eight or ten were for it, who propofed it to the houfe, and made it out to the members, that, except they embraced that fort of government, they muft be ruined. The model of it was, that the third part of the fenate or houfe fhould rote out by ballot every year, not capable of being elected again for three years to come; fo that every ninth year the fenate would be wholly altered No magiftrate was to continue above three years, and all to be chofen by the ballot, than which nothing could be invented more fair and impartial, as it was then thought, though oppofed by many for feveral reafons. This club of Commonwealthfmen, which began about Michaelmas 1659, lafted till about February 21 following; at which time, the fecluded members being reftored by general Monk, all their models vanifhed.".

After the Reftoration, he lived more privately than he had done before, but ftill was looked upon as a dangerous perfon, who maintained and propagated principles, which could never be reconciled to monarchical government. He employed himfelf now in reducing his politics into fhort and eafy aphorifms methodically digefted, and freely communicated his papers to all who vifited him. While he was putting the laft hand to his fyftem, he was, by an order from the king, feized December

cember 28, 1661, and committed to the tower of London for treasonable designs and practices. He was charged by lord chancellor Hyde, at a conference of the lords and commons, with being concerned in a plot, whereof twenty-one persons were the chief managers: "that they all met in Bow-street, Covent-garden, and in other places; that they were of seven different parties or interests, as three for the commonwealth, three for the long parliament, three for the city, three for the purchasers, three for the disbanded army, three for the independents, and three for the fifth-monarchy men; that their first consideration was how to agree on the choice of parliament-men against the ensuing session; and that a special care ought to be had about the members for the city of London, as a precedent for the rest of the kingdom to follow; whereupon they nominated the four members after chosen, and then sitting in parliament. Their next care was to frame a petition to the parliament for a preaching ministry, and liberty of conscience; then they were to divide and subdivide themselves into several councils and committees, for the better carrying on their business by themselves or their agents and accomplices all over the kingdom. In these meetings Harrington was said to be often in the chair; that they had taken an oath of secrecy, and concerted measures for levying men and money." The chancellor added, that though he had certain information of the times and places of their meetings, and particularly those of Harrington and Wildman, they were nevertheless so fixed in their nefarious design, that none of those they had taken would confess any thing, not so much as that they had seen and spoken to one another at those times or places.

But, notwithstanding these declarations of the chancellor, it is certain, that this plot was never made out; and it is not impossible but it might be imaginary. It is at least easy to account upon political principles, for Harrington's confinement, and the severity and ill usage he met with in it, when we consider not only his notions of government, which he every where enforced with the greatest zeal; but also how obnoxious he must needs have made himself to the powers then in being, by his very ill usage of the Stuart family. Nothing can be viler than the picture he has drawn of Mary queen of Scotland; he has has also painted her son James I. in the most odious colours, suggesting at the same time, that he was not born of the queen, but was a supposititious impostor, and of course had no right to the crown he inherited. His portrait of Charles I. is an abominable figure: "never was man," says he, "so resolute and obstinate in tyranny. He was one of the most consummate in the arts of tyranny that ever was; and it could be no other than God's hand, that arrested him in the height

height of his designs and greatness, and cut off him and his family." The truth is, Harrington seems in the latter end o his life to have grown fanatic in politics; and his keeping within no bounds, as such people seldom do, might make it the more expedient to put him under confinement. From the tower he was conveyed very privately to St. Nicholas's island opposite to Plymouth; and thence, upon a petition, to Plymouth, some relations obliging themselves in a bond of 5000l for his safe imprisonment. At this place he became acquainted with one Dr. Dunstan, who advised him to take a preparation of guiacum in coffee, as a certain cure for the scurvy, with which he was then troubled. He drank of this liquor in great quantities, which had probably a very pernicious effect, for he soon grew delirious; upon which a rumour prevailed at Plymouth, that he had taken some drink which would make any man mad in a month; and other circumstances made his relations suspect, that he had foul play shewn him, lest he should write any more "Oceanas." It was near a month before he was able to bear the journey to London, whither, as nothing appeared against him, he had leave from the king to go. Here he was put under the care of physicians, who could afford little help to the weakness of his body, none at all to the disorders of his mind. He would discourse of other things rationally enough; but, when his own distemper was touched upon, he would fancy and utter strange things about the operation of his animal spirits, which transpired from him, he said, in the shape of birds, flies, bees, or the like. He talked so much of good and evil spirits, that he even terrified those about him; and to those who objected to him, that these chimeras were the fruits of a disordered imagination, he would reply, that "he was like Democritus, who, for his admirable discoveries in anatomy, was reckoned distracted by his fellow-citizens." In this crazy condition he married the daughter of Sir Marmaduke Dorrel, in Buckinghamshire, a lady to whom he was formerly suitor, and with whom he spent the remainder of his life. Towards his latter end, he was subject to the gout, and enjoyed little ease; but, drooping and languishing a good while, he was at lasted seized with a palsy, and died at Westminster, September 11, 1677, and lies buried there in St. Margaret's church, on the southside of the altar, next the grave of Sir Walter Raleigh.

His writings were first collected, methodized, reviewed, and published, by Toland, 1700, in one vol. folio; but there was another edition, by Dr. Birch, set forth in 1737, which contains several articles omitted in Toland's. He made some attempts in the poetical way. Thus, in 1658, he published an English translation of two eclogues of Virgil, and two
books

books of the "Æneis," under the title of "An Essay upon two of Virgil's Eclogues, and two of his Æneis, towards the translation of the whole;" and, in 1659, was printed his translation of the four following books "of the Æneid;" but his poetry, as Wood says, gained him no reputation.

HARRIS (WILLIAM), a protestant dissenting minister of eminent abilities and character, resided at Honiton in Devonshire. September 20, 1765, the degree of D. D. was conferred on him, in the university of Glasgow, by the unanimous consent of the members of that body. " He published an historical and critical Account of the Lives of James I. Charles I. and Oliver Cromwell, in five vols. 8vo. after the manner of Mr. Bayle. He was preparing a like account of James II. He also wrote the life of Hugh Peters; besides many fugitive pieces occasionally, for the public prints, in support of liberty and virtue. All his works have been well received; and those who differ from him in principle still value him in point of industry and faithfulness." We give this character in the words of his magnificent patron, Mr. Hollis, who had presented him with many valuable books in reference to the subject of his histories and was at the expence of procuring his doctor's degree. Dr. Harris died at Honiton, February, 4, 1770.

HARRIS (JAMES) Esq.) an english gentleman of very uncommon parts and learning, was the son of James Harris, Esq. by a sister of lord Shaftsbury, author of " The Characteristics," whose elegance and refinement of taste and manners Mr. Harris seems to have inherited. He was born in the Close at Salisbury, 1709; and educated at the grammar-school there. In 1726, he was removed to Wadham-college in Oxford, but took no degree. He cultivated letters, however, most attentively, and also music, in the theory and practice of which he is said to have had few equals. He was member for Christ-church, Hants, which he represented in several successive parliaments. In 1763, he was appointed one of the lords commissioners of the admiralty, and soon after removed to the board of treasury. In 1774, he was made secretary and comptroller to the queen, which post he held to his death. He died December 21, 1780, in his 72d year, after a long illness, which he bore with calmness and resignation.

He is the author of some valuable works. 1. " Three Treatises: concerning Art; Music, Painting, and Poetry; and Happiness, 1745," 8vo. 2. " Hermes; or, a Philosophical Enquiry concerning Universal Grammar, 1751," 8vo. Of this piece bishop Lowth, in the preface to his " English Grammar," expresseth himself thus: " Those, who would enter more deeply into this subject, will find it fully and accurately

curately handled, with the greatest acuteness of investigation, perspicuity of explication, and elegance of method, in a treatise intituled, "Hermes, by James Harris, Esq;" the most beautiful and perfect example of analysis that has been exhibited since the days of Aristotle." 3. "Philosophical Arrangements." 4. "Philological Enquiries, 1782," 2 vols. 8vo. finished just before his death, and published since.

HARRIS (WALTER), an English physician and member of the college. He was in great reputation about the year 1700, and was physician to William the third. He published a treatise in much esteem on the acute diseases of children, and this he did at the earnest intreaty of Sydenham.

HARRISON (WILLIAM), a young gentleman high in esteem, and (as Swift expresses it) "a little pretty fellow, with a great deal of wit, good sense, and good nature," and fellow of New-college, Oxford; had no other income than 40l. a year as tutor to one of the duke of Queensbury's sons. In this employment he fortunately attracted the favour of Dr. Swift, whose generous solicitations with Mr. St. John obtained for him the reputable employment of secretary to ord Raby, ambassador at the Hague, and afterwards earl of Stafford. A letter of his, whilst at Utrecht, dated December 16, 1712, is printed in the Dean's works. Mr. Harrison, who did not long enjoy his rising fortune, was dispatched to London with the Barrier-treaty; and died February 14, 1712-13. See the "Journal to Stella" of that and the following day, where Dr. Swift laments his loss with the most unaffected sincerity. Mr. Tickell has mentioned him with respect, in his "Prospect of Peace," in "English Poets," Vol. XXVI. p. 113; and Dr. Young, in the beautiful close of an "Epistle to Lord Lansdown" (Vol. LII. p. 185.) most pathetically bewails his loss. Dr. Birch, who has given a curious note on Mr. Harrison's "Letter to Swift," has confounded him with Thomas Harrison, M. A. of Queen's college. In the "Select Collection," by Nichols, are some pleasing specimens of his poetry; which, with "Woodstock-Park" in Dodsley's "Collection," and an "Ode to the Duke of Marlborough, 1707," in Duncombe's "Horace," are all the poetical writings that are known of this excellent young man; who figured both as an humourist and a politician in the fifth volume of the "Tatler," of which (under the patronage of Bolingbroke, Henley, and Swift) he was professedly the editor. See the "Supplement to Swift." There was another William Harrison, author of "The Pilgrim, or the happy Convert, a Pastoral Tragedy, 1709."

HARRISON (JOHN), celebrated for the accuracy of his mechanism, and the inventor of a time-keeper to ascertain the longitude

longitude at sea; was born at Foulby, near Pontefract, in Yorkshire, in the year 1693. His father, a carpenter, was occasionally assisted by the son in his employment; to which, as was then usual among artists in the country, were added the various practices of surveying lands, and repairing clocks and watches. From his earliest youth he seems to have had a strong propensity to that kind of machinery, which is moved by wheels. In 1700, he removed with his father to Barrow, in Lincolnshire, where, with few opportunities of acquiring knowledge, he improved whatever he could attain. For assistance in the prosecution of his studies, he acknowledged himself to have been obliged to a clergyman, who officiated in the neighbourhood, and who lent him a copy of Saunderson's lectures in MS. which, with the diagrams, he carefully transcribed. But in 1726, his native genius appears to have surmounted all the disadvantages of a confined and desultory education. He had then constructed two clocks, of which the workmanship was chiefly wood. To *these time-pieces* he applied the escapement, and the compound pendulum, which he had previously invented. The accuracy of these works was thought to have surpassed that of all those of a similar kind, which had preceded them. They were said to have scarcely varied a second in a month. With expectation of being enabled by the board of longitude to execute a machine for the discovery of the longitude at sea, he arrived in London, in the year 1725. Dr. Halley, to whom he first applied, referred him to Mr. George Graham, who, discovering his extraordinary talents, advised him to construct his machine before he made application to the board. For that purpose he returned home, and, in 1735, came to London with his first machine completed. The next year he was accordingly sent to Lisbon to make a trial of its properties. In this voyage he corrected what is termed the dead reckoning, about a degree and a half. Now, having received additional encouragement to continue his labours, in 1739 he produced a second machine more simple in the construction, and more exact in its movement than the former. Though it was never tried at sea, this farther proof of the artist's talents raised him still higher in the estimation of his friends and the public. In 1749 he had completed a third machine, still less complicated, but more accurate than the second. It was said to have erred no more than three or four seconds in a week. He then imagined that his art could produce nothing more perfect; but afterwards endeavouring to improve common watches, he found his expectations so much surpassed his formed attempts, that he was encouraged to make his fourth time-keeper. It was about six inches of diameter, and formed in the shape of a watch.

watch. The utility of this laſt improvement was aſcertained by a ſon of the inventor, in two voyages, one to Jamaica, and the other to Barbadoes; both the experiments proved ſatisfactory. From them it appeared that the machine kept time within the limits required by the Act of the 12th of queen Anne. The reward of 20,000l. for the diſcovery, was accordingly adjudged to Mr. Harriſon, who received it at different times, though not without infinite trouble. The four machines delivered to the board of longitude and depoſited in the royal obſervatory at Greenwhich, where, it is ſaid, they ſtill remain totally neglected. Mr. Kendal afterwards, for the uſe of captain Cook in his circum-navigation of the world, made a time-keeper after the principles upon which Mr. Harriſon had conſtructed his fourth. This machine, during a voyage of three years, was thought to have anſwered the purpoſe, in accuracy, as well as the original could have done. Mr. Harriſon employed the latter end of his life in conſtructing another time-keeper, on the principle which he had adopted in making his fourth.

After a trial of ten weeks, which was made 1772, at the king's private obſervatory at Richmond; it was found to have erred no more than four ſeconds and a half. His conſtitution had for ſome years viſibly declined; he had been ſubject to frequent fits of the gout, which had never attacked him till his 77th year; and he died in 1776 at his houſe in Red-lion-ſquare, aged eighty. His knowledge ſeems to have been entirely confined to the mechanics, on which ſubject he could ſpeak with clearneſs and preciſion; but he could not communicate his thoughts in writing without difficulty. His language, when written, was not free from that embarraſſment and obſcurity which are ſo frequently to be obſerved in the ſtyle of thoſe who have not been accuſtomed to explain their practical acquiſitions upon paper. His deſcription concerning ſuch mechaniſm as will afford a nice or true menſuration of time, which was publiſhed in 1775, has been adduced as a proof of the preceding obſervation. In his earlier years he had been a leader of a band of church-ſingers. His experience in the various modulations of ſound, and his accuracy in keeping time in muſic, were diſplayed in a curious monochord of his conſtruction; and were reported to have been equal to the ſkill, which he afterwards diſplayed in the invention of the machine, to which he owed his fortune and his fame.

HARRISON (COLONEL), the ſon of a butcher, and one of the judges of Charles the firſt. He was an impudent and hypocritical fanatic, and fixed upon as a ſuitable perſon to delude the unwary Fairfax, with whom he continued on his knees in the affectation of prayer, till the fatal blow was ſtruck on the ill-fated monarch. On the Reftoration he was tried and executed.

HARTLEY (DAVID), an Englifh phyfician of eminence, was the fon of a clergyman, and born about 1704. He received his academical education at Jefus-college, Cambridge, of which he was fellow; and took the degree of M. A. He firft began to practife phyfic at Newark in Nottinghamfhire; removed thence to St. Edmund's Bury, in Suffolk; after this, fettled for fome time in London; and, laftly, went to live at Bath, where he died, September 30, 1757, aged fifty-three. He publifhed, in 1739, "A View of the prefent Evidence for and againft Mrs. Stevens's Medicines as a Solvent for the Stone, containing 155 Cafes, with fome Experiments and Obfervations." He was greatly inftrumental in procuring for Mrs. Stevens the 5000l. granted by parliament: her medicines were made public in the Gazette, from June 16 to June 19, 1739. Yet Dr. Hartley is faid to have died of the ftone, after having taken above 200 pounds weight of foap; and Mrs. Stevens's medicines have long been exploded, as futile and of no effect. He is faid to have written alfo in defence of inoculation; and fome letters of his are in the "Philofophical Tranfactions." But his capital work is intituled "Obfervations on Man, his Frame, his Duty, and his Expectations, in two Parts, 1749." 2 vols. 8vo.

HARTUNGUS (JOHN), born at Millenberg, in Germany, in 1505; and ftudied in the univerfity of Heidelberg. He at firft took arms againft the Turks; but foon returned to the gentler fervice of the mufes, and became greek profeffor at Heidelberg. He read lectures upon Homer, and publifhed fome prologomena and notes on the three firft books of the Odyffey. He alfo tranflated Apollonius into latin. He died in 1579.

HARVEY (WILLIAM), an eminent Englifh phyfician, who firft difcovered the circulation of the blood, was born of a good family at Folkftone, in Kent, April 2, 1578. At ten years of age he was fent to a grammar-fchool at Canterbury, and at fourteen removed thence to Caius college, in Cambridge. At nineteen, he travelled, through France and Germany, to Padua in Italy; where, having ftudied phyfic under Euftachius Radius, John Minadous, and Hieronymus Fabricius ab Aquapendente, he was created doctor of phyfic and furgery in that univerfity, 1602. He had a particular regard for his laft mafter; often quotes him in terms of the higheft refpect; and declares, that he was the more willing to publifh his book, "De Motu Cordis;" becaufe Fabricius, who had learnedly and accurately delineated in a particular treatife almoft all the parts of animals, had left the heart alone untouched. Soon after,

after, returning to England, he was incorporated M. D. at Cambridge; went to London to practife, and married. In 1604, he was admitted candidate of the college of phyficians in London; and three years after fellow. In 1615, he was appointed lecturer of anatomy and furgery in that college; and the year after read a courfe of lectures there, in which he opened his difcovery relating to the circulation of the blood. The original MS. of thefe lectures is extant in the valuable mufeum of the late Sir Hans Sloane, which was puchafed by parliament, and is intituled, "Prælectiones anatom. univerfal. per me Gulielmum Harvæium, medicum Londinenfem, anat. & chirurg. profefforem. Ann. Dom. 1616. Anno ætatis 37. Prælect. Apr. 16, 17, 18." In 1628, he publifhed his "Exercitatio anatomica de motu cordis & fanguinis;" and dedicated it to Charles I. There follows alfo another dedication to the college of phyficians, in which he obferves, that he had frequently before, in his "Anatomical Lectures," declared his new opinion concerning the motion and ufe of the heart, and the circulation of the blood; and for above nine years had confirmed and illuftrated it before the college, by reafons and arguments grounded upon ocular demonftration, and defended it from the objections of the moft fkilful anatomifts. This difcovery was of fuch vaft importance to the whole art of phyfic, that as foon as men were fatisfied, which they were in a few years, that it could not be contefted, feveral put in for the prize themfelves; a great many affirmed the difcovery to be due to others, unwilling that Harvey fhould run away with all the glory. Some afferted, that father Paul was the firft difcoverer of the circulation; but, being too much fufpected for heterodoxies already, durft not make it public, for fear of the inquifition. Honoratus Faber profeffed himfelf to be the author of that opinion; and Vander Linden, who publifhed an edition of Hippocrates, about the middle of the laft century, took a great deal of pains to prove, that this father of phyfic knew the circulation of the blood, and that Harvey only revived it. But the honour of the difcovery has been fufficiently afferted and confirmed to Harvey; and, fays Freind, "as it was entirely owing to him, fo he has explained it with all the clearnefs imaginable: and, though much has been written upon that fubject fince, I may venture to fay, his own book is the fhorteft, the plaineft, and the moft convincing, of any, as we may be fatified, if we look into the many apologies written in defence of the circulation."

In 1632, he was made phyfician to Charles I. as he had been before to king James; and, adhering to the royal caufe upon the breaking out of the civil wars, attended his majefty at the battle of Edge-hill, and thence to Oxford; where, in

1642,

1642, he was incorporated M. D. In 1645, the king got him elected warden of Merton-college, in that university; but, upon the surrendering of Oxford the year after to the parliament, he left that office and retired to London. In 1651, he published his book, intituled, "Exercitationes de generatione animalium; quibus accedunt quædam de partu, de membranis ac humoribus uteri, & de conceptione." This is a curious work, and had certainly been more so, but for some misfortune, by which his papers perished, during the time of the civil wars. For, although he had both leave and an express order from the parliament to attend his majesty upon his leaving Whitehall, yet his house, in London, was in his absence plundered of all the furniture; and his "Adversaria," with a great number of anatomical observations, relating especially to the generation of insects, were taken away by the savage hands of the rude invader This loss he lamented several years after, and the reader will be apt to lament too, when he considers the following pathetic words: "Atque hæc dum agimus, ignoscant mihi niveæ animæ, si summarum injuriarum memor levem gemitum effudero. Doloris mihi hæc causa est. Cum inter nuperos nostros tumultus, & bella plusquam civilia, serenissimum regem, idque non solum senatus permissione sed & jussu sequor, rapaces quædam manus non modo ædium mearum supellectilem omnem expilarunt, sed etiam, quæ mihi causa gravior querimoniæ, adversaria mea multorum annorum laboribus parta è musæo meo summoverunt. Quo factum est, ut observationes plurimæ, præsertim de generatione insectorum, cum reipublicæ literariæ, ausim dicere, detrimento perierint." In 1654, on Michaelmas-day, he was chosen president of the college of physicians in his absence; and, coming thither the day after, acknowledged his great obligation to the electors, for chusing him into a place of the same honour and dignity, as if he had been elected to be "Medicorum omnium apud Anglos princeps." But his age and weakness were so great, that he could not discharge the duty incumbent upon that great office; and, therefore, he requested them to chuse Dr. Prujean, who had deserved so well of the college. As he had no children, he made the college his heirs, and settled his paternal estate upon them in July following. He had three years before built them a combination-room, a library, and a museum; and, in 1656, he brought the deeds of his estate, and presented them to the college. He was then present at the first feast, instituted by himself to be continued annually, together with a commemoration-speech in latin, to be spoken on the 18th of October, in honour of the benefactors to the college; having appointed a handsome stipend for the orator, and also for the keeper of the library and museum, which are still called by his name He died June 3, 1657, and was carried to be interred at Hempsted, in Hertfordshire, where a monument is erected to his memory. Not long afterwards, a character of him was drawn up, and engraved on a copper-plate, which was put under his picture at the college, and which, though it is somewhat long, we have thought proper to subjoin here, since it not only confirms all we have said of him, but contains many particulars of his character, not to be found elsewhere.

GULIELMUS HARVÆUS,
Anglus natu, Galliæ, Italiæ, Germaniæ, hospes,
Ubique amor & desiderium.
Quem omnis terra expetisset civem,
Medicinæ Dr. Coll. Med. Lond. socius & consilarius,
Anatomes chirurgiæque professor,
Regis Jacobi familiæ Caroloque regi medicus,
Gestis clarus, omissisque honoribus,
Quorum alios tulit, oblatos renuit alios,
Omnes meruit.
Laudatis priscorum ingeniis par;
Quos honoravit maxime imitando,
Docuitque posteros exemplo.
Nullius lacessivit famam, veritatis studens magis quam gloriæ,
Hanc tamen adeptus
Industria, sagacitate, successu nobilis
Perpetuos sanguinis æstus circulari gyro
Fugientis, seque sequentis,
Primus promulgavit mundo.
Nec passus ultra mortales sua ignorare primordia,
Aureum edidit de ovo atque pullo librum,
Albæ gallinæ filium.
Sic novis inventis Apollineam ampliavit artem.
Atque nostrum Apollinis sacrarium augustius esse
Tandem voluit;
Suasu enim & cura D. D. Dni. Francisci Prujeani præsidis
Et
Edmundi Smith electoris
An. MDCLIII.
Senaculum, & de nomine suo museum horto superstruxit,
Quorum alterum plurimis libris & instrumentis chirurgicis,
Alterum omnigena supellectile ornavit & instruxit
Medicinæ patronus simul & alumnus.
Non hic anhela sustitit herois virtus, impatiens vinci
Accessit porro munificentiæ decus:
Suasu enim & consilio Dni. Drif. Edv. Alstoni præsidis
Anno MDCLVI.

Rem.

Rem noftram anguftam prius, annuo LVI. lib. reditu
   Auxit.
Paterni fundi ex affe hæredem collegium dicens;
 Quo nihil illi carius nobifve honeftius.
Unde bibliothecario honorarium fuum, fuumque oratori
   Quotannis pendi :
Unde omnibus fociis annuum fuum convivium,
Et fuum denique (quot menfes) conviviolum cenforibus parari,
   Juffit.
Ipfe etiam pleno theatro geftiens fe hæreditate exuere,
 In manus præfidis fyngrapham tradidit :
Interfuitque orationi veterum benefactorum, novorumque
   Illicis.
  Et philotefio epulo.
 Illius aufpicium, & pars maxima;
 Hujus conviva fimul, & convivator.
Sic poftquam fatis fibi, fatis nobis, fatis gloriæ,
Amicis folum non fatis, nec fatis patriæ vixerat.
  Cœlicolum atria fubiit
  Jun. iii^o MDCLVII.

 We will juft mention, that Dr. Harvey lived to fee his doctrine of the circulation of the blood univerfally received; and was obferved, by Mr. Hobbes, to be "the only perfon that ever had that happinefs." A fine edition of his works has been publifhed, fince the firft edition of this Dictionary, under the care and fuperintendency of the late Dr. Lawrence, (who hath prefixed a life of the author), in two vols. 4to, 1766.

 HARVEY (GIDEON), an Englifh phyfician alfo, was born in Surrey; acquired the greek and latin tongues in the Low Countries; and was admitted of Exeter-college, Oxford, in 1655. Afterwards he went to Leyden, and ftudied under Vanderlinden, Vanhorn, and Vorftius, all of them profeffors of phyfic, and men of eminence. He was taught chemiftry there by a german, and, at the fame place, learned the practical part of chirurgery, and the trade of an apothecary. After this he went to France, and thence returned to Holland, where he was admitted fellow of the college of phyficians at the Hague; being, at that time, phyfician in ordinary to Charles II. in his exile. He afterwards returned to London, whence he was fent, in 1659, with a commiffion to Flanders, to be phyfician to the Englifh army there; where ftaying till he was tired of that employment, he paffed through Germany into Italy, fpent fome time at Padua, Bologna, and Rome, and then returned through Switzerland and Holland to England. Here he became phyfician in ordinary to his majefty; and, after king William came over, was made phyfician of the tower. He died about 1700. He wrote a great number

number of books, which however have never been in any esteem with the faculty. He waged a perpetual war with the college of physicians, whom he endeavoured to expose in a piece intituled, "The Conclave of Physicians; detecting their Intrigues, Frauds, and Plots, against their Patients, &c. 1683," 12mo.

HARWOOD (EDWARD), born in 1729, at a village in Lancashire. He was an excellent classical scholar, and author of various works of different degrees of merit. The book which has most established his reputation as a man of learning is his "View of the various Editions of the Greek and Roman Classics." This has passed through numerous editions, and has been translated into most of the European languages. It is certainly, though an imperfect, a very useful, publication, and has had the effect of inspiring many with a taste and curiosity for matters of literature, which time and experience has improved and matured into excellence. His publications were too numerous to be here specified. He refused various overtures to conform to the established church, and died in poverty, at an advanced age, in 1794.

HASE (THEODORE DE), born at Bremen in 1682. After travelling for his improvement in Germany and Holland, he was made professor of Belles Lettres at Hanau. He was soon afterwards recalled to Bremen, to be professor of hebrew. He was a very learned man, and published some "Dissertations," which were highly esteemed. He died in 1731.

HASE (JAMES), the brother of the preceding, and a man of considerable erudition. He published many classical tracts, which were well received by the learned. He died in 1723.

HASSELQUIST (FREDERICK), was born in 1722, at Tournalla, in East Gothia. His father was a minister of the gospel, and, dying when his son was very young, left him in great distress. An uncle sent him to school, where he for some time got a scanty livelihood by teaching the younger children. In 1741, he went to the university of Upsal, where also he maintained himself by instructing others. His favourite study was physic, and, in consequence of his diligence, a royal stipend was procured him. His first publication was an "Essay on the virtue of Plants," which was well received. In consequence of what was said by Linnæus, in one of his botanical lectures, that very little was known of Palestine, Hasselquist formed the resolution of going there, and was delighted with the idea of being the first that should add the natural history of this country to the learning of Europe. He communicated his design to Linnæus, who greatly assisted him in the accomplishment of his purpose. In 1749, he went to Stockholm, where he read lectures on botany, still keeping his

voyage

voyage to Palestine in view. At length the Levant Company offered him a free passage to Smyrna: he accordingly made his intended tour. He collected an incredible quantity of the curiosities of the animal, mineral, and vegetable, kingdoms; and, after an absence of two years, was preparing to return, when, exhausted by fatigue, and overcome by the heat of the climate, he died near Smyrna in 1752, being not quite thirty years old. His creditors seized his curiosities and manuscripts; but, on the representation of Linnæus to the queen of Sweden, that princess discharged his debts. Linnæus was directed to arrange and publish the observations of Hasselquist, which has been done in a manner highly honourable to the fame of them both.

HASTINGS (ELIZABETH), daughter of Theophilus earl of Huntingdon, deserves a place in this collection, from the number of her public and private charities, which were perhaps never equalled by any of her sex. A splendid list of the charities, and a detail of this lady's character, may be found in Welford's "Memorial's;" but the "Tatler" has done the highest honour to her memory in the forty-second number of that work. She is there depictured, by Mr. Congreve, under the title of the "Divine Aspasia." See also a farther account of her private character in the forty-ninth number of the same publication. Lady Elizabeth died in the year 1740, leaving behind her the character of " an illustrious patron of all who love praise-worthy things."

HATTON (Sir CHRISTOPHER), was chancellor in the reign of Elizabeth. It is singular of this personage, that, although he had never followed the profession of the law, he was promoted to this high office. He was a great favourite with his mistress; and it is recorded of him, that, notwithstanding the expectations of the lawyers, his decisions, as chancellor, were never found deficient, either in equity or judgement. It was the artful eloquence of this man which prevailed on Mary queen of Scots to wave the claims of her royal dignity, and submit to trial.

HAVERCAMP (SIGEBERT), a celebrated critic and scholar, was born in Holland, and became an illustrious professor of history, eloquence, and the Greek tongue, at Leyden. He was particularly skilled in the science of medals, and was the author of some works in this way, that were very much esteemed. He gave good editions, as well as grand ones, of several Latin and Greek authors, of Eutropius, Tertullian's "Apologetic," Josephus, Sallust &c. and his editions of those authors are reckoned the best. He died in 1742, at Leyden, aged fifty-eight.

HAUSTEAD (PETER), a comic writer in the reign of Charles the First. He wrote a play, called the "Rival Friends," which was acted before the king and queen, when they visited the University of Cambridge. There are also according to Langbaine, some sermons with this gentleman's name, published at London in 1646.

HAUTE-FEUILLE (JOHN), an ingenious mechanic, born at Orleans in 1647. He first discovered the secret of moderating the vibration of the balance in watches, by means of a small steel spring, which has since been made use of, and these watches are, by way of distinction, called pendulum watches. The invention of Haute-Feuille was brought to perfection by Huygens. Haute Feuille wrote also many small but curious pamphlets. He died in 1724.

HAWKESWORTH (JOHN), an English writer of a very soft and pleasing cast, was born about the year 1719; though his epitaph, as we find it in the "Gentleman's Magazine for August, 1781," makes him to have been born in 1715. He was brought up to a mechanical profession; that of a watch-maker, as is supposed. He was of the sect of Presbyterians, and a member of the celebrated Tom Bradbury's meeting, from which he was expelled for some irregularities. He afterwards devoted himself to literature, and became an author of considerable eminence. In the early part of life, his circumstances were rather confined. He resided some time at Bromley in Kent, where his wife kept a boarding-school. He afterwards became known to a lady, who had great property and interest in the East-India company; and, through her means, was chosen a director of that body. As an author, his "Adventurer" is his capital work; the merits of which, if we mistake not, procured him the degree of LL.D. from Herring, archbishop of Canterbury. When the design of compiling a narrative of the discoveries in the South-seas was on foot, he was recommended as a proper person to be employed on the occasion: but, in truth, he was not a proper person, nor did the performance answer expectation. Works of taste and elegance, where imagination and the passions were to be affected, were his province; not works of dry, cold, accurate narrative. However, he executed his task, and is said to have received for it the enormous sum of 6000l. He died in 1773: some say, of high living; others, of chagrin from the ill reception of his "Narrative:" for he was a man of the keenest sensibility, and obnoxious to all the evils of such irritable natures. On a handsome marble monument at Bromley, in Kent, is the following inscription; the latter part

part of which is taken from the laft number of "The Adventurer."

To the Memory of
JOHN HAWKESWORTH, LL.D.
Who died the 16th of November,
MDCCLXXIII, aged 58 years.
That he lived ornamental and ufeful
To Society in an eminent degree,
Was among the boafted felicities
Of the prefent age;
That he laboured for the benefit of Society,
Let his own pathetic admonitions
Record and realize:

"The hour is hafting, in which whatever praife or cenfure I have acquired will be remembered with equal indifference.—Time, who is impatient to date my laft paper, will fhortly moulder the hand which is now writing it in the duft, and ftill the breaft that now throbs at the reflection. But let not this be read as fomething that relates only to another: for a few years only can divide the eye that is now reading from the hand that has written."

HAWKE (LORD HAWKE), was the fon of Edward Hawke, Efq. barrifter at law, by Elizabeth, daughter of Nathaniel Bladen, Efq. He was from his youth brought up to the fea, and paffed through the inferior ftations till, in the year 1734, he was appointed captain of the Wolf. His intrepidity and conduct were firft of all diftinguifhed in the memorable engagement with the combined fleets of France and Spain off Toulon, when the Englifh fleet was commanded by the admirals Matthews, Leftock, and Rowley. If all the Englifh fhips had done their duty on that day as well as the Berwick, which captain Hawke commanded, the honour and difcipline of the navy would not have been fo tarnifhed. He compelled the Pader, a fpanifh veffel of 60 guns, to ftrike; and, to fuccour the Princeffa and Somerfet, broke the line without orders, for which act of bravery he loft his commiffion, but was honourably reftored to his rank by the king. In 1747 he was appointed rear-admiral of the white; and on the 14th of October, in the fame year, fell-in with a large french fleet, bound to the Weft-Indies. This was a glorious day for England, and the event taught Britifh commanders to defpife the old prejudice of ftaying for a line of battle. Perceiving, fays the gallant admiral in his letters to the Admiralty,

ralty, that we loft time in forming our line, I made the signal for the whole squadron to chafe, and when within a proper diftance to engage. On October the 31ft, admiral Hawke arrived at Portfmouth with his prizes, namely, two feventy-fours, one feventy, two fixty-fours, and one fifty-gun fhip. As a reward of his bravery, he was foon afterwards made knight of the bath. In 1748 he was made vice-admiral of the blue, and elected an elder brother of the Trinity-houfe; in 1755 he was appointed vice-admiral of the white, and in 1757 commanded the squadron which was fent to co-operate with Sir John Mordaunt in the expedition againft Rochfort. In 175 , Sir Edward commanded the grand fleet, oppofed to that of the French equipped at Breft. and intended to invade thefe kingdoms. He accordingly failed from Portfmouth, and, arriving off Breft, fo ftationed his fhips that the French fleet did not dare to come out. More than this, they had the mortification of beholding their coaft infulted, and their mer- chantmen taken. The admiral, by a ftrong wefterly wind, was blown from his ftation; the French accordingly feized this opportunity and fteered for Quiberon-bay, where a fmall Englifh fquadron lay under the command of commodore Duff. Sir Edward Hawke immediately went in purfuit of them, and on the 20th of November came up with them off Bel- leifle. The wind blew exceedingly hard at the time, never- thelefs the French were engaged, and totally defeated. For thefe and fimilar fervices, the king fettled a penfion of 2000*l*. per annum on Sir Edward and his two fons, or the furvivor of them; he alfo received the thanks of the Houfe of Commons, and the freedom of the city of Cork in a gold box. In 1765 he was appointed vice-admiral of Great Britain, and firft lord of the Admiralty; and, in 1776, he was made a peer of En- gland, under the title of baron Hawke, of Towton, in the county of York. His lordfhip married Catharine the daugh- ter of Walter Brooke, of Burton Hall, in Yorkfhire, Efq. by whom he had four children. He was one of the greateft cha- racters that ever adorned the Britifh navy, but moft of all remarkable for the daring courage which induced him on many occafions to difregard thefe forms of conducting or fuf- taining an attack, which the rules and ceremonies of fervice had before confidered as indifpenfable. He died at his feat at Shepperton in Middlefex, October the 14th, 1781.

HAWKINS (SIR JOHN), a brave Englifh admiral in the reign of Elizabeth. He was rear-admiral of the fleet fent out againft the Armada, and had a principal fhare in its deftruc- tion. He alfo fignalized himfelf in feveral expeditions to the Weft-Indies, where he died in 1595.

HAWKINS (Sir John), was the son of a man, who, though defcended from Sir John Hawkins the memorable admiral and treafurer of the navy, in the reign of queen Elizabeth, followed at firft the occupation of a houfe-carpenter, which he afterwards exchanged for the profeffion of a furveyor and builder. He had married Elizabeth, daughter of Thomas Gwatkin of Townhope, in the county of Hereford, gentleman; and the iffue of this marriage were feveral children. Of thefe the prefent object of our enquiry was the youngeft, and was born in the city of London, on the 30th day of March, 1719. After having been fent firft to one fchool, and afterwards to a fecond, where he acquired a tolerable knowledge of Latin, he was placed under the tuition of Mr. Hoppus, the author of a well known and ufeful architectural compendium, publifhed in octavo in 1733, and intituled, " Proportional Architecture, or the Five Orders, regulated by equal Parts." Under this perfon he went through a regular courfe of architecture and perfpective, in order to fit him for his father's profeffion of a furveyor, for which he was at firft intended; but his firft coufin Mr. Thomas Gwatkin, being clerk to Mr. John Scott of Devonfhire-ftreet, Bifhopfgate, an attorney and folicitor in full practice, perfuaded him to alter his refolution, and embrace that of the law; which he did, and was accordingly articled as a clerk to the fame perfon, Mr. John Scott. In this fituation his time was too fully employed in the actual difpatch of bufinefs, to permit him without fome extraordinary means to acquire the neceffary knowledge of his profeffion by reading and ftudy; befides that, his mafter is faid to have been more anxious to render him a good copying clerk, by fcrupulous attention to his hand-writing, than to qualify him by inftruction to conduct bufinefs. To remedy this inconvenience, therefore, he abridged himfelf of his reft, and rifing at four in the morning, found opportunity of reading all the neceffary and moft eminent law writers, and the works of our moft celebrated authors on the fubjects of verfe and profe. By thefe means, before the expiration of his clerkfhip, he had already rendered himfelf a very able lawyer, and had poffeffed himfelf of a very accurate and elegant tafte for literature in general, but particularly for poetry, and the polite arts; and the better to facilitate his improvement, he, from time to time, furnifhed to "The Univerfal Spectator," "The Weftminfter Journal," " The Gentleman's Magazine [A]," and other pe-

[A] In fome of his vifits on thefe and fimilar occafions to Cave, the editor of " The Gentleman's Magazine," he firft became acquainted with Dr. Johnfon foon after the connection between Cave and Johnfon commenced.

riodica-
l

riodical publications of the time, essays and disquisitions on several subjects. The first of these is believed to have been an "Essay on Swearing;" but the exact time of its appearance, and the paper in which it was inserted, are both equally unknown. It was, however, re-published some years since (without his knowledge till he saw it in print) in one of the news-papers. His next production was an "Essay on Honesty,' inserted in the "Gentleman's Magazine" for March, 1739; and which occasioned a controversy, continued through the Magazines for several succeeding months, between him and a Mr. Calamy, a descendant of the celebrated Dr. Edmund Calamy, then a fellow-clerk with him.

Without friends or family-connections, or at least without such as could advance him in the profession to which he had betaken himself, he was now (his clerkship being expired, and he himself admitted an attorney and solicitor) to seek for the means of procuring business by making for himself reputable and proper connections.

About the year 1741, a club having been instituted by Mr. Immyns an attorney, a musical man, (but better known as the amanuensis of Dr. Pepusch), and some other musical persons, under the name of The Madrigal Society, to meet every Wednesday evening, he became a member of it, and continued so many years. Pursuing his inclination for music still farther, he became also a member of " The Academy of Ancient Music," which used to meet every Thursday evening at the Crown and Anchor in the Strand, but since removed to Freemasons Hall; and of this he continued a member till a few years before its removal.

Impelled by his own taste for poetry, and excited to it by his friend Foster Webb's example, who had contributed to "The Gentleman's Magazine" many very elegant poetical compositions; he had, before this time, himself become an occasional contributor in the same kind, as well to that as to some other publications. The earliest of his productions of this species, now known, is supposed to be a copy of verses "To Mr. John Stanley, occasioned by looking over some Compositions of his, lately published," which bears date 19th February, 1740, and was inserted in "The Daily Advertiser" for February 21, 1741; but, about the year 1742, he proposed to Mr. Stanley the project of publishing, in conjunction with him, six cantatas for a voice and instruments, the words to be furnished by himself, and the music by Mr. Stanley. The proposal was accepted, the publication was to be at their joint expence, and for their mutual benefit; and accordingly, in 1742, six cantatas were thus published, the five first written by Mr. Hawkins, the sixth and last by Foster Webb;

Webb; and, these having succeeded beyond the most sanguine expectations, a second set of six more, written wholly by himself, were in like manner published a few months after, and succeeded equally well.

As these compositions, by being frequently performed at Vauxhall, Ranelagh, and other public places, and at many private concerts, had become favourite entertainments, and established the author's reputation as a poet, many persons, finding him also a modest well-informed young man of unexceptionable morals, were become desirous of his acquaintance. Among these was Mr Hare of Limehouse, a brewer, who being himself a musical man, and having met him at Mr. Stanley's at musical parties, gave him an invitation to his house; and, to forward him in his profession, introduced him to a friend of his, Peter Storer of Highgate, Esq. This introduction became, from his own good conduct, the means of making Mr. Hawkins's fortune, though in a way which neither he nor Mr. Hare at that time could foresee, and different from that in which it was first intended.

In the winter of this year 1749, Dr. then Mr. Johnson was induced to institute a club to meet every Tuesday evening at the King's Head, in Ivy-lane, near St. Paul's. It consisted only of nine persons, and Mr. Hawkins was invited to become, and did become, one of the first members accordingly; and about this time, as it is supposed, finding his father's house, where he had hitherto resided, too small for the dispatch of his business now very much encreasing, he, in conjunction with Dr. Munckley, a physician, with whom he had contracted an intimacy, took a house in Clements-lane, Lombard-street. The ground floor was occupied by him as an office, and the first floor by the doctor as his apartment. Here he continued till the beginning of 1753, when, on occasion of his marriage with Sidney, the youngest of Mr. Storer's daughters, who brought him a considerable fortune, which was afterwards greatly encreased, he took a house in Austin Friers, near Broad-street, still continuing to follow his profession of an attorney.

Having received, on the death of Peter Storer, Esq. his wife's brother, in 1759, a very large addition to her fortune, he quitted business to the present Mr. Alderman Clark, who had a short time before completed his clerkship under him, disposed of his house in Austin Friers, and, an opportunity offering, he purchased that now the property of Mr. Vaillant; and soon afterwards bought the lease of one in Hatton-street, London, for a town-residence.

From a very early period of his life he had entertained a strong love for the amusement of angling; and his affection for it, tegether with the vicinity of the river Thames, was undoubtedly his motive to a refidence at this village. He had been long acquainted with Walton's " Complete Angler ;" and had, by obfervation and experience, himfelf become a very able proficient in the art. Hearing, abut this time, that Mr. Mofes Browne propofed to publifh a new edition of that work, and being himfelf in poffeffion of fome material particulars refpecting Walton, he, by letter, made Mr. Browne an offer of writing, for his intended edition, Walton's Life. To this propofal no anfwer was returned, at leaft for fome time, from which circumftance Mr. Hawkins concluded, as any one reafonably would, that his offer was not accepted; and, therefore, having alfo learnt in the mean time that Mr. B. meant not to publifh the text as the author left it, but to modernize it in order to file off the ruft, as he called it, wrote again to tell Mr. Browne that he fo underftood it ; and that, as Mr. B's intention was to fophifticate the text in the manner above mentioned, he, Mr. Hawkins, would himfelf publifh a correct edition. Such an edition, in 1760, he accordingly publifhed in octavo with notes, adding to it a " Life of Walton" by himfelf, a " Life of Cotton," the author of the fecond part, by the well-known Mr. Oldys ; and a fet of cuts defigned by Wale, and engraved by Ryland [B].

His propenfity to mufic, manifefted by his becoming a member and frequenter of the feveral mufical focieties before mentioned, and alfo by a regular concert at his houfe in Auftin Friers, had led him, at the fame time that he was endavouring to get together a good library of books, to be alfo folicitous for collecting the works of fome of the beft mufical compofers ; and, among other acquifitions, it was his fingular good fortune to become poffeffed by purchafe of feveral of the moft fcarce and valuable theoretical treatifes on the fcience itfelf any where extant, which had formerly been collected by Dr. Pepufch [c]. With this ftock of erudition, therefore, he about this time, at the inftance of fome very good judges, his friends, fet about procuring materials for a work then very much

[B] Of this work three editions, each containing a very large impreffion, were fold off before the year 1784, when, there being a demand for a fourth, he revifed and made very large additions to the " Life of Walton," and the notes to the work throughout ; and he re-wrote the " Life of Cotton," in order to comprefs it into lefs compafs, retaining, however, every fact in the former, and adding feveral others. In 1792, after his death, a fifth edition was publifhed by his eldeft fon, (in which, from his papers, were inferted his laft corections and additions,) the former impreffion of 1784 being at that time nearly difpofed of.

[c] This collection of treatifes he, after the completion of his work, gave, in 1778, to the Britifh Mufeum, where it ftill continues.

wanted,

wanted, a "History of the Science and Practice of Music," which he afterwards published.

At the recommendation of the well-known Paul Whitehead, Esq. his neighbour in the country, who, conceiving him a fit person for a magistrate, had mentioned him as such to the duke of Newcastle, then lord lieutenant for Middlesex, his name was, in 1761, inserted in the Commission of the Peace for that county; and having, besides a due attention to the great work in which he was engaged, by the proper studies, and a sedulous attendance at the sessions, qualified himself for the office, he became an active and useful magistrate in the county [D]. Observing, as he had frequent occasion to do in the course of his duty, the bad state of highways, and the great defect in the laws for amending and keeping them in repair, he set himself to revise the former statutes, and drew an act of parliament consolidating all the former ones, and adding such other regulations as were necessary. His sentiments on this subject he published in octavo, in 1763, under the title of "Observations on the State of Highways, and on the Laws for amending and keeping them in repair," subjoining to them the draught of the act before mentioned, which bill, being afterwards introduced into parliament, passed into a law, and is that under which all the highways in the kingdom are at this time kept repaired. Of this bill it is but justice to add, that, in the experience of more than thirty years, it has never required a single amendment.

Johnson, and Sir Joshua then Mr. Reynolds, had, in the winter of this year 1763, projected the establishment of a club to meet every Monday evening at the Turk's Head, in Gerrard-street, and, at Johnson's solicitation, he, Mr. H. became one of the first members. This club, since known by the appellation of The literary Club, was at first intended, like the former in Ivy-lane, to have consisted of no more than nine persons, and that was the number of the first members; but the rule was broken through to admit one who had been a member of that in Ivy-lane. Till this admission, Johnson and Mr. Hawkins were the only persons that had been members of both.

An event of considerable importance and magnitude, in the year 1764, engaged him to stand forth as the champion of the

[D] When he first began to act, he formed a resolution of taking no fees, not even the legal and authorized ones, and pursued this method for some time, till he found that it was a temptation to litigation, and that every trifling alehouse quarrel produced an application for a warrant. To check this, therefore, he altered his mode and received his due fees, but kept them separately in a purse; and at the end of every summer, before he left the country for the winter, delivered the whole amount to the clergyman of the parish, to be by him distributed among such of the poor as he judged fit.

county

county of Middlesex, against a claim then for the first time set up, and so enormous in its amount as justly to excite resistance. The city of London finding it necessary to re build the gaol of Newgate, the expence of which, according to their own estimates, would amount to 40,000 l. had this year applied to parliament, by a bill brought into the House of Commons by their own members, in which, on a suggestion that the county prisoners, removed to Newgate for a few days previous to their trials at the Old Bailey, were as two to one to the London prisoners constantly confined there, they endeavoured to throw the burthen of two thirds of the expence on the county, while they themselves proposed to contribute one third only. This attempt the magistrates for Middlesex thought it their duty to oppose, and accordingly a vigorous opposition to it was commenced and supported under the conduct of Mr. Hawkins, who drew a petition against the bill, and a case of the county, which was printed and distributed amongst the members of both houses of parliament. It was the subject of a day's conversation in the House of Lords; and produced such an effect in the House of Commons, that the city, by their own members, moved for leave to withdraw the bill. The success of this opposition, and the abilities and spirit with which it was conducted, naturally attracted towards him the attention of his fellow-magistrates; and, a vacancy not long after happening in the office of chairman of the quarter sessions, Mr. Hawkins was, on the 19th day of September, 1765, elected his successor.

In this year 1771 he quitted Twickenham, and sold his house there to Mr. Vaillant the present owner; and, in the summer of the next year, he, for the purpose of obtaining, by searches in the Bodleian and other libraries there, farther materials for his history of music, made a journey to Oxford, carrying with him an engraver from London, to make drawings from the portraits in the music-school.

On occasion of actual tumults or expected disturbances he had more than once been called into service of great personal danger. When the riots at Brentford had arisen, during the time of the Middlesex election in the year 1768, he and some of his brethren attended to suppress them; and, in consequence of an expected riotous assembly of the journeymen Spital-fields weavers in Moorfields, in 1769, the magistrates of Middlesex and he at their head, with a party of guards, attended to oppose them, but the mob, on seeing them prepared, thought it prudent to disperse. In these and other instances, and particularly in his conduct as chairman, having given sufficient proof of his activity, resolution, abilities,

ties, integrity, and loyalty, he, on the 23d of October, 1772, received from his present majesty the honour of knighthood.

Mr. Gostling of Canterbury, with whom, though they had never seen each other, he had for some years corresponded by letter, having invited him to do so, he. in this year, paid him a visit at Canterbury, and procured from him a great deal of very curious musical intelligence, which none but Mr. Gostling could have furnished; and in the month of June in the next year, 1773, he again did the same. In this latter year 1773, Dr. Johnson and Mr. Stevens published, in ten volumes octavo, their first joint edition of Shakespear, to which Sir J. H. contributed such notes as are distinguished by his name, as he afterwards did a few more on the republication of it in 1778. An address to the king from the county of Middlesex, on occasion of the American war, having, in 1774, been judged expedient, and at his instance voted, he drew up such an address, and together with two of his brethren had, in the month of October in that year, the honour of presenting it.

After sixteen years labour, he, in 1776, published, in five volumes, quarto, his " General History of the Science and Practice of Music," which, in consequence of permission obtained in 1773 for that purpose, he dedicated to the king, and presented it to him at Buckingham-house on the 14th of November 1776, when he was honoured with an audience of considerable length both from the king and queen. Few works have been attacked with more acrimony and virulence than this. Its merit, however, as containing a great deal of original and curious information, which, but for its author, would have perished, has been amply attested by the approbation of some of the very best judges of the science and of literary composition; and by that of the university of Oxford, who, in consequence of its publication, made him soon after, through the medium of a gentleman now living, a voluntary offer of the degree of doctor of Laws, which he had reasons for declining, and afterwards paid him the compliment of requesting his picture.

Not long after this publication, that is to say in November 1777, he was induced, by an attempt to rob his house, which, though unsuccessful, was made three different nights with the interval of one or two only between each attempt, to quit his house in Hatton-street; and, after a temporary residence for a short time in St. James's Place, he took a lease of one, formerly inhabited by the famous admiral Vernon, in the street leading up to Queen square, Westminster, and removed thither.

By this removal, he became a constant attendant on divine worship at the parish-church of St. Margaret, Westminster;

and having learnt, in December, 1778, that the surveyor to the board of ordnance was, in defiance of a proviso in the lease under which they claimed, carrying up a building at the East end of the church which was likely to obscure the beautiful painted-glass window over the altar there, Sir J. H. with the concurrence of some of the principal inhabitants, wrote to the surveyor, and compelled him to take down two feet of the wall, which he had already carried up above the sill of the window, and to slope off the roof of his building in such a manner as that it is not only no injury, but, on the contrary, a defence, to the window.

In the month of December, 1783, Dr. Johnson, having discovered in himself symptoms of a dropsy, sent for Sir John Hawkins, and telling him the precarious state of his health, declared his desire of making a will, and requested him to be one of his executors. On his accepting the office, he told him his intention of providing for his servant; and, after concerting with him a plan for investing a sum of money for that purpose, he voluntarily opened to him the state of his circumstances, and the amount of what he had to dispose of. Finding the doctor, however, notwithstanding his repeated solicitations from time to time, extremely averse to carrying this intention into effect by the actual execution of a Will, and thinking it might in some measure arise from the want of legal information as to the necessary form, he, Sir J. from the above communications, some time afterwards, drew and sent him a draught of a Will, with instructions how to execute it, but leaving in it blanks for the names of his executors, and for that of the residuary legatee, (for though Johnson had given no instructions on this latter head. Sir J. H. had apprized him of the absolute necessity of a bequest of the residue, that it might not become, as it would otherwise, by the silent operation of law, the property of his executors,) Johnson still procrastinated, but at length executed this draught; so carelessly, however, as to omit first filling up the blanks.

When this circumstance became known to Sir J. H. he represented this act to him (as it really was) as a meer nullity, and Johnson was prevailed upon, on the 27th of November, 1784, at Mr. Strahan's, at Islington, to give him the necessary instructions, which he, Sir J. on the spot converted into proper legal form, by dictating, conformably to them, a will to Mr. Hoole, who, with some other friends, had there called in upon Johnson, and which being completed was executed by Johnson and properly attested. In the codicil, which Johnson afterwards made, Sir J. assisted in the same manner, as to legal phraseology, and directing the proper mode of execution and attestation.

From

From so long an acquaintance with him, and from having been intimately consulted in his affairs, and as it is strongly believed, in consequence of a conversation that passed between them, Sir J. H. was induced, on the event of Johnson's death, on the 13th day of December, 1784, to undertake to write and publish a life of him, and accordingly he set himself to collect materials for that purpose, and for an edition of his works, which with his life was afterwards published.

Not three months after the commencement of the above-mentioned undertaking to write Johnson's life, he met with the severest loss of almost any that a literary man can sustain, short of that of his friends or relations, in the destruction of his library; consisting of a numerous and well-chosen collection of books, ancient and modern, in many languages, and on most subjects, which it had been the business of above thirty years at intervals to get together. This event was the consequence of a fire. Of this loss, great as it was in pecuniary value, and comprising in books, prints, and drawings, many articles that could never be replaced, he was never heard in the smallest degree to complain; but, having found a temporary reception in a large house in Orchard-street, Westminster, he continued there a short time, and then took a house in the Broad Sanctuary, Westminster.

This event, for a short time, put a stop to the progress of his undertaking. As soon, however, as he could sufficiently collect his thoughts, he recommenced his office of biographer of Johnson, and editor of his works; and completed his intention by publishing, in 1787, the life and works, in eleven volumes, octavo, which he dedicated to the king.

With this production he terminated his literary labours; and, having for many years been more particularly sedulous in his attention to the duties of religion, and accustomed to spend all his leisure from other necessary concerns in theological and devotional studies, he now more closely addicted himself to them, and set himself more especially to prepare for that event which he saw could be at no great distance; and, the better to accomplish this end, he, in the month of May, 1788, by a will and other proper instruments, made such an arrangement of his affairs as he meant should take place after his decease.

In this manner he spent his time till about the month of May, 1789, when, finding his appetite fail him in a greater degree than usual, he had recourse, as he had sometimes had before on the same occasion, to the waters of the Islington Spa. These he drank for a few mornings; but on the 14th of that month, while he was there, he was, it is supposed, seized

seized with a paralytic affection, as on his returning to the carriage which waited for him, his servants perceived a visible alteration in him. On his arrival at home he went to bed, but got up a few hours after, intending to receive an old friend from whom he expected a visit in the evening. At dinner, however, his disorder returning, he was led up to bed, from which he never rose, for, being afterwards accompanied with an apoplexy, it put a period to his life, on the 21st of the same month, about two in the morning. He was interred on the 28th in the cloisters of Westminster Abbey, in the North walk near the Eastermost door into the church, under a stone, containing, by his express injunctions, no more than the initials of his name, the date of his death and his age, leaving behind him a high reputation for abilities and integrity, united with the well-earnt character of an active and resolute magistrate, an affectionate husband and father, a firm and zealous friend, a loyal subject, and a sincere Christian, (as, notwithstanding the calumnies of his enemies, can be abundantly testified by the evidence of many persons now living;) and rich in the friendship and esteem of very many of the very first characters for rank, worth, and abilities, of the age in which he lived.

HAWKSMOOR (NICHOLAS), was the scholar of Sir Christopher Wren, but deviated a little from the lessons and practice of his master, at least he did not improve on them, though his knowledge in every science, connected with his art, is much commended, and his character remains unblemished. He was deputy-surveyor at the building of Chelsea-college, clerk of the works at Greenwich, and was continued in the same posts by king William, queen Anne, and George I. at Kensington, Whitehall, and St. James's; surveyor of all the new churches, and of Westminster-abbey, from the death of Sir Christopher, and designed many that were erected in pursuance of the statute of queen Anne for building fifty new churches: their names are: St. Mary Woolnoth, in Lombard-street; Christ-Church, in Spital-Fields; St. George, Middlesex; St. Anne, Limehouse; and St. George, Bloomsbury; the steeple of which is a masterstroke of absurdity. It consists of an obelisk: topped with the statue of George I. hugged by the royal supporters: a lion, an unicorn, and a king, on such an eminence, are very surprising.

"The things we know are neither rich nor rare,
"But wonder how the devil they got there."

He also rebuilt some part of All-Souls-college, Oxford. At Blenheim and Castle-Howard he was associated with Vanbrugh,

brugh, and was employed in erecting a magnificent maufoleum there, when he died in March, 1736, near feventy years of age. He built feveral manfions, particularly Eafton Nefton in Northamptonfhire; reftored a defect in Beverley minfter by a machine that fcrewed up the fabric with extraordinary art; repaired, in a judicious manner, the Weft end of Weftminfter-abbey; and gave a defign for the Radcliffe-library at Oxford.

HAWKWOOD (Sir John), is indebted for a place among the Britifh worthies to his actions in a foreign fervice. He has been flightly noticed by his contemporaries at home, and would not have been brought into a confpicuous point of view but for the engraved portrait of him prefented to the Society of Antiquaries, in 1775, by lord Hailes. He is faid, by the concurrent teftimony of our writers, to have been the fon of a tanner of Sible Hedingham, in Effex, where he was born in the reign of Edward II. Mr. Morant fays, the manor of Hawkwood in that parifh takes its name from Sir John. But it was holden before him by Stephen Hawkwood, probably his father, a circumftance which would lead one to doubt the meannefs of his birth as well as his profeffion. Perfons who gave names to manors were generally of more confiderable rank: and the manor appears to have have been in the family from the time of king John.

Our hero is faid to have been put apprentice to a tailor in London: "but foon," fays Fuller, "turned his needle into a fword, and his thimble into a fhield," being preft into the fervice of Edward III. for his French wars, where he behaved himfelf fo valiantly, that from a common foldier he was promoted to the rank of captain; and for fome farther good fervice had the honour of knighthood conferred on him by that king, though he was accounted the pooreft knight in the army. His genèral, the Black Prince, highly efteemed him for his valour and conduct, of which he gave extraordinary proofs at the battle of Poitiers.

Upon the conclufion of the peace between the Englifh and French by the treaty of Bretigni 1360, Sir John, finding his eftate too fmall to fupport his title and dignity, affociated himfelf with certain companies called, by Froiffart, "Les Tard Venus;" by Walfingham, "Magna Comitiva." Thefe were formed of perfons of various nations, who, having hitherto found employment in the wars between England and France, and having held governments, or built and fortified houfes in the latter kingdom which they were now obliged to give up, found themfelves reduced to this defperate method of fupporting themfelves and their foldiers by marauding

marauding and pillaging, or by engaging in the service of less states, which happened to be at war with each other. Villani, indeed, charges Edward III. with secretly authorizing these ravages in France, while outwardly he affected a strict observance of the peace. At this time in the summer, continues this historian, an English tailor, named John della Guglea, that is, John of the needle, who had distinguished himself in the war, began to form a company of marauders, and collected a number of English, who delighted in mischief, and hoped to live by plunder, surprizing and pillaging first one town, and then another. This company increased so much, that they became the terror of the whole country. All who had not fortified places to defend them were forced to treat with him, and furnish him with provision and money, for which he promised them his protection. The effect of this was, that in a few months he acquired great wealth. Having also received an accession of followers and power, he moved from one country to another, till at length he came to the Po. There he made all who came in his way prisoners. The clergy he pillaged, but let the laity go without injury. The court of Rome was greatly alarmed at these proceedings, and made preparations to oppose these banditti. Upon the arrival of certain Englishmen on the banks of the Po, Hawkwood resigned his command to them, and professed submission to the king of England, to whose servants he presented a large share of his ill-gotten wealth.

The first appearance of Hawkwood in Italy was in the Pisan service in 1364; after which period he was every where considered as a most accomplished soldier, and fought, as different occasions presented themselves, in the service of many of the Italian states. In 1387, we find him engaged in a hazardous service in defence of the state of Florence. The earl of Armagnac, the Florentine general, having been lately defeated by Venni, the governor of the Siannese, the victors marched to surprize Hawkwood, and encamped within a mile and a half of him. But this cautious general retreated into the Cremonese, and when by several skirmishes he had amused the enemy, who kept within a mile of him, and thought to force his camp, he sallied out and repulsed them with loss. This success a little discouraged them. Venni is said to have sent Hawkwood a fox in a cage, alluding to his situation: to which Hawkwood returned for answer, " the fox knew how to find his way out." This he did by retreating to the river Oglio, placing his best horse in the rear till the enemy had crossed the river, on whose opposite bank he placed 400 English archers on horseback. The rear by their assistance crossed the river and followed the
rest,

rest, who, after fording the Mincio, encamped within ten miles of the Adige. The greatest danger remained here. The enemy had broken down the banks of the river, and let out its waters swoln by the melting of the snow and mountains to overflow the plains. Hawkwood's troops, surprized at midnight by the increasing floods, had no resource but immediately to mount their horses, and, leaving all their baggage behind them, marched in the morning slowly through the water, which came up to their horses bellies. By evening, with great difficulty, they gained Baldo, a town in the Paduan. Some of the weaker horses sunk under the fatigue. Many of the foot perished with cold and struggling against the water: many supported themselves by laying hold on the tails of the stronger horses. Notwithstanding every precaution, many of the cavalry were lost as well as their horses. The pursuers, seeing the country under water, and concluding the whole army had perished, returned back. The historian observes, that it was universally agreed no other general could have got over so many difficulties and dangers, and led back his small army out of the heart of the enemy's country, with no other loss than that occasioned by the floods, which no precaution could have prevented. One of the most celebrated actions of Hawkwood's life, says Murateri, was this retreat, performed with so much prudence and art, that he deserves to be paralleled with the most illustrious Roman generals; having, to the disgrace of an enemy infinitely superior in number and in spite of all obstructions from the rivers, given them the slip, and brought off his army safe to Castel Baldo on the borders of the Paduan. Sir John Hawkwood, as soon as he found himself among his allies, employed himself in refreshing his troop and watching the enemies motions.

At the end of 1391, the Florentines made peace with Galeazzo and the rest of their enemies, though on disadvantageous terms. To reduce the expences of the state, they discharged their foreign auxiliaries, except Hawkwood, of whose valour and fidelity they had had such repeated proofs, with 1000 men under his command.

Peace being now re-established abroad, the city of Florence was, in 1393, distracted with civil feuds, which were not terminated by the execution and exile of some principal citizens. But at the close of this year they sustained a greater loss in Sir John Hawkwood, who died March 6, advanced in years, at his house in the street called Pulverosa near Florence. His funeral was celebrated with great magnificence, and the general lamentation of the whole city. His bier, adorned with gold and jewels, was supported by the first persons

sons of the republic, followed by horses in gilded trappings, banners, and other military enfigns, and the whole body of the citizens. His remains were depofited in the church of St. Reparata, where a ftatue (as Poggio and Roffi call it, though it is well known to be a portrait) of him on horseback was put up by a public decree. If the Florentine hiftorians did not diftinguifh between a ftatue and a portrait, no wonder our countryman Stowe talks of an "image as great as a mighty pillar," erected to the memory of Sir John Hawkwood at Florence; or that Weever, copying him, calls it "a ftatue."

In the reprefentation of this hero painted on the dome of the church, he appears mounted on a pacing gelding, whofe bridle, with the fquare ornament emboft on it, is covered with crimfon velvet or cloth, and the faddle is red, ftuffed or quilted. He is dreffed in armour with a furcoat flowing on from his fhoulders, but girt about his body; his greaves are covered with filk or cloth, but the knee-pieces may be diftinguifhed under them: his fhoes, which are probably part of his greaves, are pointed according to the fafhion of the times. His hands are bare: in his right he holds a yellow baton of office, which refts on his thigh; in his left the bridle. His head, which has very fhort hair, is covered with a cap not unlike our earls' coronets, with a border of wrought work.

Sir John had a cenotaph in the church of his native town, erected by his executors Robert Rokeden fenior and junior, and John Coe. It is defcribed by Weever, as "a tomb arched over, and engraven to the likenefs of hawks flying in a wood," which, Fuller fays, was "quite flown away." It is plain the laft of the writers never took any pains to vifit or procure true information about this monument, which ftill remains in good prefervation near the upper end of the fourth aile of Sible Hedingham church. The arch of this tomb is of the mixed kind, terminating in a fort of bouquet, on both fides of which, over the arch, are fmaller arches of tracery in relief. The arch is adorned with hawks and their bells, and other emblems of hunting, as a hare, a boar, a boy founding a conch-fhell, &c. The two pillars that fupport it are charged with a dragon and lion. Under this arch is a low altar-tomb with five fhields in quatrefoils, formerly painted. In the fouth window of the chantry chapel, at the eaft end of this aile, are painted hawks, hawks bells, and efcallops, which laft are part of the Hawkwood arms, as the firft were probably the creft, as well as a rebus of the name; and we find a hawk volant on Sir John's feal. In the north and weft fide of the tower are two very neat hawks on perches in relief, in rondeaux hollowed in the wall: that over the weſt door is extremely well preſerved. They probably denote that ſome of the family built the tower. Mr. Morant imagines ſome of them rebuilt this church about the reign of Edward III. but none appear to have been in circumſtances equal to ſuch munificence before our hero: and perhaps his heirs were the rebuilders.

Contemporary and ſucceeding writers agree in their praiſes of this illuſtrious general. Both friends and enemies conſidered him as one of the greateſt ſoldiers of his age. Poggio ſtyles him " rei militaris ſcientia clarus, & bello aſſuetus," " dux ſagax," " dux prudens," " tantus dux," " rei bellicæ peritiſſimus," " ad belli officia prudentiſſimus," " expertæ virtutis & fidei ;" epithets theſe which might ſerve inſtead of a particular character. Muratori calls him, " Il prode & il " accortiſſimo capitano." As he had been formed under the Black Prince, it is not to be wondered that his army became the moſt exact ſchool of martial diſcipline, in which were trained many captains, who afterwards roſe to great eminence.

The circumſtances of the times muſt make an apology for the frequent changes of his ſervice, which led him to engage as ſuited his intereſt. He was a ſoldier of fortune; and his abilities in the field occaſioned him to be courted by different rival ſtates. The Florentines offered the beſt terms, and to them he ever after adhered with an irreproachable fidelity.

His charity appears in his joining with ſeveral perſons of quality, in this kingdom, in founding the Engliſh hoſpital at Rome for the entertainment of poor travellers.

HAY (WILLIAM, Eſq.), an agreeable Engliſh writer, was born at Glenburne in Suſſex, about 1700, as is conjectured; and educated at Headley-ſchool. In 1730, he publiſhed a poem, called " Mount Caburn," dedicated to the dutcheſs of Newcaſtle; in which he deſcribes the beauties of his native country, and celebrates the virtues of his friends. When lord Hardwicke was called up to the houſe of lords in 1734, he was choſen to ſucceed him, in repreſenting the borough of Seaford among the commons; and he repreſented this borough for the remainder of his life. He defended the meaſures of Sir Robert Walpole, and was the ſuppoſed author of a miniſterial pamphlet, intituled, " A Letter to a Freeholder on the late Reduction of the Land tax to one Shilling in the Pound ;" which had been printed in 1732. In 1735, he publiſhed " Remarks on the Laws relative to the Poor, with Propoſals for their better Relief and Employment;" and at the ſame time brought in a bill for the purpoſe. He made another attempt of this kind, but without effect. May 1738, he was appointed a commissioner of the victualling-office. In 1753, appeared "Regio Philosophi; or, the Principles of Morality and Christianity, illustrated from a View of the Universe, and of Man's Situation in it." This was followed, in 1754, by his "Essay on Deformity;" in which he rallies his own imperfection, in this respect, with much liveliness and good humour. "Bodily deformity," says he, "is very rare. Among 558 gentlemen in the House of Commons, I am the only one that is so. Thanks to my worthy constituents, who never objected to my person, and I hope never to give them cause to object to my behaviour." The same year, he translated Hawkins Browne "De Immortalitate Animæ." In 1755, he translated and modernized some "Epigrams of Martial;" but survived this publication only a short time, dying June 19, the same year. A little time before, he had been appointed keeper of the records in the tower, and it is said that his attention and assiduity, during the few months he held that office, were eminently serviceable to his successors.

He left a son, who inherited the imperfect form of his father. This gentleman went into the service of the East-India company, where he acquired rank, fortune, and reputation; but, being one of those who opposed Cossim Ally Kawn, and unfortunately falling into his hands, was, with other gentlemen, ordered to be put to death at Patna, October 5, 1762. Mr. Hay's works were collected, by his daughter, in two volumes, quarto, 1794.

HAYES (CHARLES, Esq.), a very singular person, whose great erudition was so concealed by his modesty, that his name is known to very few, though his publications are many. He was born in 1678, and became distinguished in 1704 by "A Treatise of Fluxions," folio; the only work to which he ever set his name. In 1710, came out a small quarto pamphlet of nineteen pages, intituled, "A new and easy Method to find out the Longitude, from observing the Altitudes of the Celestial Bodies," and, in 1723, "The Moon, a Philosophical Dialogue;" tending to shew, that the moon is not an opaque body, but has original light of her own. During a long course of years, the management of the late Royal African Company lay in a manner wholly upon Mr. Hayes, he being annually either sub-governor or deputy-governor; notwithstanding which, he continued his pursuit after general knowledge. To a skill in the greek and latin, as well as modern languages, he added the knowledge of the hebrew; and published several pieces relating to the translation and chronology of the scriptures. The African company being dissolved in 1752, he retired to Down in Kent, where he gave himself up to study. May 1753, he began to compile, in latin, his "Chronographia Asiatica & Ægyptiaca," which he lived to finish, but not to publish; which, however, was publilhed afterwards. August 1758, he left his house in Kent, and took chambers in Gray's inn, where he died, December 18, 1760, in his 82d year. The title of his posthumous work runs thus: "Chronographiæ Asiaticæ & Ægyptiacæ Specimen, in quo, 1. Origo Chronologiæ LXX Interpretum investigatur. 2. Conspectus totius operis exhibetur," 8vo.

HAYNES (HOPTON), assay-master of the Mint near fifty years, and principal tally-writer of the Exchequer for above forty years, in both which places he always behaved himself highly worthy of the great trust reposed in him, being indefatigable and most faithful in the execution of his offices, was a most loyal subject, an affectionate husband, a tender father, a kind master, and a sincere friend; charitable and compassionate to the poor, a complete gentleman, and consequently a good christian. He died at his house in Queen-Square, Westminster, November 19, 1749. In the next year appeared a miscellaneous work of his [A], under the title of "The Scripture Account of the Attributes and Worship of God: and of the Character and Offices of Jesus Christ. By a candid Enquirer after Truth. Published at the desire of the deceased Author. Lond. 1750."

HAYNES (SAMUEL), M. A. son to the above, was tutor to the earl of Salisbury, with whom he travelled, and who rewarded him, in June 1737, with the valuable rectory of Hatfield, Herts. In 1740 he published "A Collection of State Papers," folio; in March, 1743, on the death of Dr. Snape, succeeded to a canonry at Windsor; and, in May 1747, he was presented also by his noble patron to the rectory of Clothall (the parish in which the earl of Salisbury's seat, called Quickswood, is situated). He was an amiable man and a chearful companion; and died June 9, 1752.

HAYWARD (Sir JOHN), an English historian, was educated at Cambridge, where he took the degree of LL.D In 1599, he published, in 4to, "The first Part of the Life and Raigne of King Henrie IV. extending to the End of the first Yeare of his Raigne;" dedicated to Robert earl of Essex; for which he suffered a tedious imprisonment, on account of having advanced something in defence of hereditary right. We are informed, in lord Bacon's "Apophthegms," that queen Elizabeth, being highly incensed at this book, asked Bacon, who was then one of her council learned in the law, "whether there was any treason contained in it?" who an-

[A] See Lindsey's Sequel to his Apology," pp. 18. 23; and Baron's "Preface to his Cordial for Low Spirits," p. xviii.

swered,

fwered, "No, madam; for treason, I cannot deliver my opinion there is any; but there is much felony." The queen, apprehending it, gladly afked, "How and wherein?" Bacon anfwered, "becaufe he had ftolen many of his fentences and conceits out of Cornelius Tacitus." Camden tells us, that this book being dedicated to the earl of Effex, when that nobleman and his friends were tried, the lawyers urged, that "it was written on purpofe to encourage the depofing of the queen;" and they particularly infifted on thefe words in the dedication, in which our author ftyles the earl "Magnus & præfenti judicio, & futuri temporis expectatione." In 1603, he publifhed, in quarto, "An Anfwer to the firft Part of a certaine Conference concerning Succeffion, publifhed not long fince under the Name of R. Doleman." This R. Doleman was the jefuit Parfons. In 1610, he was appointed by king James one of the hiftoriographers of Chelfea-college, near London. This college was intended, fays Fuller, for a fpiritual garrifon, with a magazine of all books for that purpofe, where learned divines fhould ftudy and write in maintenance of all controverfies againft the papifts. Befides the divines, at leaft two able hiftorians were to be maintained in the college, to record and tranfmit to pofterity all memorable paffages in church and ftate. This fcheme was pufhed by the king and other confiderable perfonages, and was in agitation for fome years; but dropped at length, nobody knows how. In 1613, he publifhed, in 4to. "The Lives of the Three Normans, Kings of England; William I. William II. Henry I." and dedicated them to Charles prince of Wales. In 1619, he received the honour of knighthood from his majefty at Whitehall. In 1624, he publifhed a difcourfe, intituled, "Of Supremacie in Affaires of Religion;" dedicated to prince Charles. It is written in the manner of a converfation held at the table of Dr. Toby Matthews, bifhop of Durham, in the time of the parliament, 1605; and the propofition maintained is, that fupreme power in ecclefiaftical affairs is a right of fovereignty. He wrote likewife "The Life and Raigne of King Edward VI. with the Beginning of the Raigne of Queen Elizabeth, 1630," 4to. but this was pofthumous; for he died June 27, 1627. He was the author of feveral works of piety.

For the judgements that have been paffed upon him, Wood tells us, that "he was accounted a learned and godly man, and one better read in theological authors, than in thofe belonging to his profeffion; and that, with regard to his hiftories, the phrafe and words in them were in their time efteemed very good; only fome have wifhed, that, in his 'Hiftory of Henry IV.' he had not called Sir Hugh Lynne by fo light a word as

Mad-

Mad-cap, though he were such; and that he had not changed his historical style into a dramatical, where he introduceth a mother uttering a woman's passion in the case of her son." Nicolson observes, that " he had the repute, in his time, of a good clean pen and smooth style; though some have since blamed him for being a little too dramatical." Strype says, that our author " must be read with caution; that his style and language is good, and so is his fancy; but that he uses it too much for an historian, which puts him sometimes on making speeches for others, which they never spake, and relating matters which perhaps they never thought on:" In confirmation of which censure, Kennet has since affirmed him to be " a professed speech-maker through all his little History of Henry IV."

HEARNE (THOMAS), an English antiquary, and indefatigable collector and editor of books and MSS. was the son of George Hearne, parish clerk of White-Waltham in Berkshire, and born there in 1680. For some time he received no other instruction than from his father, who kept a writing-school at Waltham; but, in 1693, Francis Cherry, of Shottesbrooke, Esq; took him under his own patronage, and put him to the free-school of Bray, in Berks. Here he made so extraordinary a progress in the greek and latin tongues, and was withal so remarkable for his sobriety and good manners, that Mr. Cherry, by the advice of his friend Mr. Dodwell, who then lived at Shottesbrooke, took him into his family, and provided for him as if he had been his own son. He instructed him every day in religion and classical learning; as did Mr. Dodwell, when he was absent. Mr. Cherry, pleased with cultivating an understanding so susceptible of improvement, determined to bestow on him a liberal education; and accordingly, in December 1695, entered him of Edmund-hall, Oxford. That foundation was then governed by Dr. Mill, who had under him as vice-president Dr. White Kennet, afterwards bishop of Peterborough, then one of the most eminent tutors in the university, and at the same time vicar of Shottesbrooke, to which cure he had been presented by Mr. Cherry. Happily for Hearne, both the head of his college and his tutor were votaries of antiquity, to which he himself had a natural and even violent propensity. This was conspicuous in him, even while a boy; when he was observed to be continually plodding over the old tomb-stones of his own parish-churchyard, as soon almost as he was master of the English alphabet. This disposition, joined with his unwearied industry, recommended him particularly to Dr. Mill; who being then busy about an Appendix to his " Greek Testament," and finding him to be well versed in MSS. got him to examine several he had

had occasion to make use of in that work. When he was no more than three years standing, he went, at Dr. Mill's request, to Eton, to collate a MS. of Tatian and Athenagoras in the library there. The copy of the variations he had noted, written by his own hand, is in the Bodleian library, and was used by Mr. Worth in his edition of Tatian, and by Mr. de Chaire in that of Athenagoras, though neither of these editors have made any mention of it. He was likewise of great service to Dr. Grabe, at that time resident in Edmund-hall, for whom he compared many MSS. and made considerable collections.

In act term, 1699, he took the degree of B. A. and soon after was offered very advantageous terms to go a missionary to Maryland; but, being unwilling to leave Oxford, and the valuable acquaintance he had contracted there, he declined the offer. After he had taken his degree, he became a constant student in that noble repository of antiquities, the Bodleian library; and was so noted for the length and frequency of his visits, that Dr. Hudson, soon after he was chosen keeper thereof, took him for a coadjutor, having first obtained the consent of the curators. He became M. A. in 1703, was afterwards made janitor of the public library, and, in 1712, second librarian of the Bodleian. In January 1714-15, he was elected archetypographus of the university, and esquire-beadle of the civil law; which post he held, together with that of underlibrarian, till November following; but then, finding they were not tenable together, he resigned the beadleship, and very soon after the other place also, by reason of the oaths, with which he could not conscientiously comply. He continued a nonjuror to the last, much at the expence of his worldly interest; for, on that account, he refused several preferments, which would have been of great advantage and very agreeable to him. He died at Oxford, and was buried in St. Peter's church-yard, where there is a tomb erected for him, with this inscription written by himself: " Here lyeth the Body of Thomas Hearne, M. A. who studied and preserved Antiquities." He died June 10, 1735, aged 55 Years. Deut. xxxii. 7. " Remember the days of old, consider the years of many generations; ask thy father, and he will shew thee, thy elders, and they will tell thee,—Job. viii. 8, 9, 10. Enquire, I pray thee." He had with great parsimony saved about 1300l. which his relations, who were poor, found after his death among his books and papers.

A list of the books he published, for he was rather an editor than an author, may be acceptable to the curious; and therefore we will enumerate them as briefly as possible. They are as follow: 1. " Reliquiæ Bodleianæ; or, some genuine Remains

Remains of Sir Thomas Bodley, &c. 1703." 2. "Plinii Epiſtolæ & Panegyricus, &c. 1703." 3. "Eutropius. Meſſala Corvinus. Julius Obſequens, &c. 1703." 4. "Ductor Hiſtoricus," 2 vols. They did not come out together; a ſecond edition of the firſt was publiſhed in 1705, and the ſecond volume was publiſhed in 1704. Our author was not ſolely concerned in this work, ſome parts of it being written by another hand, as was the preface. He had made great collections for a third volume, but laid aſide this deſign, upon the appearance of the Engliſh tranſlation of Puffendorf's introduction, which begins where the ſecond volume of the "Ductor Hiſtoricus" ends, and continues the hiſtory to the preſent times. 5. "Juſtini Hiſtoria, 1705." 6. "Livy, 1708," 6 vols. 7. "A Letter, containing an Account of ſome Antiquities between Windſor and Oxford, with a Liſt of the ſeveral Pictures in the School gallery adjoining to the Bodleian Library," printed in 1708, in the "Monthly Miſcellany, or Memoirs for the Curious:" and reprinted at the end of the fifth volume of Leland's "Itinerary," but without the liſt of the pictures; which, however, being greatly ſought by the curious cauſed him to reprint 100 copies of the whole in 1725. 8. "The Life of Ælfred the Great, by Sir John Spelman, from the original MS. in the Bodleian Library, 1710." 9. "The Itinerary of John Leland the Antiquary, intermixed with divers curious Diſcourſes, written by the Editor and others, 1710," 9 vols. A new edition was printed in 1744. 10. "Dodwelli de Parma equeſtri Woodwardiana diſſertatio, &c. 1713." 11 "Lelandi de rebus Britannicis collectanea, 1715," 6 vols. 12. "Acta Apoſtolorum, Græco-Latine, literis majuſculis. E codice Laudiano, &c. 1715." 13. "Joannis Roſſi antiquarii Warwicenſis hiſtoria regum Angliæ, 1716." It was printed again with the ſecond edition of Leland's "Itinerary," and now goes along with that work. 14. "Titi Livii Foro Julienſis vita Henrici V. regis Angliæ. Accedit ſylloge epiſtolarum à variis Angliæ principibus ſcriptarum, 1716." 15. Aluredi Beverlacenſis annales; ſive hiſtoria de geſtis regum Brittanniæ, &c. 1716." 16. "Gulielmi Roperi vita D. Thomæ Mori equitis aurati, lingua Anglicana contexta, 1716." 17. "Gulielmi Camdeni Annales rerum Anglicarum & Hibernicarum. regnante Elizabetha, 1717," 3 vols. 18. "Gulielmi Neubrigenſis hiſtoria ſive chronica rerum Anglicarum, 1719." 19. "Thomæ Sprotti chronica, &c. 1719." 20. "A Collection of curious Diſcourſes written by eminent Antiquaries upon ſeveral Heads in our Engliſh Antiquities, 1720." 21. "Textus Roffenſis, &c. 1720." 22. "Roberti de Aveſbury hiſtoria de mirabilibus geſtis Edwardi III. &c. Appendicem etiam ſubnexuit, in qua inter alia continentur

tinentur Letters of King Henry VIII. to Anne Boleyne, 1720."
23. Johannis de Fordun Scotichronicon genuinum, una cum
ejufdem fupplemento ac continuatione, 1722." 24. " The
Hiftory and Antiquities of Glaftonbury, &c. 1722. 25.
" Hemingi Chartularium ecclefiæ Wigornienfis, &c. 1723."
26. " Robert of Gloucefter's Chronicle, 1724, &c." in 2 vols.
27. " Peter Langtoft's Chronicle, as illuftrated and improved
by Robert of Brune, from the Death of Cadwalader to the
End of King Edward the Ift's Reign, &c. 1720," in 2 vols.
28. " Johannis, confratris & monachi Glaftonienfis, chronica:
five hiftoria de rebus Glaftonienfibus, &c. 1726." 29.
" Adami de Domerham hiftoriæ de rebus geftis Glaftonienfibus, &c. 1727, in 2 vols. 30 " Thomæ de Elmham vita &
gefta Henrici V. Anglorum regis, &c. 1727." 31. " Liber
niger Scaccarii, &c. 1728," 2 vols. 32. " Hiftoria vitæ &
regni Richardi II. Angliæ regis, à monacho quodam de Eve-
fham confignata, 1729." 33. " Joannis de Trokelowe annales
Edwardi II. &c. 1729." 34. " Thomæ Caii vindiciæ antiquitatis academiæ Oxonienfis, &c. 1730," 2 vols. 35. "Wal-
teri Hemingforde, canonici de Giffeburne, hiftoria de rebus
geftis Edvardi I. II. III. &c. 1731," in 2 vols. 36. " Duo rerum Anglicarum fcriptores veteres, videlicet, Thomas Otter-
bourne & Johannes Wethamftade, ab origine gentis Britan-
nicæ ufque ad Evardum IV &c. 1733," in 2 vols. 37. " Chronicon five annales prioratus du Dunftable &c. 1733." 38.
" Benedictus, abbas Petroburgenfis, de vita & geftis Henrici
II. Richardi I. &c. 1735," in two vols. The reader will be
apt to fancy that Mr. Hearne had laboured pretty fufficiently,
having probably publifhed more than would be ever read ;
however, he was going on in the fame way, and was got to
the eve of another publication in two vols. 8vo, when death
very cruelly withheld his hand. He was an editor of a very peculiar caft ; for he fcarcely ever publifhed an old writer, without intermixing with or adding to him a parcel of papers,
which had little or perhaps no relation at all to the principal
work. Thefe odd farragoes are generally introduced by long
and elaborate prefaces, fome in latin, others in englifh, as
mifcellaneous as their following collections. The capriciouf-
nefs of the man's genius, and the oddity of his tafte, are indeed fufficiently obvious; yet, without doubt, there are many
readers, to whom his compofitions will afford entertainment.
All his works, except the firft, were printed at Oxford; all
in 8vo.

We have obferved above, that he lived and died a nonjuror ;
yet it appears, that he was not thus rigid in the beginning of
his life, from a pamphlet afcribed to him, and faid to be
written

written in his 22d year. The title is, " A Vindication of those who take the Oath of Allegiance to his prefent Majefty, from Prejudice, Injuftice, and Difloyalty, charged upon them by fuch as are againft it." It is addreffed to Mr. Cherry, from whom it came with many other MSS. exprefsly by will to the Bodleian library. It is dated from Edmund-hall in Oxford, June 11, 1730. In 1731, it was printed by an anonymous editor, who prefixed to it a print of the author and a preface, containing a fatyrical account of him. The piece itfelf is fo wretched a compofition in all refpects, as to be a real curiofity ; fo that it is not a wonder, that it did not convert the gentleman to whom it was addreffed. Befides the Herculean labours already mentioned, he made indexes to feveral works ; and, among the reft, to the folio edition of "·Lord Clarendon's Hiftory of the Rebellion," in 1704.

HEATH (JAMES), an Englifh hiftorian, was born, 1629, in London, where his father, who was the king's cutler, lived. He was educated at Weftminfter-fchool, and became a ftudent of Chrift-church, Oxford, in 1646. In 1648, he was ejected thence, by the parliament-vifitors, for his adherence to the royal caufe; lived upon his patrimony, till it was almoft fpent; and then, foolifhly marrying, was obliged to write books and correct the prefs, in order to maintain his family. He died, of a confumption and dropfy, at London, in Auguft 1664, and left feveral children to the parifh. He publifhed, 1. " A brief Chronicle of the late inteftine War in the three Kingdoms of England, Scotland, and Ireland, &c. 1661," 8vo. afterwards enlarged by the author, and completed from 1637 to 1663, in four parts, 1663, in a thick 8vo. To this was again added a continuation from 1663 to 1675 by John Philips, nephew by the mother to Milton, 1676, folio. 2. " Elegy upon Dr. Thomas Fuller, 1661." 3. " The Glories and Magnificent Triumphs of the bleffed Reftoration of King Charles II. &c. 1662," 8vo. 4. " Flagellum; or, the Life and Death, Birth and Burial, of Oliver Cromwell, the late Ufurper, 1663" The third edition came out with additions in 1655, 8vo. 5. " Elegy on Dr. Sanderfon, Bifhop of Lincoln, 1662." 6. " A new Book of loyal Englifh Martyrs and Confeffors, who have endured the Pains and Terrors of Death, Arraignment, &c. for the Maintenance of the juft and legal Government of thefe Kingdoms both in Church and State, 1663," 12mo. 7. " Brief but exact Survey of the Affairs of the United Netherlands, &c." 12mo. The reafon why fuch writers as our author continue to be read, and will probably always be read, is, not only becaufe " Hiftoria quoquo modo fcripta delectat;" but alfo becaufe in

the

the meaneſt hiſtorian there will always be found ſome facts, of which there will be no cauſe to doubt the truth, and which yet will not be found in the beſt. Thus Heath, who perhaps had nothing but pamphlets and newſpapers to compile from, frequently relates facts that throw light upon the hiſtory of thoſe times, which Clarendon, though he drew every thing from the moſt authentic records, has omitted.

INDEX

TO THE

SEVENTH VOLUME.

	Page		Page
GESNER, Conrad	1	Gillespie, George	19
——— Solomon	2	Gilpin, Bernard	ib.
——— Solomon	ib.	——— Richard	24
——— John Matthew	ib.	Giolito, Del Farrari	ib.
Geta, Septimius	3	Gioia, Flavio	ib.
Gethin, Lady Grace	ib.	Giorgione	25
Gething, Richard	5	Gioseppino	26
Gevartius, John Gaspar	ib.	Giotto	27
Ghilini, Jerome	6	Giraldi, Lilio Gregorio	29
Ghirlandaio, Domenico	ib.	——— John Baptist Cintio	30
Giannoni, Peter	ib.	Giraldus, Silvester	31
Gibbon, Edward	ib.	——— Cambrensis	33
Gibbs, James	8	Giron, D. Pierre	ib.
Gibalyn, Le Compte de	ib.	Giry, Louis	ib.
Gibson, Edmund	ib.	Giselinus	ib.
——— Richard	12	Glain, N. Saint	ib.
——— William	13	Glandorp, Matthias	34
——— Edward	ib.	Glanvil, Joseph	ib.
——— William	ib.	Glapthorne, Henry	40
——— Thomas	ib.	Glass, John	ib.
——— Thomas	ib.	——— John *son*	41
Gisanius, Hubertus or Obertus	ib.	Glaphyra	ib.
		Glaser, Christopher	ib.
		Glaubert, Rodolphus	ib.
Gilberr, William	14	Glen, John	ib.
——— Thomas	16	Glicas or Glycas	ib.
——— Sir Humphrey	17	Glisson, Francis	ib.
——— Jeffery	ib.	Gloucester, Robert of	43
Gildas	ib.	Glover, Richard	44
Gildon, Charles	ib.	Gmelin, Samuel Gottlieb	ib.
Giles, John	18	——— John George	ib.
Gill, Alexander	ib.	Goar, James	ib.
——— Alexander (*son*)	19	Gobier, Charles	ib.
——— Dr. John	ib.	Goclenius, Conrad	ib.

Vol. VII.   C c   Goclenius,

INDEX.

	Page		Page
Goclenius, Rodolphus	44	Gordon, Thomas	75
Goddard, Jonathan	45	——— Alexander	76
Godeau, Anthony	48	——— James	77
Godfrey, Sir Edmund Bury	49	——— Robert	ib.
——— of Boulogne	ib.	Gore, Thomas	ib.
Godiva	50	Gorelli	78
Godolphin, John	ib.	Gorgias, Leontinus	ib.
Godwin, Thomas	51	———	ib.
——— Francis	52	Gorius, Antonius Francifcus	ib.
——— Dr. Thomas	54	Gorlæus, Abraham	ib.
Goeree, William	55	——— David	79
Goertzs, John Baron of	ib.	Goropius, John	ib.
Goefius, William	56	Gorreus	ib.
Goez, Damian de	ib.	Goffelini, Julian	ib.
Goff, Thomas	57	Godefchale	80
Gogava, Antonius Hennannus	ib.	Goffelin, Antony	ib.
		Gothofred	ib.
Goguet, Antony-Yves	ib.	——— Theodofius *eldeſt ſon*	ib.
Goldaſt, Melchior Haiminsfield	ib.	——— James, *another ſon*	ib.
Goldhagen, John Euftachius	59	——— Dennis, ſon of Theodoſius	81
Goldman, Nicolas	ib.	——— John, *the ſon*	ib.
Goldſmith or Gouldſmith, Francis	ib.	Gotti, Vincent Louis	ib.
——— Oliver	ib.	Gottleber, John Chriſtopher	ib.
Golius, James	62	Gaudelin or Goudouli	ib.
——— Peter	66	Goudimel, Claudius	ib.
Goltzius, Henry	ib.	Govea, Martial	ib.
——— Hubert	67	——— Andrew	ib.
Gomar, Francis	68	——— Antony	ib.
Gombauld, John Ogier de	ib.	Gouge, William	82
Gomerſal, Robert	70	——— Thomas	ib.
Gomez, De Cividad, near Alvarez	ib.	Goujet, Claude Peter	ib.
		Goujon, John	ib.
——— De Caſtro Alvarez	ib.	Goulart, Simon	ib.
——— Magdeline Angelica Poiſſon de	ib.	Goulſton, Theodore	83
		Gould, Robert	ib.
Gondi, John Paul	ib.	Goulu, John	ib.
Gongora, Lewis de	71	Goupy, Joſeph	84
Gondiin, Louis Antoine	72	Gournay, Mary de Jars Lady of	ib.
Gonet, John Baptiſt	ib.		
Gonnelli, John	ib.	Gounville, John Herauld	85
Gonſalva	ib.	Gouſſet, James	ib.
Gonthier	73	Guthieres, James	ib.
——— John and Leonard	ib.	Gower, John	ib.
Gonzaga, Lucretiz	ib.	Gouye, John	87
Gouzalez, Thyrſius	74	Graaf, Regnier de	88
Gool, John Var.	ib.	Grabe, John Erneſt	ib.
Goodall, Walter	ib.	Gracian, Balthaſar	94
Goodwin, John	ib.	Gracchus, Tiberius and Caius	ib.
——— Thomas	ib.		
——— Thomas	75	Gradenigo, Peter	ib.
Gordianus, the elder	75		
			Græme,

INDEX.

	Page		Page
Græme, John	94	Green, Matthew	134
Graffio	97	Greene, Dr. Maurice	135
Grafigny, Frances	ib.	Greenhill, John	ib.
Grafton, Richard	ib.	Greenville, Sir Richard	136
Graham, George	ib.	Gregory	ib.
Grain, John Baptist le	98	——— James	145
Graindorge, Andrew	99	——— David	149
Gramaye, John Baptist	ib.	——— John	151
Grammond, Gabriel, lord of	ib.	——— Edmund	153
Gramont, Antony, duke of	100	——— Nazianzen	ib
——— Philibert, count of	ib.	——— Nyssen	159
Grancolas, John	ib.	——— Theodorus	160
Grand, Antony le	ib.	Gregorius, Georgius Floren-	
——— Joachim le	ib.	tius, or Gregory of Tours	162
——— Marc Antony le	ib.	Gregory, Peter	ib.
——— Louis	101	Grenan, Benignus	ib.
Grandet, Joseph	ib.	Grenee	163
Grandier, Urban	ib.	Gresham, Sir Thomas	ib.
Grandin, Martin	103	Gresset, John Bap. Louis	169
Granduar, Charles	ib.	Gretser, James	ib.
Grandius, Guido	ib.	Grevenbroeck	170
Granet, Francis	ib.	Greville, Fulk or Foulk	ib.
Grange, Joseph de Chancel	ib.	Grevin, James	174
——— N.	ib.	Grevius or Grævius, John	
Granger or Grainger, James	104	George	175
Grant, Francis	105	Greuze	177
——— Patrick	108	Grew, Obadiah	ib.
Granville, George	109	——— Nehemiah	178
Grapaldus, Francis Marius	115	Grey, Lady Jane	179
Gras, Antony le	ib.	——— Dr. Zachary	189
Graswinckel, Theodore	ib.	——— Dr. Richard	ib.
Gratarolus, William	ib.	Gibaldus, Matthew	190
Gratian	116	Gribner, Michael Henry	191
	ib.	Grierson, Constantia	ib.
Gratiani, Jerome	ib.	Griffet, Henry	192
Gratius, Faliscus	ib.	Grisslier, John	ib.
——— Ortuinus	117	Griffin, Prince of Wales	ib.
Gravelot, Henry Francis Bourguignon		Grignon, Jaques	ib.
		Grimaldi, John Francis	ib.
Graverol, Francis	ib.	Grimarest, Leonard	193
Gravesande, William James	ib.	Grimoux	164
Gravina, Peter	118	Grindal, Edmund	ib.
——— John Vincent	119	Gringonneur, Jacquemin	198
Graunt, Edward	121	Gringore, Peter	ib.
——— John	ib.	Grisaunt, William	ib.
Gray, Thomas	125	Grive, John de la	ib.
Grazzini, Antony Francis	126	Grocyn, William	ib.
Grearakes, Valentine	ib.	Grednius, Stanislaus	200
Greaves, John	ib.	Gronovius, John Frederic	ib.
Green, Robert	132	——— James	ib.
——— John	133	Gropper, John	202
——— Edward Burnaby	134	Grose, Peter	ib.

Cc 2 Grose,

INDEX.

	Page		Page
Grose, Nicolas	202	Guicciardini, Louis	244
——— Francis	ib.	Gucheron, Samuel	ib.
Grosley, Peter John	203	Guidi, Alexander	ib.
Grosseteste, Robert	ib.	Guido, Reni	247
Grosseste, Claude	204	Guidotti, Paul	248
Grotius, Hugo	ib.	Guignard, John	ib.
——— William	218	Guild, Dr. William	ib.
——— Peter	ib.	Guillandius, Melchior	249
Grove, Henry	219	Guillemeau, James	ib.
Gruchius, Nicolas	221	Guillet, De St. George	ib.
Grudius, Nicolas Everard	ib.	Guillim, John	ib.
Grue, Thomas	ib.	Guise, Henry	250
Gruget, Claude	ib.	Guiscard, Robert	ib.
Grüner, John Frederic	ib.	——— Charles	251
Gruterus, Janus	ib.	Guise, William	ib.
——— Peter	224	Guitton, d'Arezzo	ib.
Grynæus, Simon	ib.	Gulvenstaedt, John Antony	ib.
——— Thomas	225	Gundling, Nicolas Jerome	252
Gryphiander, John	ib.	Gunning, Peter	ib.
Gryphius, Sebastian	226	Gunter, Edmund	254
——— Andrew	ib.	———	256
——— Christian	227	Gurtler, Nicolas	ib.
Guadagnolo, Philip	ib.	Gutman, Lewis	257
Guagnin, Alexander	228	Guttauvillan, Peter	ib.
Gualbert, S. John	ib.	Gustavus, Vasa	ib.
Gualdus, Prioratus, alias Galeazzo	ib.	——— Adolphus	258
		Guthrie, William	259
Gualterus, Rodolphus	ib.	Guttemberg, John	ib.
Guarin, Peter	ib.	Guy, Thomas	260
Guarini	ib.		261
——— John Baptist	229	Guyard, Dr. Berville	ib.
	231	Guyet, Francis	ib.
Guasco, Octavian	ib.	Guyon, Johanna-Mary Bouviers de la Mothe	263
Guazzi, Stephen	ib.		
——— Mark	232	——— Claude	ib.
Gudius, Marquard	ib.	Gwynn, Eleanor	ib.
——— Gottlob Frederic	235	Gwynne, Matthew	264
Guercheville, Antoinette de Pons Marchioness of	ib.	H.	
Guercino	ib.		
Gueret, Gabriel	236	Habakkuk	265
Guerin, Francis	ib.	Habert, Henry Louis	ib.
Gueriniere, Francis Robichon		Habicot, Nicolas	ib.
	ib.	Habington, William	ib.
Guesclin, Bertrand du	ib.	Hacker, William	266
Guettard, John Stephen	ib.	——— John	267
Guevara, Antony de	237	Hackspan, Theodore	269
——— Louis Velez de	238	Haddock, Sir Richard	ib.
Gueuletie, Thomas Simon	ib.	Haddon, Dr. Walter	ib.
Guglielmini	ib.	Hadrian VI.	270
Guichard, Claude de	ib.	Haen, Antony de	273
Guicciardini, Francesco	ib.	Hagedorn	ib.
			Haggai,

INDEX.

	Page		Page
Haggai	273	Harding, Thomas	324
Haquenier, John	274	Hardinge, Nicolas	324
Hahn, Simon Frederic	ib.	Hardion, James	325
Haillan, Bernard de Girard, lord of	ib.	Hardouin, John	ib.
Haines, Joseph	275	Hardwicke, Philip Yorke, earl of	329
Hakem	276	Hardy, Alexander	333
Hakewill, George	ib.	——— Charles	ib.
Hakluyt, Richard	277	Hare, Francis	ib.
Halde, John Baptist du	281	Hariot, Thomas	ib.
Hale, Sir Matthew	ib.	Harley, Robert	335
Hales, John	286	Harmer, Thomas	338
——— Stephen	289	Harmodius	ib.
Halh-beigh	291	Harold	339
Hall, Joseph	ib.	Harpalus	ib.
——— John	294	Harpocration, Valerius	ib.
——— Henry	ib.	Harrington, Sir John	ib.
——— John	295	——— James	341
——— Jacob	ib.	Harris, William	347
——— Richard	296	——— James	ib.
Halle, Peter	ib.	——— Walter	348
——— Antony	297	Harison, William	ib.
——— Claude Guy	ib.	——— John	ib.
Haller, Albert	ib.	——— Colonel	350
Halley, Edmund	ib.	Hartley, David	351
Hallifax, Samuel	303	Hartungus, John	ib.
Hamberger, George Albart	ib.	Harvey, William	ib.
——— George Christopher	ib.	——— Gideon	355
		Harwood, Edward	356
Hamel, John Baptiste du	ib.	Hase, Theodore de	ib.
Hamilton, Antony Count	305	——— James	ib.
——— George	ib.	Hasselquist, Frederic	ib.
Hamlet	ib.	Hastings, Elizabeth	357
Hammond, Dr. Henry	306	Hatton, Sir Christopher	ib.
——— Antony	309	Havercamp, Sigebert	ib.
——— James	310	Haustead, Peter	358
Hamon, John	ib.	Haute-feuille, John	ib.
Hampden, John	311	Hawkesworth, John	ib.
Hamsa	314	Howe, Lord Hawke	359
Handel, George Frederic	ib.	Hawkins, Sir John	360
Hankins, Martin	321	——— Sir John	361
Hanmer, Sir Thomas	ib.	Hawksmoor, Nicolas	370
Hanneken, Mennon	ib.	Hawkwood, Sir John	371
——— Philip Louis	322	Hay, William	375
Hannibalianus, Flavius Claudius	ib.	Hayes, Charles	376
Hanno	ib.	Haynes, Hopton	377
	ib.	——— Samuel	ib.
Hannsachs	ib.	Hayward, Sir John	ib.
Hanneman, Adrian	ib.	Hearne, Thomas	379
Hanway, Jonas	ib.	Heath, James	383

END OF THE SEVENTH VOLUME.

www.ingramcontent.com/pod-product-compliance
Lightning Source LLC
Chambersburg PA
CBHW032020220426
43664CB00006B/312